Emotion in Education

This is a volume in the Academic Press
EDUCATIONAL PSYCHOLOGY SERIES

Critical comprehensive reviews of research knowledge, theories, principles, and practices

Under the editorship of Gary D. Phye

Emotion in Education

Paul A. Schutz
University of Texas at San Antonio (UTSA)

Reinhard Pekrun
University of Munich

AMSTERDAM • BOSTON • HEIDELBERG • LONDON
NEW YORK • OXFORD • PARIS • SAN DIEGO
SAN FRANCISCO • SINGAPORE • SYDNEY • TOKYO

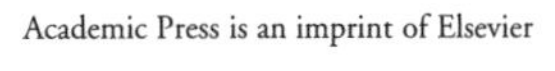

Academic Press is an imprint of Elsevier
30 Corporate Drive, Suite 400, Burlington, MA 01803, USA
525 B Street, Suite 1900, San Diego, California 92101-4495, USA
84 Theobald's Road, London WC1X 8RR, UK

This book is printed on acid-free paper. ♾

Library of Congress Cataloging-in-Publication Data
Emotion in education / [edited by] Paul A. Schutz, Reinhard Pekrun.
p. cm.
ISBN 0-12-372545-3
1. Educational psychology. 2. Affective education. 3. Emotions in children.
I. Schutz, Paul A. II. Pekrun, Reinhard, 1952-
LB1073.E457 2007
370.15'34–dc22 2006101551

British Library Cataloguing-in-Publication Data
A catalogue record for this book is available from the British Library.

ISBN 13: 978-0-12-372545-5
ISBN 10: 0-12-372545-3

For information on all Academic Press publications
visit our Web site at www.books.elsevier.com

Printed in the United States of America
07 08 09 10 11 9 8 7 6 5 4 3 2 1

To Sonja, Petra, Isaac, and Susanna

Contents

Introduction

Theoretical Perspectives on Emotions in Education

Students' Emotions in Educational Contexts

Implications and Future Directions

Contributors

Mary Ainley (9), Psychology Department, University of Melbourne, Victoria 3010, Australia

Monique Boekaerts (3), Center for the Study of Education and Instruction, Leiden University, 2300 RB Leiden, The Netherlands

Dionne I. Cross (13), Department of Educational Psychology and Instructional Technology, University of Georgia, Athens, GA 30602-7143 USA

Erik De Corte (11), Center for Instructional Psychology and Technology, University of Leuven, B-3000 Leuven, Belgium

Jessica T. DeCuir-Gunby (12), Department of Curriculum and Instruction, North Carolina State University, Raleigh, NC 27695-7801 USA

Kathleen deMarrais (16), College of Education, University of Georgia, Athens, GA 30602 USA

Andrew J. Elliot (4), Department of Clinical and Social Psychology, University of Rochester, Rochester, NY 14627-0266 USA

Anne C. Frenzel (2), Institute of Educational Psychology, University of Munich, D-80802 Munich, Germany

Thomas Goetz (2), Institute of Educational Psychology, University of Munich, D-80802 Munich, Germany

Ji Y. Hong (13), Department of Educational Psychology and Instructional Technology, University of Georgia, Athens, GA 30602-7143 USA

Anna Liljestrom (16), Department of Educational Psychology and Instructional Technology, University of Georgia, Athens, GA 30602-7143 USA

Elizabeth A. Linnenbrink (7), Department of Psychology and Neuroscience, Duke University, Durham, NC 27708-0085 USA

Debra K. Meyer (14), Department of Education, Elmhurst College, Elmhurst, IL 60126 USA

Peter Op 't Eynde (11), Center for Instructional Psychology and Technology, University of Leuven, B-3000 Leuven, Belgium

Jennifer N. Osbon (13), Department of Educational Psychology and Instructional Technology, University of Georgia, Athens, GA 30602-7143 USA

Reinhard Pekrun (1, 2, 4, 18), Institute of Educational Psychology, University of Munich, D-80802 Munich, Germany

Raymond P. Perry (2), Department of Psychology University of Manitoba, Winnipeg, Manitoba R3T 2N2 Canada

Carl Ratner (6), Institute for Cultural Research and Education, Trinidad, CA 95570 USA

Kathryn Roulston (16), Department of Lifelong Education and Policy, University of Georgia, Athens, GA 30602 USA

Paul A. Schutz (1, 13, 18), Department of Counseling, Educational Psychology, and Adult and Higher Education, University of Texas at San Antonio (UTSA), San Antonio, TX 78207 USA

Rosemary E. Sutton (15), Department of Curriculum and Foundations, Cleveland State University, Cleveland, OH 44115 USA

Jeannine E. Turner (8), Educational Psychology and Learning Systems, Florida State University, Tallahassee, FL 32306 USA

Julianne C. Turner (14), Department of Psychology, University of Notre Dame, Notre Dame, IN 46556 USA

Lieven Verschaffel (11), Center for Instructional Psychology and Technology, University of Leuven, B-3000 Leuven, Belgium

Ralph M. Waugh (8), Department of Educational Psychology, University of Texas at Austin, Austin, TX 78712 USA

Bernard Weiner (5), Department of Psychology, University of California Los Angeles (UCLA), Los Angeles, CA 90095-1563 USA

Meca R. Williams (12), Curriculum Foundations and Reading, Georgia Southern University, Statesboro, GA 30460 USA

Moshe Zeidner (10), Faculty of Education-Counseling Program, University of Haifa, Mt. Carmel, Haifa 31905, Israel

Michalinos Zembylas (17), Open University of Cyprus, 2002 Strovolos, Nicosia, Cyprus

Preface

This book examines some of the current inquiries related to the study of emotions in educational contexts. There has been a notable increase in interest in educational research on emotions. In fact, 2005 was the first year in which the term *emotions and emotional regulation* was included in the list of descriptors for proposals submitted for the American Educational Research Association's annual meeting. In other words, there are a growing number of paper submissions to the program for this annual meeting that relate to emotions in education. This growing interest in emotions in education can also be seen in the increasing number of journals that have devoted special issues to the topic (e.g., *Educational Psychologist*, 2002; *Learning and Instruction*, 2005; *Teaching and Teacher Education*, 2006; *Educational Psychology Review*, in press).

The increase in inquiries on emotions can be traced to the growing importance of the emotional nature of educational contexts. A notable example is the push for accountability in the school system that has brought with it an increase in the use of high-stakes testing (Nichols & Berliner, 2007). This accountability movement brings with it the emotions that are associated with high-stakes testing in both students and teachers. In addition, in countries around the world, including, for example, the United States and many European nations, there are high attrition and early retirement rates of teachers. In the United States, many teachers stop teaching as early as 3 years into their teaching careers. A number of studies identified unpleasant affective states such as anger, anxiety, subjective stress, and burnout as core factors influencing teachers' decisions to dropout (Wilhelm, Dewhurst-Savellis, & Parker, 2000; Wisniewski & Gargiulo, 1997). According to Hughes (2001), teacher burnout is often associated with the coping strategy of escape or the desire to escape an affectively stressful situation. Thus, the educational context is an emotional place, and emotions have the potential to influence teaching and learning processes (both positively and negatively). By implication, there is a great need to study emotions in education.

As such, *Emotion in Education* represents some of the most exciting research on emotions and education and will have the potential to impact research in this area. This combination of uniqueness, variety, timeliness, and potential for transformation of the field will make this a very important book. The chapters have been written for scholars in the area, but we also encouraged the authors to write with graduate students in mind. Therefore, the book should also be of great interest for graduate seminars.

A book such as this can only be compiled with the help of a number of people. We first would like to thank the authors, who agreed to participate and who sent their chapters to us in a timely manner. Every author we asked agreed to participate and the book was delivered to the publisher on the contract date. We would also like to thank Nikki Levy and Gary Phye, who agreed to take on this project at Academic Press. In addition, Sonja Lanehart, Dionne Cross, and Ji Yeon Hong read and critiqued a number of chapters from the book.

Finally, on a personal note, we want to thank Sonja Lanehart, Isaac Schutz, Petra Barchfeld, and Susanna Pekrun, who have made it clear that "Sorry, I'm working on the book" is no longer an acceptable excuse!

Paul A. Schutz
Reinhard Pekrun

References

Hughes, E. (2001). Deciding to leave but staying: teacher burnout, precursors and turnover. *International Journal of Human Resource Management, 12,* 288–298.

Nichols, S. L., & Berliner, D. C. (2007). Collateral damage: How high-stakes testing corrupts America's schools. Cambridge, MA: Harvard Education Press.

Wilhelm, K., Dewhurst-Savellis, J., & Parker, G. (2000). Teacher stress? An analysis of why teachers leave and why they stay. *Teachers and Teaching, 6,* 291–304.

Wisniewski, L., & Gargiulo, R. (1997). Occupational stress and burnout among special educators: A review of the literature. *Journal of Special Education, 31,* 325–346.

PART

I

Introduction

Introduction to Emotion in Education

PAUL A. SCHUTZ
University of Texas at San Antonio (UTSA)

REINHARD PEKRUN
University of Munich

In spite of the emotional nature of classrooms, inquiry on emotions in educational contexts, outside of a few notable exceptions (attribution theory: Weiner, 1985; research on test anxiety: Zeidner, 1998) has been slow to emerge (Pekrun, Goetz, Titz, & Perry, 2002). Students' test anxiety has been the only emotion in this field that strongly and continuously attracted researchers' interest. From more than 1,000 empirical studies conducted over a span of more than five decades, we have evidence on the structures, antecedents, and effects of this emotion, as well as on measures suited to prevent excessive test anxiety by changing education, and to treat this emotion once it occurred. However, what about student emotions other than test anxiety? And what about teachers' emotions? What do we know about students' and teachers' unpleasant emotions, other than anxiety, such as anger, hopelessness, shame, or boredom; and what do we know about pleasant emotions, such as enjoyment, hope, or pride in educational settings? Until recently, the answer to this question had to be "next to nothing."

The lack of inquiry on emotions in education has been noted by a variety of scholars. For example, motivation scholar Martin Maehr (2001) suggested that we needed to "rediscover the role of emotions in motivation" (p. 184). In addition, the editors of the *Handbook of Self-Regulation* (Boekaerts, Pintrich, & Zeidner, 2000) posed the following question in their concluding chapter: "How should we deal with emotions or affect?" (Zeidner, Boekaerts, & Pintrich, 2000, p. 754). This question came after 22 chapters that spanned 744 pages of some of the most current theory and research on motivation and self-regulated learning, and it reflected the state of the art in educational research on emotions and emotional regulation at that time. It is clear that research on emotions in education was and is currently needed.

The call for research on emotions has been heard, and recently there has been a discernable increase in the number of scholars investigating emotions in educational contexts. This edited volume represents the accumulation of some of that current interest in inquiry on emotions. As such, the book showcases some of the most contemporary, informative, and formative research in the areas of emotions and emotional regulation in education. The goal of the book will be to provide some perspectives on how to answer the question posed by Zeidner et al. (2000) cited above.

To do so, *Emotion in Education* features a number of important scholars from around the world (Australia, Belgium, Canada, Cyprus, Germany, Israel, The Netherlands, and the United States) who represent a variety of disciplines (e.g., emotion psychology, educational psychology, cultural psychology, sociology, and teacher education), scientific paradigms (e.g., nomothetic social-cognitive perspectives, idiographic case studies, phenomenological approaches, critical race theory, poststructuralist, and postpositivistic perspectives), and inquiry methods (e.g., quantitative, qualitative, ethnographic, and multimethods). In addition, the chapters in the edited volume will deal with a variety of populations (e.g., teachers; K-12 and university students) and educational activity settings (e.g., classrooms and independent study).

The book is organized into five sections. The first section, this introductory chapter, raises some of the issues related to inquiry on emotions and sets the stage for the remainder of the book. The second section, "Theoretical Perspectives on Emotions in Education," focuses on some theoretical foundations of research into emotions in education. This section features such scholars as Reinhard Pekrun, Anne C. Frenzel, Thomas Goetz, and Raymond P. Perry; Monique Boekaerts; Andrew J. Elliot; Bernard Weiner; and Carl Ratner. The work of these international authors represents the main theoretical foundations for much of the current work on emotions in education.

In Chapter 2, Reinhard Pekrun, Anne Frenzel, Thomas Goetz, and Raymond P. Perry provide an overview of control-value theory of academic and achievement emotions and its implications for education. The control-value theory provides a social-cognitive perspective on students' and teachers' academic emotions. The theory integrates assumptions of attributional and expectancy-value approaches. It is assumed that control and value appraisals relating to learning, teaching, and achievement are of primary importance for students' and teachers' emotions, and that different emotions are predicted by different types and combinations of control-value appraisals. Furthermore, the theory implies that specific features of classroom and social environments contribute to the development of academic emotions, and that emotions influence students' learning and achievement as well as teachers' instructional behavior and professional development. The authors' empirical research on these assumptions is summarized in the chapter, and implications for theory and practice are outlined.

In Chapter 3, Monique Boekaerts discusses her perspective on how students' emotions relate to her theory of self-regulation and social-constructivist learning environments. From this perspective, traditional classrooms do not provide much room for self-regulated learning. Students are cognitively, emotionally, and socially dependent on their teachers, who formulate classroom goals as well as determine which type of interaction is allowed. Several researchers described the destructive dynamics of classrooms that are set up in terms of a competitive learning game, where teachers are the gatekeepers of success and approval. In social-constructivist learning environments, students are invited to self-regulate their motivation and learning processes and to learn from and with each other. Teachers scaffold the learning process and withdraw external regulation when students are ready to fly solo.

Chapter 4, presents a theoretical account of the implications of goal theory for achievement-related emotions. Andrew J. Elliot is one of the most renowned contemporary authors in the field of achievement-related personality, and achievement goal theory more specifically. Elliot and Pekrun outline the implications of Elliot's four-fold approach to achievement goals for achievement-related emotions. They discuss the available empirical evidence linking students' goals to their emotions, including their own studies on these linkages.

In Chapter 5, Bernard Weiner, who for a long time has been one of the lone voices in educational research on emotions by discussing emotions as a main component of his attributional theory, lays out his views about "moral" classroom emotions from an attributional theory perspective. As the originator of attributional theory, Weiner discusses the diversity of these emotions within the classroom context, including emotions such as envy and scorn as well as admiration and gratitude. Many of these emotions have not yet been systematically explored to date. Weiner calls for more research on these emotions and their implications for the classroom.

In Chapter 6, Carl Ratner discusses his perspective of macro-cultural psychology related to emotions in education. This perspective was developed in his book, *Cultural Psychology: A Perspective on Psychological Functioning and Social Reform* (2006). It draws upon anthropology, history, and sociology. Ratner develops the notion that culture consists of macro-cultural factors, such as cultural institutions (e.g., educational organizations), artifacts, and cultural concepts. Psychological phenomena develop in order to facilitate macro-culture. They, therefore, originate in, take on characteristics of, and function to support macro-cultural factors. Emotions in education are a case in point. Ratner discusses the cultural function of emotions and historical variations in emotions.

The third section, "Students' Emotions in Educational Contexts," focuses on researchers who foreground students' emotions and their relations to academic motivation, learning, and achievement. This section features chapters that represent the work of Elisabeth Linnenbrink; Jeannine Turner and Ralph Waugh; Mary Ainley; Moshe Zeidner; Peter Op 't Eynde, Erik De Corte, and Lieven Verschaffel; and Jessica DeCuir-Gunby and Meca Williams. This group of international scholars focuses their chapters on students and their emotions as they transact in educational settings.

In Chapter 7, Elizabeth Linnenbrink discusses her program of research, which focuses on understanding the role of affect in students' motivation and learning using a multidimensional model of affect. This model is based on a broad definition of affect as a state consisting of both short-lived, intense emotions and longer-lived, less intense moods. Based on this model, affect can be defined along the two intersecting dimensions of valence (pleasant, unpleasant) and activation/arousal (high, low) within circumplex models of affect. A bidirectional, asymmetrical model linking motivation, affect, and cognitive processing is presented based on a series of studies examining upper elementary-school and middle-school students' affect during mathematics instruction. As such, affect is considered as a predictor and outcome of achievement goal orientations in mathematics. Affect is also examined in relation to cognitive engagement and learning as well as to broader social interactions in the classroom. Finally, the possibility that affect might mediate the relation between motivation and cognitive processing is discussed.

In Chapter 8, Jeannine Turner and Ralph Waugh discuss the dynamic patterns of students' emotional reactions to exam feedback relative to their academic goals, values, self-regulation, and achievement. Turner and Waugh use a nonlinear dynamical systems theoretical approach to help increase our understanding of students' unfolding motivation-emotion-behavior-achievement sequences that occur throughout a semester. Sustained learning is a complex phenomenon comprising a myriad of processes, such as those involved in perceptual-cognitive appraisals, affective responses, fulfilling motivational goals, striving for future goals, and self-regulation. A nonlinear dynamical systems approach offers a promising conceptualization of these complexly interwoven, qualitatively different, emergent interrelationships (Lewis, 2000). As students strive to create personal meaning with respect to their learning and personal goals, each student may be thought of as a self-organizing system that adaptively reacts to external and internal informational signals relative to their personal goals and values.

In Chapter 9, Mary Ainley presents the findings from a range of studies that have tracked students' reports of how they feel at critical points in learning tasks. She explores the findings, testing them against a model proposing that emotion processes 'alert' students to specific features of their environment and provide direction for the decisions and choices students make about processing information. Her analyses look at the emotion processes implicated in a range of phenomena, including personal dispositions (such as curiosity), mood, and reports on how students feel at critical points during the course of specific achievement activities. For each of these levels of experience (disposition, mood, feeling states), she presents what her research shows about the nature of the affective processing that occurs and how it contributes to processing task information, the decisions students make, and their achievement outcomes.

Chapter 10, features the work of Moshe Zeidner, who has written extensively on test anxiety. For this chapter, he discusses the evidence on test anxiety that was accumulated by researchers over many years. Specifically, Zeidner discusses the consistent findings that emerged on the assessment of test anxiety, its antecedents and consequences, as well as selected modes of intervention, with a focus on classroom settings. Because of the increasing emphasis on testing in the United States and other countries, the discussion of test anxiety in educational contexts is incredibly timely and helpful in understanding this issue.

In Chapter 11, starting from a short discussion of current theories on self-regulated learning, Peter Op 't Eynde, Erik de Corte, and Lieven Verschaffel introduce a meta-emotion perspective that conceptualizes the self-regulation of emotional processes in learning. Rather than being an entirely individual and autonomous process, the self-regulation of emotional processes is perceived here as highly situated within a classroom context. They also present an overview of empirical research that allows for a more detailed understanding of the role of emotions in mathematical problem solving and students' regulation of these emotional processes in the classroom context. Finally, they discuss the affordances and constraints of the developed meta-emotion perspective for research on learning and instruction, taking into account the empirical findings presented.

Chapter 12 provides the opportunity for Jessica DeCuir-Gunby and Meca Williams to examine how issues of race and racism impact emotions within the school context. Using a Critical Race Theory framework, they explore how African American students at a predominantly White, elite, and private school express and regulate their emotions when faced with a racially-charged situation. They do so through counterstorytelling, with a focus on Critical Race Theory's tenet of the permanence of racism. The racially-charged situation they examine involves a school assembly where a civil rights leader discussed the history of race and racism in the United States and how it impacts present-day society, including their school. The analysis focuses on how African American students reacted to the speaker, how they reacted to their White classmates' reactions, and how the incident impacted their views of race and racism. Recommendations are given regarding regulating emotions within a racially-charged environment.

The fourth section, "Teachers' Emotions in Educational Contexts," focuses on researchers that foreground teacher emotions and the role of emotions in classroom interactions. The section features chapters that represent the research programs of Paul A. Schutz, Dionne I. Cross, Ji Yeon Hong, and Jennifer N. Osbon; Debra K. Meyer and Julianne C. Turner; Rosemary Sutton; Anna Liljestrom, Kathryn Roulston, and Kathleen deMarrais; and Michalinos Zembylas. This group of scholars focuses their chapters on teachers and their emotions as they transact in educational contexts.

In Chapter 13, Paul A. Schutz, Dionne Cross, Ji Hong, and Jennifer Osbon build on their research that focuses on teachers' thoughts and beliefs about emotions and emotional relationships in the classroom. They explicate how they see the nature of teacher identities and beliefs influencing and are influenced by emotional experiences in classroom contexts. Using a multi-methods approach to their inquiry, they investigate understandings and beliefs related to constructing relationships with their students, developing their classrooms' emotional climate, and negotiating their role as a teacher.

In Chapter 14, Debra Meyer and Julianne Turner examine the existing literature on "classroom climate" and how this research is related to the co-regulation or emotional scaffolding in classroom settings. They focus on the teacher's role in creating an "affective climate," and on the relationships among student emotion, motivation, and higher-level learning in those climates. They integrate findings and brief case descriptions from their extensive research to exemplify major themes that are emerging on classroom research involving student and teacher emotions.

In Chapter 15, Rosemary Sutton reviews some of the limited literature on teachers' frustration and anger, summarizes findings from an interview and diary study she conducted on teachers' anger and frustration, and analyzes this work in the context of the current social-psychological research on these emotions. Because individuals frequently attempt to self-regulate their unpleasant emotions, research on self-regulation is an essential facet of work on teachers' anger and frustration. Sutton reviews studies on teachers' self-regulation of unpleasant emotions, including her own research, which began with a set of interviews and now focuses on the development of a questionnaire.

In Chapter 16, Anna Liljestrom, Kathryn Roulston, and Kathleen deMarrais focus on women teachers' experiences of anger in their professional lives in schools. Through a sociological framework, the authors examine how professional roles and teachers' moral beliefs about these roles, school structures, and experiences with administrators constrain teachers' expressions of anger in the workplace. Less-experienced young women teachers describe a socialization process where "appropriate" emotional expressions have to be acquired. More experienced teachers report undertaking elaborate and extensive steps to suppress or control their anger and other difficult emotions. The authors' findings show the problems arising from simplified "cultures of care" associated with traditionally female professions that stand in direct opposition to expressing or even having difficult emotions in the workplace. The chapter suggests implications for teacher education and teachers' professional development.

In Chapter 17, Michalinos Zembylas analyzes the politics of emotions in education through an investigation of emotion discourses in teaching and learning. Through several ethnographic studies that he conducted in the last five years, he offers examples of some of these discourses and how they are implicated in the play of power relations. His aim is to sketch a theory that has a place for ambivalence and change, and that shows the relations between teacher emotions and personal and institutional enactments of power. Thus his ethnography of emotions in education goes beyond an exploration of the emotional rules that teachers accommodate or resist in expressing, suppressing, or neutralizing emotions. Zembylas suggests that emotions play an important role in enabling change, something that is currently missing from many accounts of emotions in education.

The final section, Chapter 18, synthesizes the themes that emerged from the other chapters. For this section, Reinhard Pekrun and Paul A. Schutz focus on discussing future directions for inquiry on emotions in education. They discuss the need to build more comprehensive theoretical frameworks and enrich these frameworks with theoretical perspectives regarding the dynamic, multilevel, and contextualized nature of students' and teachers' emotions. They then address the advantages of multimethod research paradigms integrating diverse methodologies, such as qualitative and quantitative, experimental and nonexperimental, and nomothetic and idiographic approaches. Also, they address the need for more exploration of facets of emotions in education, of the dynamics of these emotions, and of their interlinkages with institutional and socio-historical contexts. In closing, Pekrun and Schutz call for better interdisciplinary collaboration of researchers in the many different scientific disciplines that are, or should be, involved in studying emotions in education.

As educational researchers, the authors in this volume seek to identify and understand factors within the educational process that have the potential to influence the effectiveness of learning environments. Emotions have surfaced as an important contributing factor to the success of both students and teachers in academic settings. The authors featured in this book suggest that emotions are a significant part of schooling and the daily lives of those involved in the educational process. Considering the importance of education, it is crucial to not only understand the causes or antecedents of emotional events and how these affect classroom transactions, but also to better understand how these events influence students' and teachers' success in the classroom. This book helps us move in that direction.

References

Boekaerts, M., Pintrich, P. R., & Zeidner, M. (Eds.) (2000). *Handbook of Self-regulation*.

Lewis, M. (2000). Emotional self-organization at three time scales. In M.D. Lewis & I. Granic (Eds.), *Emotion, development, and self-organization: Dynamic systems approaches to emotional development* (pp. 125-152). Cambridge, UK: Cambridge University Press.

Maehr, M. L. (2001). Goal theory is not dead—not yet anyway: A reflection on the special issue. *Educational Psychology Review, 13*, 177-185.

Pekrun, R., Goetz, T., Titz, W., & Perry, R.P. (2002). Academic emotions in students' self-regulated learning and achievement: A program of quantitative and qualitative research. *Educational Psychologist, 37*, 91-106.

Ratner, c. (2006). Cultural psychology: A perspective on pyschological functioning and social reform. Mahwah, NJ: Lawrence Erlbaum Associates.

Weiner, B. (1985). An attributional theory of achievement motivation and emotion. *Psychological Review, 52*, 548-573.

Zeidner, M. (1998). *Test anxiety: The state of the art*. New York: Plenum.

Zeidner, M., Boekaerts, M., & Pintrich, P. R. (2000). Self-regulation: Directions and challenges for future research. In M. Boekaerts, P. R. Pintrich, & M. Zeidner (Eds.), *Handbook of self-regulation* (pp. 750-768). San Diego, CA: Academic Press.

PART II

Theoretical Perspectives on Emotions in Education

CHAPTER

2

The Control-Value Theory of Achievement Emotions: An Integrative Approach to Emotions in Education

REINHARD PEKRUN, ANNE C. FRENZEL, & THOMAS GOETZ

University of Munich

RAYMOND P. PERRY

University of Manitoba

For students and teachers alike, educational settings are of critical importance. Over the years, many hours are spent in the classroom, social relationships are created there, and the attainment of important life goals depends on individual and collective agency in educational institutions. Because of their subjective importance, educational settings are infused with intense emotional experiences that direct interactions, affect learning and performance, and influence personal growth in both students and teachers (Pekrun, Goetz, Titz, & Perry, 2002a, 2002b).

The significance of emotions experienced in educational settings has been recognized by researchers in different fields, including personality research that has analyzed students' test anxiety since the 1930s (Zeidner, 1998), research on achievement motivation (Heckhausen, 1991), and more recent educational studies focusing on a variety of emotions in education (as evidenced in the chapters of this volume). Emanating from these different research traditions, various theoretical accounts of students' and teachers' emotions have evolved, but to date, these different traditions and their allied theoretical accounts have operated in relative isolation. As such, research on emotions in education, and on achievement emotions more generally, is in a state of fragmentation today. More integrative frameworks seem to be largely lacking, thereby limiting theoretical and empirical progress.

The control-value theory of achievement emotions (Pekrun, 2000, in press a) described here is an attempt to provide such an integrative framework. It is based on the premise that current approaches to achievement emotions share a number of common basic assumptions, and can be regarded as being complementary rather than mutually exclusive. More specifically, the theory builds on assumptions from expectancy-value theories of emotions (Pekrun, 1984, 1988, 1992a; Turner & Schallert, 2001), transactional theories of stress appraisals and related emotions (Folkman & Lazarus, 1985), theories of perceived control (Patrick, Skinner & Connell, 1993; Perry, 1991, 2003), attributional theories of achievement emotions (Weiner, 1985), and models addressing the effects of emotions on learning and performance (Fredrickson, 2001; Pekrun, 1992b; Pekrun et al., 2002a; Zeidner, 1998, 2007).

In this chapter, we first provide a brief overview of the theory, including a definition of the term *achievement emotion*. We then address the assumptions of the theory regarding the appraisal antecedents of achievement emotions. Next, conceptual corollaries and extensions of the theory are outlined. Specifically, we discuss implications for the multiplicity of achievement emotions, and for their more distal individual and social antecedents. Furthermore, we address assumptions of the theory regarding the effects of achievement emotions on learning and performance; the reciprocal relations between achievement emotions, antecedents, and effects; the regulation of these emotions; and their relative universality across socio-historical contexts, genders, and individuals. In closing, implications for educational practice are outlined.

In describing implications of the theory for emotions in education, we will primarily address emotions as experienced by students. Also, the related empirical evidence gathered so far primarily pertains to students' emotions. However, it should be noted that the theory applies to the achievement emotions experienced by other participants in educational settings as well, such as teachers (Frenzel, Goetz, Pekrun, & Wartha, 2006), principals, administrators, school employees, and parents. For example, many of the emotions experienced by teachers pertain to achievement-related occupational goals of increasing students' competences and fostering their development. The theory aims at explaining these emotions experienced by teachers, in similar ways as it explains the achievement emotions experienced by students.

OVERVIEW OF THE CONTROL-VALUE THEORY

Definition and Dimensions of Achievement Emotions

In the control-value theory, *achievement emotions* are defined as emotions tied directly to achievement activities or achievement outcomes. Achievement can be defined simply as the quality of activities or their outcomes as evaluated by some standard of excellence (Heckhausen, 1991). By implication, most emotions pertaining to students' academic learning and achievement are seen as achievement emotions, since they relate to behaviors and outcomes that are typically judged according to standards of quality—by students themselves and by others. However, not all of the emotions in educational settings are achievement emotions. Specifically, social emotions are frequently experienced in these same settings, as for example, a student's caring for a friend in the classroom. Achievement and social emotions can overlap, as in emotions directed towards the achievement of others (e.g., contempt, envy, empathy, or admiration instigated by the success or failure of others; see Weiner, 2007).

In past research, studies on achievement emotions typically focused on emotions relating to achievement outcomes (e.g., research on test anxiety, Zeidner, 2007; studies on emotions following success and failure, Weiner, 1985). The perspective used here implies that emotions pertaining to achievement-related activities are also considered to be achievement emotions (see Table 1). Examples of outcome-related achievement emotions are the joy and pride experienced by students when academic goals are met, and the frustration and shame when efforts fail. The excitement arising from learning, boredom experienced in classroom instruction, or anger about task demands are but a few examples of activity-related emotions. Activity emotions have traditionally been neglected by research on achievement emotions. The present perspective implies that the scope of existing research should be broadened to include this important class of emotions as well.

The differentiation of activity vs. outcome emotions pertains to the *object focus* of achievement emotions. In addition, as emotions more generally, achievement emotions can be grouped according to their *valence* (positive vs. negative; or pleasant vs. unpleasant), and to the degree of *activation* implied (activating vs. deactivating; see also Linnenbrink, 2007). Using these three dimensions, achievement emotions can be organized in a three-dimensional taxonomy (Table 1; Pekrun et al., 2002a).

Structure of the Theory: Overview of Assumptions and Implications

Figure 1 provides an overview of the different elements of the theory. Assumptions regarding the arousal of achievement emotions are at the heart of the theory. It is assumed that appraisals of ongoing achievement activities,

TABLE 1
A Three-Dimensional Taxonomy of Achievement Emotions

	Positive[a]		Negative[b]	
Object Focus	Activating	Deactivating	Activating	Deactivating
Activity Focus	Enjoyment	Relaxation	Anger Frustration	Boredom
Outcome Focus	Joy Hope Pride Gratitude	Contentment Relief	Anxiety Shame Anger	Sadness Disappointment Hopelessness

[a]Positive, pleasant emotion; [b]Negative, unpleasant emotion.

and of their past and future outcomes, are of primary importance in this respect (Figure 1, link 1). Succinctly stated, this key element of the theory stipulates that individuals experience specific achievement emotions when they feel in control of, or out of control of, achievement activities and outcomes that are subjectively important to them, implying that *control appraisals* and *value appraisals* are the proximal determinants of these emotions.

To the extent that this is true, more distal individual antecedents should affect these emotions by influencing control and value appraisals in the first place (Figure 1, link 2). Examples of such antecedents are individual achievement goals as well as achievement-related control and value beliefs. However, the theory acknowledges that emotions are also influenced by non-cognitive factors, including genetic dispositions and physiologically bound temperament (Figure 1, link 3). Concerning determinants in classroom interaction, social environments, and the broader socio-historical context, the theory implies that factors influencing individual control-value appraisals should affect the individual's achievement emotions (Figure 1, link 4).

The theory also addresses the effects of achievement emotions on students' academic engagement and performance. Specifically, it is posited that emotions influence cognitive resources, motivation, use of strategies, and self-regulation vs. external regulation of learning (Figure 1, link 5). The effects of emotions on achievement are posited to be mediated by these processes (Figure 1, link 6). Furthermore, processes of learning as well as their achievement outcomes are expected to act back on students' emotions (Figure 1, link 7), and on the environment within, and outside of, the classroom (Figure 1, link 8). By implication, antecedents, emotions, and their effects are thought to be linked by reciprocal causation over time (see the chain of links 1 to 8 in Figure 1), in line with dynamic systems accounts of emotions in education (Turner & Waugh, 2007). Assumptions on reciprocity have implications for the

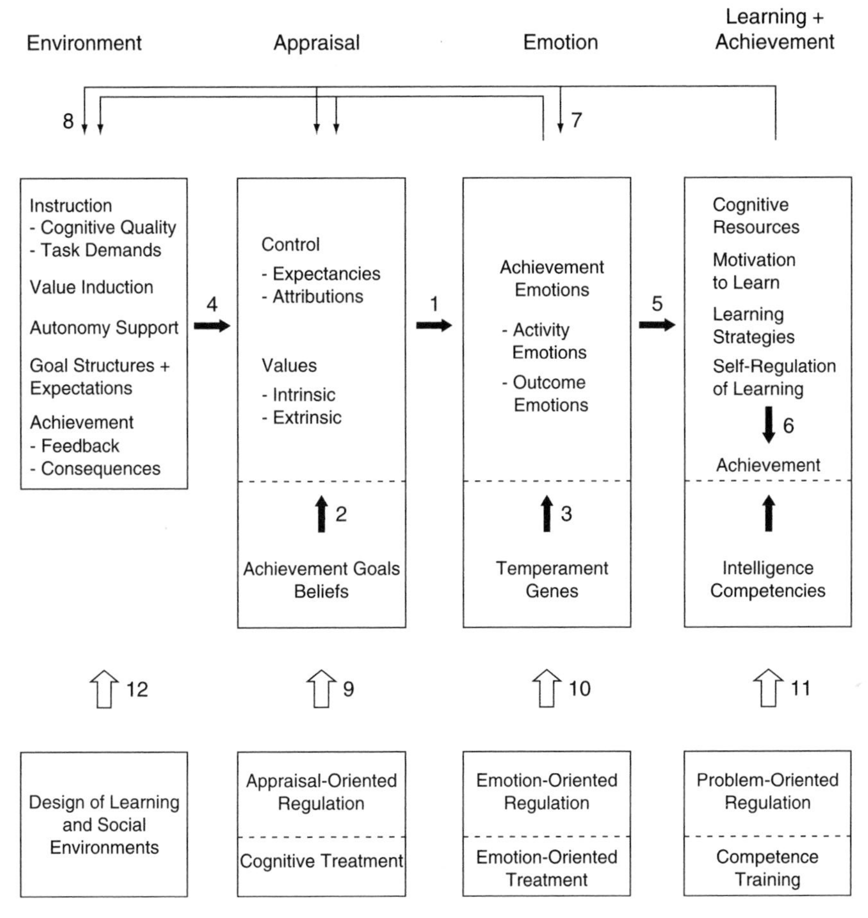

FIGURE 1

Overview of the control-value theory of achievement emotions.

regulation and treatment of achievement emotions (Figure 1, links 9 to 11), and for the design of "emotionally sound" (Astleitner, 2000) learning environments (Figure 1, link 12). Finally, there are some additional features of the theory that are not displayed in Figure 1, including assumptions on the multiplicity of achievement emotions and on their relative universality.

In the following sections, these elements of the theory are addressed in turn. A more complete treatment, however, is beyond the scope of the present chapter (see Goetz, Frenzel, Pekrun, & Hall, 2006; Pekrun, 1988, 1992a, 1992b, 2000, in press a; and Pekrun et al., 2002a, 2002b, for more elaborate discussions of facets of the theory).

CONTROL, VALUES, AND EMOTIONS: LINKAGES BETWEEN APPRAISALS AND AFFECT

Generally, emotions can be influenced by a host of proximal factors, such as situational perceptions, cognitive appraisals, physiological processes, or feedback from facial expression. For the emotions arising from achievement activities and performance outcomes, however, appraisals relating to these activities and outcomes can be assumed to be most important. Among the different appraisals addressed by appraisal theories of emotions (Scherer, Schorr, & Johnstone, 2001), subjective control over activities and outcomes and the subjective values of these activities and outcomes are held to be most relevant by the control-value theory, as noted previously.

Subjective control over achievement activities and their outcomes is assumed to depend on causal expectancies and causal attributions that imply appraisals of control. Three types of causal expectancies are relevant (Pekrun, 1988, in press a; see also Skinner, 1996): *action-control* expectancies that an achievement activity can successfully be initiated and performed ("self-efficacy expectations"; Bandura, 1977); *action-outcome* expectancies that these activities lead to outcomes one wants to attain; and *situation-outcome* expectancies that these outcomes occur in a given situation without one's own action. Examples would be a student's expectation that he will be able to invest sufficient effort in learning some material (action-control expectancy); the expectation that he will, because of his efforts, attain a good grade (action-outcome expectancy); and the expectation that he will get a good grade even if he does *not* act at all (situation-outcome expectancy). Expectations of the latter type, however, will typically be low in achievement situations. The attainment of success and prevention of failure are normally contingent on one's own efforts. By implication, expectations that success can be attained, or failure prevented, presuppose to perceive sufficient internal control over activities and their achievement outcomes, as implied by positive action-control and action-outcome expectancies.

Regarding the *subjective values* of activities and outcomes, the theory makes a distinction between intrinsic and extrinsic values. *Intrinsic values* of activities relate to appreciating an activity per se, even if it does not produce any relevant outcomes. For example, being interested in mathematics, a student can value dealing with math problems, irrespective of the contribution this activity might have to getting good grades in math. *Extrinsic values* pertain to the instrumental utility of activities to produce outcomes, and of outcomes to produce further outcomes (Heckhausen, 1991). An example would be a student who values academic studying because it helps in getting good grades, and who values good grades because they contribute to the attainment of future goals like getting the job she wants (Husman & Lens, 1999).

The theory next makes predictions about how different patterns of these appraisals instigate different achievement emotions (Figure 1, link 1; see Table 2 for a summary of assumptions). These predictions can be grouped according to the type of emotion addressed. Using the object focus dimension of the taxonomy of achievement emotions described above, three types of achievement emotions are distinguished: prospective outcome emotions, retrospective outcome emotions, and activity emotions.

Prospective Outcome Emotions

Prospective, anticipatory outcome emotions are experienced when positively valued success or negatively valued failure are to be expected. If perceived control is high and the focus is on success, *anticipatory joy* is assumed to be instigated. For example, if a student expects to be able to master an upcoming exam, she may simply look forward to the good grade that will result. If the focus is on failure, on the other hand, and there is high subjective control implying the expectation that failure can be avoided, *anticipatory relief* will be experienced. For example, if a student notices that she will be able to prevent an anticipated failure on an exam because her preparation for the exam was successful, she will feel relief upon noticing that she likely worried needlessly, even if the exam has not started yet.

If there is partial control only, implying that success and failure are subjectively uncertain, *hope* will be instigated if the focus is on success, and *anxiety* if the focus is on failure. Since uncertainty implies both chances for success and the threat of failure, mixed feelings comprising both hope and anxiety will be quite typical for many outcome-focused achievement situations. For example, a student who wants to pass an important exam, but does not know if he will be able to do so, can hope for success and can at the same time be afraid of failure (see also Folkman & Lazarus, 1985). Finally, if success is perceived as not being attainable and failure to be certain, hope and anxiety are assumed to be replaced by *hopelessness*. Hopelessness is posited to occur whenever a positive achievement outcome cannot be attained or a negative outcome is subjectively certain. As such, hopelessness is experienced both when cognitions focus on the nonattainability of success, and when the focus is on the nonavoidability of failure.

Retrospective Outcome Emotions

For retrospective outcome emotions following subjectively important successes and failures, subjective control as implied by causal attributions of these outcomes is of importance. More specifically, in line with Weiner's (1985) assumptions on attribution-independent emotions, it is assumed that some of the immediate affective reactions to success or failure do not depend on subjective control (*control-independent* emotions), in contrast to emotions

TABLE 2
The Control-Value Theory: Basic Assumptions on Control, Values, and Achievement Emotions

	Appraisals		
Object Focus	Value	Control	Emotion
Outcome / Prospective	Positive (Success)	High Medium Low	Anticipatory joy Hope Hopelessness
	Negative (Failure)	High Medium Low	Anticipatory relief Anxiety Hopelessness
Outcome / Retrospective	Positive (Success)	Irrelevant Self Other	Joy Pride Gratitude
	Negative (Failure)	Irrelevant Self Other	Sadness Shame Anger
Activity	Positive Negative Positive/Negative None	High High Low High/Low	Enjoyment Anger Frustration Boredom

involving more elaborate, control-dependent cognitive mediation (*control-dependent* emotions). Concerning control-independent emotions, success is posited to induce *joy* and *contentment*, and the nonoccurrence of expected success is posited to induce *disappointment*. Failure is expected to instigate *sadness* and *frustration*, and the nonoccurrence of expected failure is expected to instigate *relief*.

The emotions *pride*, *shame*, *gratitude*, and *anger* are assumed to be control-dependent (Table 2). These emotions are instigated by causal attributions of success and failure implying that the self, other persons, or situational factors produced the achievement outcome. Pride and shame are posited to be induced by attributions of success and failure to the self, and gratitude and anger by attributions to other persons. These assumptions reflect Weiner's (1985) attributional theory of achievement emotions, but there also are differences. Specifically, the controllability of the perceived *causes* of success and failure as addressed by Weiner's theory is not held to be critical for the instigation of outcome-related emotions. Rather, it is the perceived controllability of the *outcome* itself that is posited to determine which emotion is instigated.

As one implication, the antecedents of pride and shame are seen as being symmetrical. Both of these emotions are self-related affects triggered by success and failure that are appraised as being caused by oneself. Both of them can be induced by any self factors that are perceived as influencing achievement outcomes, typical examples being ability (or lack of ability) and effort (or lack of effort). For example, if a student performs well, and she attributes this performance to her abilities or successful efforts at learning, she will be proud of her accomplishments. Similarly, if she fails an exam and attributes this failure to a perceived lack of abilities, or to insufficient effort at studying, shame about the failure will be experienced.[1]

Activity Emotions

Emotions relating to achievement activities are assumed to depend on the perceived controllability of the activity and on its value. If the activity is seen as being controllable and valued positively, *enjoyment* is instigated. For example, if a student is interested in some learning material and feels capable of dealing with this material, he will enjoy studying. Enjoyment of achievement activities can take different shades, including excitement at challenging tasks, as well as more relaxed states when performing pleasant routine activities. If there is controllability, but the activity is negatively valued, *anger* is posited to be experienced. Examples are activities that can be performed, but are subjectively aversive (e.g., because they require much mental or physical effort). In contrast, if the activity is valued, but there is no sufficient control and obstacles inherent in the activity cannot be handled successfully, *frustration* will be experienced.

Finally, if the activity is valued neither positively nor negatively, *boredom* is induced. For example, if demands are too low, as in monotonous routine activities, there may be insufficient challenge and a lack of intrinsic value, thus producing boredom. Conversely, if demands exceed capabilities and cannot be met, it may also be difficult to detect meaning in the activity, thus reducing its value. Furthermore, subjectively devaluing material that is too difficult may serve to cope with the threat implied by high demands. By implication, as a consequence of lack of value, boredom can be experienced under both low- and high-demand conditions.

Multiplicative Relations of Control and Value

The control-value theory implies that appraisals of both control and value are necessary for an achievement emotion to be instigated. More precisely, the

[1] In Weiner's attributional theory, shame is seen as being primarily linked to attributions of failure to lack of ability, whereas attributions to lack of effort are assumed to arouse guilt. In contrast, pride is held to be both ability- and effort-linked (see, e.g., Weiner, 1985, p. 561 ff.). Weiner's theory thus implies cognitive asymmetry of pride and shame. In the control-value theory, successes and failures perceived as being self-caused are assumed to instigate the self-emotions of pride and shame, respectively, including the arousal of shame by failure that is attributed to lack of effort. This assumption seems to be in line with the empirical evidence from attributional research (including some of Weiner's own studies; e.g., Brown & Weiner, 1984). Guilt, on the other hand, is seen to be aroused by violations of moral norms, implying that guilt is expected to be induced by failure when failure avoidance is seen as a moral obligation. In this case, both shame *and* guilt would be expected to be aroused.

intensity of achievement emotions is assumed to be a multiplicative function of appraisals of controllability, on the one hand, and value, on the other (see Pekrun, 1988, in press a, for formalized versions of this assumption). For most emotions, emotional intensity increases with increasing controllability (in positive emotions) or uncontrollability (in negative emotions), and with increasing subjective value. If one of the two is lacking, the emotion will not be induced.

Positive, pleasant achievement emotions are posited to be a multiplicative function of the perceived controllability and positive values of activities or outcomes. For example, if a student values some learning material and believes she will be able to master it, she will enjoy learning the material. In contrast, if she is not interested in the material or perceives a lack of control over how to learn it, the learning activity will not be enjoyable. Similarly, negative, unpleasant achievement emotions are assumed to be a joint function of perceived lack of controllability and negative values. For example, if a student perceives failure at an upcoming exam to be possible and not sufficiently controllable, and judges the exam to be important because of its consequences for attaining career goals, he will be afraid of the exam. In contrast, if there is no anticipation of failure, or the exam is irrelevant to the student's goals, no anxiety will be experienced.

These assumptions imply that subjective value moderates the effects of perceived control on achievement emotions. More specifically, the theory implies that values influence both the *type* of emotion experienced and its *intensity*. If an activity or outcome is valued positively, as implied by approach goals, positive emotions are assumed to be instigated. If the subjective value of the activity or outcome is negative, as implied by goals to avoid the activity or outcome, negative emotions are thought to be aroused. The intensity of the emotion experienced is posited to be a function of the degree of subjective value, such as the degree of interest in some learning material or the perceived importance of success and failure on an exam. If no value is perceived, implying that the activity or outcome is subjectively irrelevant, no emotion is instigated, with the exception of boredom that is aroused by nonvalued activities.

Subconscious Appraisals and Habitualized Achievement Emotions

In everyday classroom situations, an elaborate processing of control- and value-related cognitions likely is the exception rather than the rule. With repeated experience, appraisals need no longer be the focus of conscious attention. Also, recurring sequences of situational perceptions, appraisals, and emotions can habitualize over time. Habitualized emotions are based on procedural schemes that short-circuit perceptions and emotions, meaning that perceptions alone are sufficient to induce an emotion, without any need for intervening appraisals (Pekrun, 1988; Reisenzein, 2001). For example, a student who has had many positive experiences in a class can experience anticipatory enjoyment before entering the classroom, without any need for an elaborate processing of expectancies and values relating to attending the class. With habitualized emotions, appraisals can be part of the activation of procedural emotion schemes, but appraisals are an epiphenomenon of emotion induction in this case, rather than its cause. However, whenever new experiences are made that contradict existing schemes, these schemes can be broken up again and be replaced by renewed appraisal processes, such that emotions can be modified and adapted to new situational circumstances.

Empirical Studies on Appraisals and Achievement Emotions

While assumptions for some emotions addressed by the control-value theory have yet to be tested, evidence from a number of sources corroborates many of the predictions for enjoyment, hope, pride, anger, anxiety, shame, hopelessness, and boredom. In our own research, we used a multimethod paradigm including both qualitative and quantitative studies to test assumptions of the theory. Our qualitative data gathered in 11 interview studies using samples of K-12 and higher education students showed that students' reports about their emotions experienced with regard to learning in the classroom and taking exams were systematically connected to their thinking about control and values in these situations (e.g., Pekrun, 1992c; Titz, 2001). In addition to testing assumptions, we also learned from our interview data how to refine our hypotheses. For example, when initially considering boredom, we focused on the assumption that low-demand conditions implying insufficient challenge induce boredom (Csikszentmihalyi, 2000). However, in our participants' reports about boredom, academic settings characterized by very high perceived demands were also reported to instigate boredom (Titz, 2001), motivating us to enlarge our original hypothesis on relations between demands and this emotion.

In much of our quantitative research, we used scales of the *Achievement Emotions Questionnaire* (AEQ) (Pekrun, Goetz, & Perry, 2005) to explore the relations between emotions and subjective control and values. The AEQ is a multidimensional 24-scale instrument assessing students' class-related, learning-related, and exam-related achievement emotions. Emotions measured within these categories include enjoyment, hope, pride, relief, anger, anxiety, shame, hopelessness, and boredom (alpha >.80 for 20 of the 24 scales in the normative investigation reported by Pekrun et al., 2005). In a number of studies, the enjoyment, hope, and pride scales of the AEQ showed positive correlations with students' achievement-related action-control, action-outcome, and overall success expectations, whereas correlations for the anxiety, shame, and hopelessness scales were negative (for overviews of some of these studies, see Pekrun, Goetz, Perry, Kramer, & Hochstadt, 2004; Pekrun et al., 2002a; Titz, 2001). The scales also revealed consistent relations with indicators of the subjective value of success and failure, including students' achievement approach and avoidance goals (Pekrun, Elliot, & Maier, in press; see Elliot & Pekrun, 2007).

In addition to main effects, we also examined the interactive effects as implied by our assumptions on multiplicative relations. For example, in a study on university students' enjoyment, anxiety, and boredom experienced in statistics courses, we found that perceived control over and value of course achievement significantly interacted in producing these emotions. In line with assumptions, enjoyment was highest when both control and value were high, and anxiety was highest when control was low, but value was high (Pekrun, Barrera, Goetz, & Maier, 2003).

COROLLARIES AND EXTENSIONS OF THE THEORY

In this section, we provide an overview of the implications of the theory regarding the multiplicity of achievement emotions and their more distal individual and social antecedents. Also, we address assumptions of the theory on the effects of emotions on learning and achievement; on reciprocal linkages between antecedents, achievement emotions, and effects; and on the regulation and treatment of achievement emotions. Finally, the relative universality of the functional mechanisms of achievement emotions is discussed.

Multiplicity and Domain Specificity of Achievement Emotions

The control-value theory implies that specific, discrete achievement emotions arise from different combinations of appraisal antecedents and show qualitative differences in terms of their components. By implication, a full account of these emotions presupposes to acknowledge their multiplicity. Furthermore, control-related and value-related variables have been shown to be organized in domain-specific ways (e.g., students' academic self-concepts and interests; Bong, 2001). Therefore, it follows from the assumptions of the theory that the emotions determined by control and values should be domain-specific as well, in contrast to more traditional conceptions regarding achievement emotions as generalized personality traits (e.g., test anxiety; Zeidner, 1998). This can be assumed not only for the emotions experienced by students, but also for teachers' domain-related emotions pertaining to different subjects they teach. For teachers' emotions, evidence on domain specificity still seems to be lacking. Assumptions on the domain specificity of students' emotions, however, were corroborated in recent studies (Goetz, Pekrun, Hall, & Haag, 2006; Goetz, Frenzel, Pekrun, & Hall, in press).

Goals and Beliefs as Antecedents of Achievement Emotions

Since appraisals of control and values are regarded as proximal antecedents of achievement emotions, it follows from the theory that any individual variables that affect these appraisals can influence resulting emotions as well. Two important groups of such variables are individual achievement goals and enduring control and value beliefs (Figure 1, link 2). As to goals, Pekrun, Elliot and Maier (2006) presented a theoretical model and related empirical evidence implying that different achievement goals help to focus attention on specific sets of activity-related and outcome-related appraisals, thus affecting achievement emotions mediated by these appraisals (see Elliot & Pekrun, 2007).

Similarly, control-related beliefs (e.g., self-concepts of ability) and value-related beliefs (e.g., individual interests) can be assumed to affect appraisals and resulting achievement emotions, in addition to physiologically based temperament directly affecting individual propensities to experience certain emotions. For example, if a student holds favorable control beliefs regarding her achievement in an academic domain like mathematics, an activation of these beliefs will lead to appraisals of challenging tasks as being manageable, and to related positive emotions.

Classroom Instruction and Social Environments as Antecedents of Achievement Emotions

In line with assumptions of social-cognitive learning theories, the control-value theory implies that the impact of environments on individual achievement emotions is also largely mediated by control-value appraisals (Figure 1, link 4). By implication, environmental factors affecting students' appraisals should be important for their emotions. Since all of these factors are of immediate practical relevance, they are discussed in the final section on implications for educational practice.

The Effects of Emotions on Learning and Achievement

Of the three dimensions used by the taxonomy of achievement emotions introduced at the outset, the two dimensions of *valence* and *activation* are posited to be most important to describe the performance effects of emotions. Using these two dimensions, achievement emotions can be grouped into four basic categories: *positive activating* emotions, such as enjoyment, hope, and pride; *positive deactivating* emotions, such as relief and relaxation; *negative activating* emotions, like anger, anxiety, and shame; and *negative deactivating* emotions, like boredom or hopelessness (see Table 1). The theory makes the following predictions regarding the effects of these emotions on cognitive resources, motivation, strategy use, self-regulation, and resulting achievement (Figure 1, links 5 and 6; see Pekrun, 1992b; Pekrun et al., 2002a).

Cognitive Resources

Emotions help focus attention on the object of the emotion. Therefore, enlarging assumptions of the resource allocation model put forward by Ellis and Ashbrook (1988), it can be assumed that positive or negative emotions that do not relate to an ongoing achievement activity distract attention away from the activity, so that they reduce cognitive resources available for task purposes and impair performance needing such resources. For example, if a student is angry about failure, or worries about an upcoming exam, she will experience difficulties in concentrating on learning. Positive emotions relating to the activity, on the other hand, are assumed to focus attention on the activity, thus benefiting performance.

In line with these assumptions, we found that students' enjoyment of learning correlated positively with their flow experiences (which imply focusing cognitive resources on learning), and negatively with their task-irrelevant thinking at learning (Pekrun et al., 2002a). In contrast, anxiety, shame, and hopelessness arising from negative achievement outcomes were negatively related to flow experiences and positively to task-irrelevant thinking (Pekrun et al., 2004). We also used experimental procedures to analyze the effects of extra-task emotions on task-related attention, and found that emotional states induced by affective pictures or recollections of critical life events reduced the cognitive resources available for task purposes, as indicated by event-related brain potentials (Meinhardt & Pekrun, 2003). In line with theoretical assumptions, this was true not only for negative emotional states, but also for positive states related to task-irrelevant stimuli.

Interest and Motivation

Positive activating emotions such as enjoyment of learning are assumed to increase interest and strengthen motivation. Negative deactivating emotions, such as hopelessness and boredom, are held to be detrimental for motivation. In contrast, the effects of positive deactivating emotions like relief, as well as negative activating emotions like anger, anxiety, and shame, are posited to be more complex and ambivalent. Failure-related anxiety, for example, can reduce interest and intrinsic motivation, but can also strengthen motivation to invest effort to avoid failure. If a student is afraid of failing an upcoming exam, intrinsic motivation to learn the material will be reduced, while motivation to avoid failure can be strengthened.

In line with these assumptions, we found that students' enjoyment of learning and instruction related positively to their intrinsic and extrinsic motivation, whereas relations for hopelessness and boredom were negative, and relations for anxiety and shame ambivalent (e.g., Pekrun et al., 2002a; Pekrun et al., 2004). The following excerpts from qualitative interviews with college students demonstrate the motivationally ambivalent nature of achievement-related anxiety (Titz, 2001). Being asked about the motivational impulses triggered by experiencing anxiety before an important university exam, the answers given by three students were: "I'd rather avoided the exam"; "...no motivation anymore"; and "I just wanted it to be over." However, other participants responded: "I want to pass it...I don't want to fail the exam"; "I wanted to solve the test as well as possible"; and "...it [the feeling] has motivated me to see the exam as a challenge." Finally, one student was able to express the ambivalent nature of his anxiety in a succinct way in the following statement: "...you would rather run away, but on the other hand, you want to fulfil your obligations—overall truly ambivalent feelings."

Strategies of Learning and Problem Solving

Mood research has shown that positive affective states tend to facilitate holistic, flexible, and creative ways of solving problems, whereas negative states can facilitate more rigid and analytical ways of thinking (e.g., Isen, 2000). In line with these findings, it is assumed that positive activating emotions help using flexible learning strategies, such as elaboration of learning material, whereas negative activating emotions can facilitate the use of more rigid strategies, such as simple rehearsal. If a student enjoys learning mathematics, for example, it may be easier for her to engage in creative mental modeling of mathematical problems, whereas anger or anxiety can lead her to resort to exercising algorithmic procedures. For deactivating emotions, it is assumed that these emotions are detrimental to any more elaborate processing of task-related information.

In our field studies, we found positive relations for students' enjoyment, hope, and pride, and their use of flexible learning strategies (elaboration and organization of learning material). The evidence for beneficial effects of anger, anxiety, and shame on rehearsal was weaker (Pekrun et al., 2002a). Similarly, a recent study on the emotions experienced by teachers in math classrooms showed that these teachers' enjoyment of teaching related positively to their use of creative teaching methods oriented towards mental modelling of mathematical problems (Frenzel et al., 2006).

Self-Regulation vs. External Regulation of Learning and Problem-Solving

Self-regulation of behavior requires flexible use of meta-cognitive, meta-motivational, and meta-emotional strategies, making it possible to adapt behavior to goals and environmental demands. It is assumed that positive activating emotions, such as enjoyment of learning, enhance self-regulation, whereas negative emotions, such as anxiety or shame, facilitate reliance on external guidance. In line with this assumption, we found that students' enjoyment of learning related positively to their perceived self-regulation of academic learning, whereas achievement-related anxiety and shame related to perceived external regulation of learning by teachers and parents (Pekrun et al., 2002a; Pekrun et al., 2004).

Academic Achievement

The effects of emotions on achievement are assumed to be a joint product of the four mechanisms described above, and any interactions between these mechanisms and task demands. By implication, the overall effects of emotions on achievement are inevitably complex. For most task conditions, however, it can reasonably be assumed that positive activating emotions, such as activity-related enjoyment, exert positive overall effects, and negative deactivating emotions, such as hopelessness and boredom, exert negative effects. The effects of positive deactivating emotions, such as relaxation, and of negative activating emotions, such as anger, anxiety, and shame, can be assumed to be more complex, because of the ambivalence of the effects of these emotions on motivation and cognitive processing. If a student is able, for example, to use the motivational energy implied by exam-related anxiety to increase his efforts, and if task demands are congruent to a more rigid processing of information as facilitated by anxiety, exam performance can be enhanced instead of being impaired (see Turner & Waugh, 2007, for related assumptions on shame).

Our empirical findings are largely in line with these assumptions. In a number of studies, we consistently found that students' enjoyment, hope, and pride as assessed by scales of the AEQ related positively to their academic achievement, whereas their hopelessness as well as boredom related negatively to achievement (Pekrun et al., 2002a). For anger, anxiety, and shame, overall sample correlations were negative as well, suggesting that negative effects of these emotions outweigh positive effects across individuals. However, as expected, we also found that there are individual students who can profit, in terms of motivation and achievement, from their anxiety. Specifically, in a diary study investigating students' individual trajectories of achievement emotions experienced before and during their final university exams, we found that exam anxiety correlated negatively with achievement-related agency over time in many students, but showed positive correlations in others (Pekrun & Hofmann, 1996).

Feedback Loops of Emotions, Antecedents, and Effects: The Individual and Social Dynamics of Emotion Systems

Emotions are assumed to influence learning, but learning and achievement outcomes are among the antecedents of students' appraisals and emotions, thus implying that emotions, their effects, and their antecedents are linked by reciprocal causation over time within individuals (Figure 1). Furthermore, the relationship between appraisals and emotions is conceived to be bidirectional as well, with appraisals triggering emotions, and emotions acting on appraisals by mechanisms of emotion-congruent activation of memory networks. Beyond the individual level, the assumptions of the theory imply that teachers' and students' emotions also reciprocally influence each other, implying that their emotions are closely and often inextricably intertwined in classroom settings (also see Meyer & Turner, 2007). Teachers' enjoyment and enthusiasm, for example, can induce enjoyment of classroom instruction in students, and students' enjoyment can in turn enhance teachers' positive affect, one important mechanism being emotional contagion (Hatfield, Cacioppo, & Rapson, 1994) transmitting emotions between teachers and students. Reciprocal causation implies that there can be co-development of emotions in teachers and students that can extend over months and years, and can take beneficial as well as detrimental forms.

In line with perspectives of dynamical systems theory (see Turner & Waugh, 2007), it is assumed that reciprocal causation can take different forms, and can extend over fractions of seconds (e.g., in linkages between appraisals and emotions), days, weeks, months, or years. Positive feedback loops likely are quite typical (e.g., enjoyment of learning and success on exams reciprocally reinforcing each other), but negative feedback loops can also be important (e.g., failure inducing anxiety in a student, and anxiety motivating the student to successfully avoid failure on the next exam).

In our empirical studies, we found evidence for feedback loops within students, and preliminary evidence for relations between teachers' and students' affect (Frenzel et al., 2006). Specifically, in structural equations modeling of longitudinal data on students' academic development from grade five to ten, we found that students' emotions and their achievement were reciprocally linked over the years, implying that academic success and failure were important antecedents of students' emotional development and that their emotions reciprocally affected their academic achievement (e.g., Pekrun, 1992a). Typically, these feedback loops were positive, with success and positive emotions as well as failure and negative emotions reinforcing each other over the years.

Regulation and Treatment of Achievement Emotions

Since emotions, antecedents, and effects are assumed to be reciprocally linked over time, the control-value theory implies that emotions can be regulated and changed by addressing any of the elements involved in these cyclic feedback processes (Figure 1, links 9 to 12). Regulation and treatment of achievement emotions can target the emotion itself (*emotion-oriented* regulation and treatment; e.g., using relaxation techniques or taking drugs); the control and value appraisals underlying emotions (*appraisal-oriented* regulation and treatment; e.g., cognitive restructuring and therapy); the academic competences determining students' agency (*competence-oriented* regulation and treatment; e.g., training of learning skills); and the environment within educational institutions, including classroom instruction (*design of academic environments*). A more complete analysis of emotion regulation and the treatment of achievement emotions is beyond the scope of this chapter (see Goetz, et al., 2006; Zeidner, 1998; and see below for implications regarding the design of academic environments).

Relative Universality of Achievement Emotions Across Socio-Historical Contexts, Genders, and Individuals

The control-value theory is based on the assumption that general functional mechanisms of human emotions are bound to universal, species-specific characteristics of our mind. In contrast, specific contents of emotions as well as specific values of process parameters (e.g., the intensity of emotions) may be specific to different cultures, genders, and individuals. This assumption implies that the basic structures and causal mechanisms of emotions follow general nomothetic principles, whereas contents, intensity, and duration of emotions can differ.

Concerning gender differences, for example, it follows from the theory that relations between control and value appraisals, on the one hand, and achievement emotions, on the other, should be structurally equivalent for males and females: Emotions depend on control and value appraisals in both genders. However, to the extent that perceived control and academic values differ between the genders, resulting emotional experiences can differ as well. Corroborating this assumption, we found, for example, that the relationships between girls' and boys' control and value appraisal in mathematics, on the one hand, and their mathematics emotions, on the other, were structurally equivalent across genders (Frenzel, Pekrun, Goetz, & vom Hofe, 2006). However, mean scores for perceived control were substantially lower in girls. As a consequence, girls reported less enjoyment in mathematics as well as more anxiety and shame. Corroborating assumptions of the theory, these differences in emotions proved to be mediated by the gender differences of appraisals.

Similar arguments can be made for different countries and cultures. For example, in a cross-cultural comparison of Chinese and German middle school students' achievement emotions, we found structurally equivalent relations of appraisals and emotions. Mean levels of emotions, however, differed between cultures. The Chinese students reported significantly more achievement-related enjoyment, pride, anxiety, and shame and significantly less anger than the German students (Frenzel, Thrash, Pekrun, & Goetz, in press).

IMPLICATIONS FOR EDUCATIONAL PRACTICE

The control-value theory implies that students' and teachers' achievement emotions can be influenced by changing subjective control and values relating to achievement activities and their outcomes. This can be achieved by shaping the learning environments of students and the occupational environments of teachers in "emotionally sound" ways (Astleitner, 2000). Important features of these environments likely affecting students' and teachers' emotions are described in the following sections (Pekrun, in press a, b).

Cognitive Quality of Academic Environments

The cognitive quality of learning environments and the implied task demands are assumed to influence students' valuing of learning material, as well as their competences and perceived control. Clearly structured, cognitively activating material and challenging task demands that match students' capabilities likely benefit students' competences and interest, thus positively affecting their appraisals and emotions. If task demands are too low or too high, boredom can result instead, as argued previously. Similar assumptions can be made for the emotional effects of teachers' academic environments. For example, if students contribute to a cognitively stimulating classroom environment by asking challenging questions and giving competent answers, teachers' enjoyment of teaching likely is enhanced.

Motivational Quality of Academic Environments: Induction of Values

By a number of different mechanisms, including both direct verbal messages and more indirect messages conveyed by the behavior of significant others, environments shape students' and teachers' interests and values underlying their emotions. For example, matching learning tasks to students' needs, and occupational tasks to teachers' needs, is posited to be beneficial. Examples are authentic learning tasks that meet students' interests, and teaching assignments that meet teachers' motivation to educate students, instead of spending hours in serving administrative duties. Also, by way of observational learning and emotional contagion, teachers' and parents' own enthusiasm can induce enthusiasm in students (see Meyer & Turner, 2007, on teachers' "emotional scaffolding" of students' emotions). Conversely, teachers' enjoyment can be fostered by students' positive classroom emotions.

Support of Autonomy and Cooperation

To the extent that students are capable and motivated to self-regulate their learning, environments supporting self-regulated learning are held to increase students' sense of control, valuing of learning, and resulting emotions. Also, cooperative learning can be beneficial, on condition that students are provided with the social competences to make use of collaboration. Similarly, teachers' perceived control, values, and emotions likely are strongly influenced by chances for autonomy and cooperation within the faculty of their school.

Goal Structures and Expectations

Institutional goal structures probably exert profound effects on subjective control, values, and emotions of the institution's members (Johnson & Johnson, 1975; Pekrun, in press a). Competitive structures as defined by social-comparison performance goals (Elliot & Pekrun, 2007) imply negative contingencies between different members' chances for success, thus likely reducing perceived control over success in many individuals, and instigating negative emotions, such as anger, anxiety, or hopelessness. Individualistic goal structures pertaining to mastery of goals, as well as cooperative goal structures, probably are more beneficial in terms of mean levels of perceived control. This can be assumed to be true both for the goal structures provided for students in their classrooms and for the occupational goals set for teachers. Goal structures can be influenced by the reference norms used to evaluate students' and teachers' achievements (e.g., social comparison norms vs. criterion-referenced and individual norms).

The expectations of significant others can exert similar effects, on condition that they are adopted by the teachers and students themselves. For example, parents and administrators often expect teachers to have control over students' discipline and classroom learning, but teachers typically have only partial influence over their students' behavior. If teachers take over expectations that are too high, a loss of subjective internal control will be experienced that can trigger feelings of anger and frustration, with burnout and quitting the job being possible long-term consequences (see Sutton, 2007, as well as Liljestrom, Roulston, & deMarrais, 2007).

Feedback and Consequences of Achievement

Feedback of success and failure at learning affects students' outcome-related achievement emotions. Also, feedback shapes the expectancies and perceived values of future performance that determine students' prospective emotions. Similarly, the feedback given by administrators, students, or parents influences teachers' performance-contingent emotions. Information about the controllability and values of performance, as implied, for example, by teachers' messages about the causes of students' performance, are especially critical for ensuing appraisals and emotions pertaining to future performance. Furthermore, contributing to the extrinsic value of achievement, the long-term consequences of achievement are of critical importance. For example, if a student can expect that she will not get employment after high school, irrespective of any academic grades, academic attainment is devalued, thus reducing related emotions as well as achievement-related motivation.

Treatment of Appraisals and Emotions

Appraisal theories like the control-value theory imply that educators can make an attempt to change students' emotions by changing their appraisals. For example, attributional retraining has been proven to be effective in changing students' motivation, and can probably be used to change their emotions as well (Perry & Penner, 1990; Perry, Hall, & Ruthig, 2005). In similar ways, college professors could help students in pre-service teacher education, and instructors in continuing education could help teachers later on, to deal with the affective aspects of their professional development.

Fostering Self-Regulation of Emotions

Finally, educators can assist students in developing regulatory skills enabling them to self-regulate their control and value appraisals, and resulting achievement emotions (Goetz et al., 2006). Similarly, it should prove possible to develop programs helping teachers to enhance their competences for regulating the appraisals and emotions they experience in the classroom and when interacting with colleagues, administrators, and parents.

CONCLUSION

In this chapter, we provided an overview of the assumptions and corollaries of the control-value theory of achievement emotions, as well as some of its implications for educational practice. On a conceptual level, the theory makes an attempt to provide a theoretical framework making it possible to integrate constructs and assumptions from a variety of theoretical approaches to emotions in education and to achievement emotions more generally. Empirically, many facets of the theory have consistently been corroborated in qualitative and quantitative investigations. Other facets, however, still await empirical analysis (the assumptions on activity emotions, for example, have not yet been tested directly in experimental studies). Also, some parts of the theory have been tested in pilot investigations, but the evidence collected so far is too preliminary to warrant firm conclusions (e.g., the assumptions on relations between achievement goals and students' emotions; Pekrun et al., 2006).

Perhaps most importantly, the assumptions provided by the theory on how to design emotionally sound learning environments for students, and occupational environments for teachers, have yet to be tested in empirical intervention studies. There is evidence that educational interventions can reduce students' test anxiety (e.g., Ruthig et al., 2004; Zeidner, 1998, 2007). The control-value theory implies that shaping educational environments in adequate ways can help to change achievement emotions other than anxiety as well. Future research should systematically explore measures to help both students and teachers to develop adaptive achievement emotions, prevent maladaptive emotions, and use their emotions in productive and healthy ways (Pekrun & Schutz, 2007).

References

Astleitner, H. (2000). Designing emotionally sound instruction: The FEASP-approach. *Instructional Science, 28,* 169-198.

Bandura, A. (1977). Self-efficacy: Toward a unifying theory of behavioral change. *Psychological Review, 84,* 191-215.

Bong, M. (2001). Between- and within-domain relations of motivation among middle and high school students: Self-efficacy, task value and achievement goals. *Journal of Educational Psychology, 93,* 23-34.

Brown, J., & Weiner, B. (1984). Affective consequences of ability versus effort ascriptions: Controversies, resolutions, and quandaries. *Journal of Educational Psychology, 76,* 146-158.

Csikszentmihalyi, M. (2000). *Beyond boredom and anxiety.* San Francisco: Jossey-Bass.

Elliot, A. J., & Pekrun, R. (2007). Emotion in the hierarchical model of approach-avoidance achievement motivation. In P. A. Schutz & R. Pekrun (Eds.), *Emotion in education* (pp. 53-69). San Diego: Elsevier Inc.

Ellis, H. C., and Ashbrook, P. W. (1988). Resource allocation model of the effect of depressed mood states on memory. In K. Fiedler and J. Forgas (Eds.), *Affect, cognition, and social behavior.* Toronto: Hogrefe International.

Folkman, S., & Lazarus, R. S. (1985). If it changes it must be a process: Study of emotion and coping during three stages of a college examination. *Journal of Personality and Social Psychology, 48,* 150-170.

Fredrickson, B. L. (2001). The role of positive emotions in positive psychology: The broaden-and-build theory of positive emotions. *American Psychologist, 56,* 218-226.

Frenzel, A. C., Goetz, T., Pekrun, R., & Wartha, S. (2006, April). *Antecedents and effects of teacher enjoyment and anger.* Paper presented at the annual meeting of the American Educational Research Association, San Francisco, CA.

Frenzel, A. C., Pekrun, R., Goetz, T., & vom Hofe, R. (2006). *Girls' and boys' emotional experiences in mathematics.* Manuscript submitted for publication.

Frenzel, A. C., Thrash, T. M., Pekrun, R., & Goetz, T. (in press). A cross-cultural comparison of German and Chinese emotions in the achievement context. *Journal of Cross-Cultural Psychology.*

Goetz, T., Frenzel, A. C., Pekrun, R., & Hall, N. C. (in press). The domain specificity of academic emotional experiences. *Journal of Experimental Education.*

Goetz, T., Frenzel, A., Pekrun, R., & Hall, N. C. (2006). Emotional intelligence in the context of learning and achievement. In R. Schulze & R. D. Roberts (Eds.), *Emotional intelligence: An international handbook* (pp. 233-253). Cambridge, MA: Hogrefe & Huber Publishers.

Goetz, T., Pekrun, R., Hall, N. C., & Haag, L. (2006). Academic emotions from a socio-cognitive perspective: Antecedents and domain specificity of students' affect in the context of Latin instruction. *British Journal of Educational Psychology, 76,* 279-308.

Hatfield, E., Cacioppo, J. T., & Rapson, R. L. (1994). *Emotional contagion.* New York: Cambridge University Press.

Heckhausen, H. (1991). *Motivation and action.* New York: Springer.

Husman, J., Lens, W. (1999). The role of the future in student motivation. *Educational Psychologist, 34,* 113-125.

Isen, A. M. (2000). Positive affect and decision making. In M. Lewis & J. M. Haviland-Jones (Eds.), *Handbook of emotions* (pp. 417-435). New York: Guilford Press.

Linnenbrink, E. A. (2007). The role of affect in student learning: A multi-dimensional approach to considering the interaction of affect, motivation, and engagement. In P. A. Schutz & R. Pekrun (Eds.), *Emotion in education* (pp. 101-118). San Diego: Academic Press.

Johnson, D. W., & Johnson, R. T. (1975). *Learning together and alone: Cooperation, competition, and individualization*. Englewood Cliffs, NJ: Prentice-Hall.

Meinhardt, J., & Pekrun, R. (2003). Attentional resource allocation to emotional events: An ERP study. *Cognition and Emotion, 17,* 477-500.

Meyer, D. K. & Turner, J. C. (2007). Scaffolding Emotions in Classrooms. In P. A. Schutz & R. Pekrun (Eds.), *Emotion in education* (pp. 235-249). San Diego: Academic Press.

Patrick, B. C., Skinner, E. A., & Connell, J. P. (1993). What motivates children's behavior and emotion? Joint effects of perceived control and autonomy in the academic domain. *Journal of Personality and Social Psychology, 65,* 781-791.

Pekrun, R. (1984). An expectancy-value model of anxiety. In H. M. van der Ploeg, R. Schwarzer & C. D. Spielberger (Eds.), *Advances in test anxiety research* (Vol. 3, pp. 53-72). Lisse, The Netherlands: Swets & Zeitlinger.

Pekrun, R. (1988). *Emotion, Motivation und Persönlichkeit* (Emotion, motivation and personality). Munich/Weinheim: Psychologie Verlags Union.

Pekrun, R. (1992a). The expectancy-value theory of anxiety: Overview and implications. In D. G. Forgays, T. Sosnowski, & K. Wrzesniewski (Eds.), *Anxiety: Recent developments in self-appraisal, psychophysiological and health research* (pp. 23-41). Washington, DC: Hemisphere.

Pekrun, R. (1992b). The impact of emotions on learning and achievement: Towards a theory of cognitive/motivational mediators. *Applied Psychology: An International Review, 41,* 359-376.

Pekrun, R. (1992c). Kognition und Emotion in studienbezogenen Lern- und Leistungssituationen: Explorative Analysen (Achievement-related cognition and emotion in higher education: An exploratory analysis). *Unterrichtswissenschaft, 20,* 308-324.

Pekrun, R. (2000). A social cognitive, control-value theory of achievement emotions. In J. Heckhausen (Ed.), *Motivational psychology of human development*. Oxford, UK: Elsevier Science.

Pekrun, R. (in press a). The control-value theory of achievement emotions: Assumptions, corollaries, and implications for educational research and practice. *Educational Psychology Review*.

Pekrun, R. (in press b). Emotions in students' scholastic development. In R. Perry & J. Smart (Eds.), *The scholarship of teaching and learning in higher education: An evidence-based perspective*. New York: Springer.

Pekrun, R., Barrera, A., Goetz, T., & Maier, M. (2003, April). *Control-value theory of academic emotions: Implications for the motivational determinants of students' emotions in the domain of mathematics and statistics*. Paper presented at the annual meeting of the American Educational Research Association, Chicago, IL.

Pekrun, R., Elliot, A. J., & Maier, M. A. (2006). Achievement goals and discrete achievement emotions: A theoretical model and prospective test. *Journal of Educational Psychology 98,* 583–597.

Pekrun, R., Goetz, T., & Perry, R. P. (2005). *Achievement Emotions Questionnaire (AEQ). User's manual*. Department of Psychology, University of Munich, Munich, Germany.

Pekrun, R., Goetz, T., Perry, R. P., Kramer, K., & Hochstadt, M. (2004). Beyond test anxiety: Development and validation of the Test Emotions Questionnaire (TEQ). *Anxiety, Stress and Coping, 17,* 287-316.

Pekrun, R., Goetz, T., Titz, W., & Perry, R. P. (2002a). Academic emotions in students' self-regulated learning and achievement: A program of quantitative and qualitative research. *Educational Psychologist, 37,* 91-106.

Pekrun, R., Goetz, T., Titz, W, & Perry, R. P. (2002b). Positive emotions in education. In E. Frydenberg (Ed.), *Beyond coping: Meeting goals, visions, and challenges* (pp. 149-174). Oxford, UK: Elsevier.

Pekrun, R., & Hofmann, H. (1996, April). *Affective and motivational processes: Contrasting interindividual and intraindividual perspectives*. Paper presented at the annual meeting of the American Educational Research Association, New York.

Pekrun, R., & Schutz, P. A. (2007). Where do we go from here? Implications and future directions for inquiry on emotions in education. In P. A. Schutz & R. Pekrun (Eds.), *Emotion in education* (pp. 303-321). San Diego: Academic Press.

Perry, R. P. (1991). Perceived control in college students: Implications for instruction in higher education. In J. Smart (Ed.), *Higher education: Handbook of theory and research* (Vol. 7, pp. 1-56). New York: Agathon.

Perry, R. P. (2003). Perceived (academic) control and causal thinking in achievement settings. *Canadian Psychologist, 44,* 312-331.

Perry, R. P., & Penner, K. S. (1990). Enhancing academic achievement in college students through attributional retraining and instruction. *Journal of Educational Psychology, 82,* 262-271.

Perry, R. P., Hall, N. C., & Ruthig, J. C. (2005). Perceived (academic) control and scholastic attainment in higher education. In J. Smart (Ed.), *Higher education: Handbook of theory and research* (Vol. 20, pp. 363-436). New York: Springer.

Reisenzein, R. (2001). Appraisal processes conceptualized from a schema-theoretic perspective. In K. R. Scherer, A. Schorr, & T. Johnstone, T. (Eds.), *Appraisal processes in emotion* (pp. 187-201). Oxford, UK: Oxford University Press.

Ruthig, J. C., Perry, R. P., Hall, N. C., & Hladkyj, S. (2004). Optimism and attributional retraining: Longitudinal effects on academic achievement, test anxiety, and voluntary course withdrawal in college students. *Journal of Applied Social Psychology, 34,* 709-730.

Scherer, K. R., Schorr, A., & Johnstone, T. (Eds.). (2001). *Appraisal processes in emotion*. Oxford, UK: Oxford University Press.

Skinner, E. A. (1996). A guide to constructs of control. *Journal of Personality and Social Psychology, 71,* 549-570.

Sutton, R. E. (2007). Teachers' anger, frustration, and self-regulation. In P. A. Schutz & R. Pekrun (Eds), *Emotion in education* (pp. 251-266). San Diego: Academic Press.

Titz, W. (2001). *Emotionen von Studierenden in Lernsituationen* [Students' emotions at learning]. Münster: Waxmann.

Turner, J. E., Schallert, D. L. (2001). Expectancy-value relationships of shame reactions and shame resiliency. *Journal of Educational Psychology, 93,* 320-329.

Turner, J. E. & Waugh, R. M. (2007). A dynamical systems perspective regarding students' learning processes: Shame reactions and emergent self-organizations. In P. A. Schutz and R. Pekrun, (Eds.), *Emotion in education*. San Diego: Academic Press.

Weiner, B. (1985). An attributional theory of achievement motivation and emotion. *Psychological Review, 92,* 548-573.

Weiner, B. (2007). Examining emotional diversity in the classroom: An attribution theorist considers the moral emotions. In P. A. Schutz & R. Pekrun (Eds.), *Emotion in education* (pp. 55-84). San Diego: Academic Press.

Zeidner, M. (1998). *Test anxiety. The state of the art*. New York: Plenum.

Zeidner, M. (2007). (2007). Test anxiety in educational contexts: Concepts, findings, future directions. In P. A. Schutz & R. Pekrun (Eds.), *Emotion in education* (pp. 159-177). San Diego: Elsevier Inc.

CHAPTER

3

Understanding Students' Affective Processes in the Classroom

MONIQUE BOEKAERTS
Leiden University

Affect is a generic term that refers to both feelings, experiences of valence, affective states, emotions, and moods (Carver, 2003; Forgas, 1992; Russell & Carroll, 1999). To date, an extensive body of literature documents that individuals' emotions arising from an experience influence the content of their cognitions, motivations, and actions, as well as the way they act and process information. The article is organized into three sections. First, I present my own theory of self-regulation in which emotions and appraisals play a prominent role. Second, I summarize results from classroom studies. In the third section, I summarize findings from our recent studies examining the mediating effect of positive and negative emotions. These studies were designed within the framework of control-value theory. Finally, I will suggest some directions for future research on the effect of emotions in the classroom.

THE DUAL PROCESSING SELF-REGULATION MODEL

Over the years I have tried to tailor the concept of self-regulation to the appraisals and emotions that are elicited during the learning process. It was necessary to considerably refine the appraisal and emotion constructs over the years to accommodate the many self-regulation strategies that students need to use in the classroom. I will describe the current version of the dual processing self-regulation model (previously referred to as the Model of

Adaptable Learning) to highlight what makes appraisals and emotions key constructs in any self-regulation theory.

Adapting the Model to Fit New Theoretical Insights

In previous versions of the model (e.g., Boekaerts, 1996; Boekaerts & Niemivirta, 2000) a distinction was made between two parallel self-regulatory pathways, the mastery or growth pathway, and the well-being pathway (see Figure 1). It was assumed that students have two priorities in the classroom, namely (1) increasing their assets by improving their competence, and (2) keeping their well-being within reasonable bounds. Students who appraise a learning activity favorably—meaning that positive cognitions and emotions are dominant—start activity in the mastery or growth pathway. The assumption is that the former students perceive the task as congruent with their personal goals, values, and needs, and therefore, the learning activity is energized from the top down. By contrast, students who perceive environmental or internal cues during the learning process that signal a mismatch between the learning activity and their personal goals, needs and interest, experience negative cognitions and emotions, which prompt them to switch to the well-being route.

It was necessary to elaborate the model with a pathway that connects the two original processing routes to explain why some students do and others do not make the switch to the well-being route and back again to the mastery track during the learning process. As can be seen in Figure 1 the two pathways are connected with dotted lines. It is assumed that the level of arousal increases when students perceive incongruity of goals or obstacles on their way to the goal during the learning process. This acts as a warning signal that something is wrong and that they need to appraise the situation in terms of potential threat, harm, or loss to well-being (for further discussion, see Boekaerts, 2002a).

Students interpret the increased level of arousal in different ways. Students who interpret it as benign and those who perceive a potential threat to well-being but feel that they have sufficient resources to deal with the potential threat will continue on the mastery track. They might not even be aware that they have experienced an emotion. However, those students who label the increased level of arousal negatively (e.g., I feel irritated, tense, disappointed, worried) will be aware that they have experienced a negative emotion, and they may interpret it as a potential harm, loss, or threat to well-being. Lazarus (1991) argued that negative emotions automatically call forth a secondary appraisal process that concerns coping options; the individual judges whether (s)he has access to sufficient resources to prevent harm and loss, or reduce it without producing additional harm.

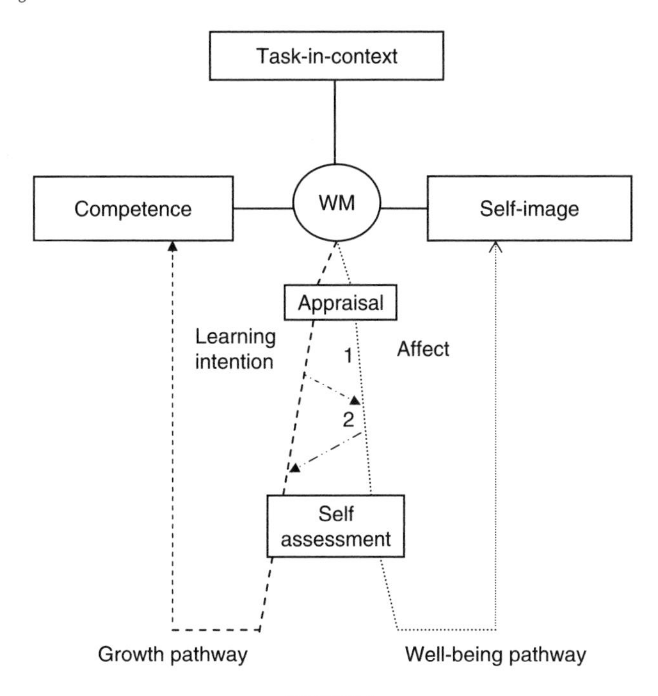

FIGURE 1
Dual processing self-regulation model (Boekaerts, 2006)

Volitional Strategies Act as a Switching Track Between the Two Pathways

There is accumulating evidence that students' efforts during the learning process to stay focused on the learning task—despite perception of incongruity and obstacles and the concomitant increase in the level of arousal—are located in the self-regulation system at the level of volitional strategies (e.g., Boekaerts, 2006; Boekaerts & Corno, 2005; Corno, 2004; Wolters & Rosenthal, 2000). These researchers argued that students who have easy access to volitional strategies use these strategies appropriately to stay focused on the task. Corno (2004) described volitional strategies as when-where plans or good work habits. Examples of such strategies are clearing away distracting objects before starting work, telling oneself to stop thinking about negative outcomes, looking for signals that show one's strength instead of negative signals, and using a stop-and-think strategy when one meets obstacles.

Please note that students who experience negative emotions during the learning process will not switch automatically to the well-being pathway. Boekaerts and Corno (2005) argued that students will stay on the mastery pathway when learning goals are dominant *and* they have access to sufficient resources to deal with perceived obstacles. However, chances are high that students will redirect their attention when nonlearning goals gain dominance in their goal network (e.g., belongingness, tranquility, or entertainment goals). At such a point, cue-driven, bottom-up processing takes over, and students begin to explore the friction they feel.

Essentially, students who can monitor their progress to the goal and who have access to volitional strategies also know how to handle obstacles and frustration en route to the goal in a particular domain of study (i.e., they have acquired the necessary scripts). These students realize that some problems are controllable and others are not. By contrast, students who cannot make an adequate mental representation of a learning situation that they find complex, unfamiliar, difficult, or ambiguous, will define many learning situations in that domain as "stressful."

Emotion-Focused Coping vs. Problem-Focused Coping

Stress researchers (e.g., Compas, 1987; Lazarus, 1991) explained that coping with stressful situations is a form of adaptation that takes place when individuals are confronted with situations that are taxing or exceeding their resources. They view coping as an inherent aspect of a student's functioning, explaining that it is not associated with successful strategies as such, but includes all purposeful efforts to manage the stressful situation.

Coping strategies have been categorized into problem-focused strategies, which refer to attempts to alter the stressor and are, as such, akin to the volition strategies described previously; and emotion-focused strategies, which refer to attempts to regulate negative emotional reactions to the stressor. Emotion-focused coping strategies, or ego-defensive coping, refer to self-protective strategies, such as ignoring salient cues, handicapping oneself, using avoidance behavior, entering a state of denial, becoming aggressive, crying, and using cognitive and behavioral distraction (for further discussion see Boekaerts, 1993; Boekaerts & Röder, 1999; Elliot & Church, 2003).

Boekaerts (2002a) compared the coping strategies that children (10 to 12 years old) and adolescents (14 to 15 years old) use to cope with stressors. Both groups used more problem-focused strategies (volitional strategies) in response to academic stressors than to interpersonal stressors. In general, primary school students did not differ from secondary school students in the type of problem-focused coping strategies they used. Compas, Malcarne, and Fondacaro (1988) reported an increase in the use of emotion-focused coping strategies between 12 and 14 years of age, with girls using more emotion-focused coping in response to academic stressors than boys. Interestingly, these authors found that the intensity of reported stress (i.e., reported negative emotions) was low when there was a match between the students' *perception of control in the situation* and the selected coping strategy. Students who used problem-focused coping (volitional strategies) when they had perceived control over the stressor and those who reported emotion-focused coping when they had low perceived control, reported low levels of negative emotions. The reverse was true for a mismatch. The lesson we might learn from these findings is that students, particularly girls, need to learn (1) how to pick up cues that inform them whether a learning task is controllable, and (2) how to select the most appropriate self-regulation strategy for the current learning conditions.

Over the years some students have encoded highly effective action programs and scripts to deal with obstacles in a school subject, including meta-cognitive strategies (e.g., trying to understand why the problem-solving process went wrong, (re)designing a plan of action, looking, for bugs in the current strategy), social scripts (e.g., asking questions, listening more intensively, role taking, anticipating the consequences of personal and social acts), and volitional strategies (e.g., restructuring the environment to avoid distraction, disengaging from the task with the intention to come back to it later). Other students may have learned to react with emotion-focused coping strategies when they encounter obstacles during problem solving that they find taxing or exceeding their resources (e.g., giving up, cheating, taking a deep breath, counting to seven, swearing, soliciting emotional support).

It is highly likely that the students' preferential strategies to deal with obstacles in a domain are activated quasi-automatically when they are working on familiar tasks and assignments, either in the classroom or at home (homework). Clearly, students who do not have easy access to the needed meta-cognitive, volitional, or coping strategies might experience more negative and fewer positive emotions than students who have easy access. Boekaerts and Niemivirta (2000) argued that lack of access to appropriate strategies affects students' perception of the task as well as their appraisal and assessment of progress.

Affect Regulation: Acquiring the Skill to Deal with Obstacles

Using emotion-focused coping strategies to control one's emotions and volitional strategies to redirect attention to the learning activity comes at a cost. Chances are high that students who perceive a learning situation as "uncontrollable" will focus on their emotions and will want to come to terms with their emotions *before* they continue with the learning task. In other words, they use emotion-focused coping strategies to reduce, relabel, or control their emotions. Graham, Hudley, and Williams (1992) pointed out that some students have the inner strength to tolerate negative emotions and hide them while they are trying to make a mental representation of an ambiguous situation and design a plan of action. Other students, particularly those who want to attain a performance goal (e.g., I want to be better than my peers on this task), may not be prepared to tolerate negative affect associated with the difficult tasks. In addition, they may not be prepared to pay the mental (e.g., effort) and social costs of admitting difficulty (e.g., embarrassment; see Butler, 1992).

Social psychologists Baumeister and Eppes (2005) showed that students who had to focus on their emotions and make an attempt to control them tended to give up faster on a successive task than those who were not requested to control their emotions. The coping strategies that were needed to control emotions interfered with task engagement and persistence, unless the spent energy was restored. In another experiment Baumeister and Eppes (2005) told students to resist eating chocolates during the experiment. The students who had to fight temptation tended to give up faster on solving a difficult puzzle compared to the students who either had not been tempted with chocolates or had been allowed to eat the chocolates. Hence it seems that the effort needed to resist temptation interacted with the effort needed to persist on the task. Therefore, students evidently needed a clear sense of purpose to use volitional strategies to redirect their attention to the learning task after experiencing negative affect. In the third section of this chapter, I will refer to some of our recent studies in which we explored the connections between having a sense of purpose in relation to a domain of study and two important outcome variables, namely reported effort after a homework session and self-assessment. Before I detail the results of these studies, I will highlight some of the findings of classroom studies on the effect of emotions on learning and achievement. These results will act as a backdrop for interpreting our own results.

STUDYING AFFECT IN THE CLASSROOM

Several researchers have shown that young children possess an insatiable need to acquire new knowledge and skills. Harter (1986) explained that increments in learning come as a product of student engagement in tasks that they enjoy. Deci and Ryan (1985) and later publications by the same researchers (e.g., Ryan & Deci, 2000) documented that children and adolescents who are attracted to a learning activity for the sake of experiencing the positive feelings of progress and competence that it generates are intrinsically motivated. Other researchers referred to flow (Csikzentmihalyi, 1990) and individual interest (Krapp, 2003) to describe the outcome of engagement characterized by experiencing positive emotions related to the activity itself.

Academic Emotions Are Part of Everyday School Life

Pekrun (2000) presented a social-cognitive perspective on academic emotions, assuming that control and value-related appraisals are key antecedents of students' emotional experiences in the classroom. Arguing that academic emotions are part of everyday school life, Pekrun showed that some of these emotions are directly linked to learning tasks and achievement goals. He described six main achievement-related emotions; hope and joy are referred to as positive academic emotions; and anxiety, anger, boredom, and hopelessness are among the negative academic emotions. These emotions can be measured reliably in a classroom context. In line with research from self-efficacy theory, Pekrun reported high correlations between self-efficacy in mathematics and joy (.62), anger (−.53), anxiety (−.63), hopelessness (−.67), and boredom (−.42). These emotions were also associated with values of achievement in mathematics but the associations were much weaker than those with self-efficacy.

Emotions Are Not Independent of the Domain and the Classroom Context

Pekrun and his colleagues further argued that students' academic histories shape their academic emotions and appraisals and that these processes are not independent of the domain of study and the classroom context. Goetz, Pekrun, Hall, and Haag (2006) showed that emotional experiences are organized in a domain-specific way, in line with findings reported in the literature that other psychosocial constructs are domain specific (self-concept of ability, self-efficacy, causal attributions, value judgments, and goal orientations). These researchers further reported that emotions differ in their degree of domain-specificity, with enjoyment being the most domain-specific, followed by boredom and anxiety. Granted, many researchers have argued that for interest and motivation to develop in relation to a school subject, it is important that students experience positive affect related to the activity itself (mastery goals, Dweck, 1986; Pintrich, 2000). But Pekrun's recent findings shed new light on the development of positive affect in relation to the mathematics domain (Pekrun, 2006).

Pekrun, Frenzel, Goetz, and Perry (2006) examined the effect of context on the experience of academic emotions. They predicted that the gender composition and the average achievement level of the classroom influence students' experiences of success and failure, appraisals, and emotions. They found that girls experience a more negative emotional development than boys, and that in classrooms with a high percentage of boys, a less positive emotional development is observed. The debilitating female mathematics profile had previously been reported by Seegers and Boekaerts (1993) and Vermeer, Boekaerts, and Seegers (2000). These researchers showed that at the primary school level, girls who possessed excellent mathematics problem-solving strategies (as evidenced by their performance on the task) expressed less confidence and positive emotions than boys in the initial stages of mathematics problem solving. Vermeer et al. (2000) suggested that under similar learning conditions, boys' chances to experience enjoyment are higher, because mathematics tasks elicit a higher initial feeling of self-efficacy in male than in female students.

Goetz, Pekrun, Hall, and Haag's (2006) recent results show that the classroom achievement level affects the development of enjoyment and pride in mathematics unfavorably, whereas the students' individual achievement level has the opposite effect. The reverse pattern was noted for hopelessness, shame, and anxiety. What these findings suggest is that students are painfully aware of the social context in which they acquire mathematics skills. Seeing other students shine in mathematics decreases positive and increases negative feelings, whereas watching them suffer increases positive and decreases negative affect.

GOAL PURSUIT, PERCEPTION OF OBSTACLES, AND GOAL FRUSTRATION

To date, social psychologists assume that students' *current goals* are crucially important to understand their emotions. Frijda (2000) and Lazarus (1991) maintained that emotional experiences change from moment to moment, as the situation unfolds in line with the person's short-term and long-term goals. Accordingly, I assumed that students' emotions (positive and negative) experienced in the classroom depend largely on whether they judge their current goals to be congruent or incongruent with the learning activities that somebody else (i.e., the teacher, parents, or peers) wants them to pursue. In turn, the emotions serve as inputs to their goal system.

It is important in this respect to understand that students bring many different goals to the classroom, and not only achievement goals. Following Ford (1992) and Wentzel (1998), Boekaerts, De Koning, and Vedder (2006) clarified that all students pursue person goals alongside person-environment goals, and that their multiple goals may be in harmony or in conflict. For example, a specific student may want to master a new skill, may want to be liked by her classmates, may want to provide and receive social support from peers, and may simultaneously pursue an avoidance goal, such as "I do not want to lose face in front of my friends." To achieve their multiple goals, students must have access to different action programs and scripts, and they need to be able to coordinate their actions.

Boekaerts, De Koning, and Vedder (2006) argued that students have established multiple connections between the goals in their goal network to self-regulate the learning process, and at the same time comply with the social expectations and rules that are prominent in their peer group. Goal frustration and its concomitant negative emotions may arise when students discover that it is impossible to achieve all their salient goals simultaneously. For example, irritation may be triggered when a student needs to make a choice between academic goals (e.g., doing one's homework) and belongingness goals (e.g., consoling a friend who has just failed a test). Students try to combine academic and social goals and discover that it is often difficult to align these goals.

Although many studies have been set up to study the achievement and social goals that students pursue in the classroom, little is known about the conflicts that may arise when students strive to attain multiple goals *simultaneously*, such as achievement and tranquility goals, or belongingness and mastery goals. Ford (1992) showed that lower-order goals that have an isolated position in a person's goal hierarchy have less of a chance to be attained than goals that are well integrated into the goal network. Multiple connections between a lower-order goal (e.g., I want to finish my French talk before the end of the week) and several higher-order goals (e.g., I want to please my parents, I want to be successful in school, I want to show my classmates that I am the best) ensure that good intentions will be realized because the individual has more than one reason for preparing his talk.

Environmental conditions influence goal-directed behavior directly (Austin & Vancouver, 1996). Perception of environmental pressure (e.g., my parents force me to do my homework; my peers want to complete this task but I want to go home early) and perceived obstacles en route to the goal (e.g., this task is too difficult: I do not know how to tackle it; there is nobody to turn to for help) may turn a desired end-state into an undesired one (e.g., I do not want to be controlled by my parents; I do not want to do such a difficult task). Several researchers (e.g., Carver & Scheier, 2000; Kuhl, 2000) described the mechanism through which goal frustration and perception of unfavorable environmental cues trigger emotions.

Mild Negative Emotions in the Presence of Positive Emotions Might Be Beneficial for Learning

Students recollect events where they did poorly or events where they lost face in front of their fellow students. This activated information reminds them that they might (again) fall short of (their own) standards and lose face. Such a recollection might produce uneasy feelings in most students, reported as negative affect. However, it is important to note that the same student's reaction to a particular configuration of stimuli may be dissimilar on different occasions, depending on his or her current goals. Students who pursue recognition and reward goals might give up more easily than students who pursue mastery goals when negative emotions are elicited during the learning process. As Pintrich (2000) explained, a performance goal creates a mindset that prompts students to interpret negative emotions as signals that the task is too complex to look smart in front of class. By contrast, students who want to master a difficult task because they value its content have created a mindset that helps them to interpret negative emotions as signals that more effort needs to be invested to achieve a learning goal or that social support needs to be solicited.

Although the latter type of students might also experience negative affect when they face obstacles during the learning process, the presence of positive cognitions and feelings in relation to the learning task might mitigate the effect of the negative emotions. For example, a student might feel tension and anxiety rise when she is provoked by her group mates to present the group's solution. At the same time she may anticipate excitement and pride because she recollects the pleased look on the teacher's face when she presented her previous solution. This favorable recollection elicits positive affect, which acts as a signal that she is capable of doing the task well and that the chances of negative outcomes are minimal. One should note that a student who tries to attain multiple goals simultaneously (e.g., wanting to understand a text and not wanting to look dumb) has to reconcile two conflicting goals.

Following the argument put forward by Fredrickson and Losada (2005), I would like to suggest that a good ratio of positive to negative emotions is beneficial for learning in the classroom. Admittedly, enjoyment of a learning activity for its own sake is the optimum learning experience in the classroom because it boosts self-efficacy judgments and reduces ego-protective behaviors. Unfortunately, not all learning activities elicit enjoyment, and students might sometimes feel entrapped in learning activities by the teacher or their peers. Under such sub-optimal learning conditions, mild negative emotions in the presence of positivity might reduce the chances that students show relaxation, which is associated with overestimation and coasting behavior.

EXPLORING THE EFFECT OF POSITIVE AND NEGATIVE AFFECT ON SELF-ASSESSMENT AND EFFORT

Based on the literature and our own findings on mathematics learning (Boekaerts, 2002b), we assumed that students who value mathematics learning and those who feel competent in that domain have a clear sense of purpose and differ from students who lack this sense of purpose on many grounds. Evidently, students have different learning histories in a domain of study; they have accumulated positive and negative experiences in relation to that domain, and this information has been integrated in an extensive network of mental representations involving the self. When triggered, these mental representations create a mind-set that directs attention towards specific cues in the learning environment. Students, who have a positive mind-set at the time of learning will observe mainly favorable cues in the learning environment, whereas students who have established a negative mind-set will pay attention to unfavorable cues in the environment. In the next sections we will look more closely at the effect of two appraisals, namely competence and value judgments, on the students' self-assessment and reported effort. The role that positive and negative emotions play during the learning process will be explored as well.

Study 1: Formulating Hypotheses About the Effect of Positive and Negative Emotions

As discussed previously, Goetz, Pekrun, Hall, and Haag (2006) investigated the direct links between students' control and value judgments and their academic emotions. In one of my recent studies, I went one step further and examined the tenability of an affect mediation effect, students' competence and value judgments in mathematics on the one hand and two important aspects of their task-related cognitions on the other (self-assessment and effort reported after a homework task had been completed). Two rival hypotheses were identified: (1) affect serves as input to the feedback system (Carver & Scheier, 2000), and (2) affect is a resource (Fredrickson, 2001).

Carver and Scheier (2000) conceptualized *affect as input to specific feedback systems*. In their view, affect predicts that individuals who feel competent that they can do a task experience positive affect while doing that task, but tend to overestimate their performance and reduce their effort. By contrast, individuals who perceive the task as exceeding their capacity experience negative affect while doing the task; they are likely to underestimate their performance and report increased effort. Models of affect that view *affect as a resource* (e.g., Aspinwell & Taylor, 1997; Fredrickson, 2001) predict that individuals who experience positive affect during goal pursuit interpret experienced positive emotions as a resource and tend to invest more effort in growth goals (e.g., extending their knowledge and skills). By contrast individuals who experience negative affect during goal pursuit will interpret that as a threat signal and redirect their attention and effort to explore the nature of the threat to deal with it (volition and coping strategies). In other words, Carver and Scheier's theory predicts that positive emotions reduce effort, whereas resource theory predicts the opposite tendency. Likewise, Carver and Scheier's theory predicts that negative emotions increase effort, whereas resource theory predicts that they decrease effort.

We hypothesized that during mathematics homework sessions, students' initial sense of competence has a direct positive effect on their self-assessment, but that this effect is also mediated by positive and negative emotions. More specifically, students who feel competent before they initiate their mathematics homework will experience many positive and few negative emotions. Presence of positive emotions and relative absence of negative emotions might increase their self-assessment in that context and might also facilitate effort investment. In other words, it is hypothesized that experiencing positive affect opens an extra window, allowing students to take advantage of learning opportunities, whereas the presence of negative affect temporarily closes that window, redirecting attention to the regulation of affect.

TESTING OUR HYPOTHESES

Students were in their second year of junior high school when the study began and in their third year at the end of the study. The sample that pertains to the mathematics domain consisted of 357 students who attended one of the regular tracks in junior high school. They were asked to keep a diary of their mathematics homework during three two-week periods. They were also requested to complete the online motivation questionnaire (Boekaerts, 2002b) before beginning their mathematics homework assignments and after they had finished them. The diary periods were selected together with the participating teachers so that they did not coincide with regular exams or holidays. Each period was separated from the previous or successive period by at least three months. In each two-week diary period (January, May, and November), students provided information about at least two mathematics homework tasks.

The study that is referred to here focused on the students' appraisals, emotions, self-assessment of homework, and reported effort registered in relation to two mathematics homework sessions at the first data collection point (January). The students' competence judgment at the second data collection point (May) served as an outcome variable. Students completed the pre-task version of the online motivation questionnaire when they were about to begin their mathematics homework tasks, indicating among other things their competence (six items) and value judgment (two items). When they had finished the mathematics homework task, they completed the post-task version of the online motivation questionnaire, reporting on the emotions they had experienced during their math homework (eight items), the effort they had invested (two items), and their assessment of the outcome of the homework tasks (two items). Eight feeling states are included in the online motivation questionnaire, namely joy; contentment; feeling at ease and feeling secure (measuring positive emotions); and feeling worried, tense, irritated, and displeased (measuring negative affect). Students had to rate on four-point scales the extent to which they endorsed statements, such as "I felt tense while doing the task," or "I felt at ease while doing the task." In previous research the eight affective states had been (factor analytically reduced) to two basic affective states, namely a subjective sense of positivity and a negative affective state, each including four items. The same two-factor solution fitted the data from this study.

The overall pattern of correlations found in this study confirmed our predictions: positive affect was positively linked to competence judgment (.42), value judgment (.16), self-assessment (.36), and effort (.22); and negative affect was inversely related to the same variables, namely competence judgment (−.14), value judgment (−.05), self-assessment (−.32), and effort investment (−.15). Structural equation modeling was used to test the direct and indirect effects of value and competence appraisals—measured before starting the homework tasks—on self-assessment and reported effort measured after doing the task and on value and competence appraisals reported later in the course. We competitively compared mediated and less restricted models and found that the model depicted in Figure 2 had the best fit ($\chi^2 = 15, 82$; df = 6; χ^2/df = 2.63; GFI = .99; RMSEA = .06).

Conclusion 1: Positive affect leads to overshooting and negative affect to undershooting. The first part of Carver and Scheier's (2000) prediction was confirmed; positive affect signals that one is exceeding one's standards for satisfactory performance and leads to favorable judgments of one's performance (overshooting), whereas negative affect signals that one is falling short of one's standards and decreases self-assessment (undershooting).

Conclusion 2: Affect signals that (in)sufficient resources are available. The second part of Carver and Scheier's (2000) prediction was not confirmed. These researchers hypothesized that experienced affect provides critical information about whether additional effort is needed to achieve a

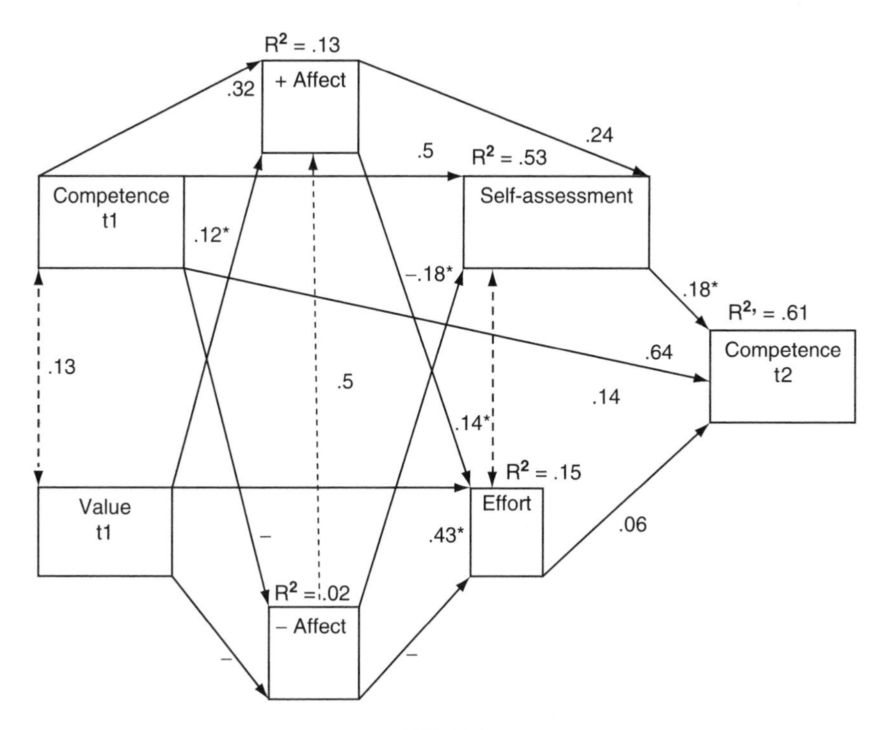

FIGURE 2
Structural equation modeling was used to test the direct and indirect effects of value and competence appraisals on self-assessment and reported effort. Affect amplified or reduced this effect.

specific goal. They predicted that negative affect motivates individuals to engage in exploration (increased effort) and that positive affect leads to coasting behavior. In fact, our results show that positive affect increases and negative affect decreases effort. This finding provides support for Fredrickson's "affect is a resource" hypothesis: students who are doing their mathematics homework in the presence of positive affect seem to interpret their affective state as a signal that they have enough resources to accept the challenge, which is reflected in effort investment, and vice versa.

Conclusion 3: Affect amplifies or moderates the effect of the appraisals. Examination of Figure 2 reveals that 53% of the variance in self-assessment could be explained. As predicted, students' *competence judgment* to do mathematics homework and their value judgment were linked directly to self-assessment, but these effects were also mediated by positive and negative affect. This implies that students who felt competent to do their math homework and those who valued doing it assessed the homework outcomes more favorably than students who did not express such judgments. However, students who experienced negative affect *during* the actual mathematics homework session tended to underestimate their performance, whereas those who reported positive affect were inclined to perceive their outcomes more positively.

Fifteen percent of the variance in reported effort could be explained and was in line with our predictions; students' competence judgment to do mathematics homework was not directly linked to their reported effort but *perceived value* was. The latter effect seems to be amplified or reduced by the students' emotions. A word of caution is in order here, because affect was measured at the same time as reported effort and self-assessment. An affect mediation effect would imply that students who valued the mathematics homework tasks were likely to experience many positive emotions during their homework sessions and few negative emotions. This may have produced a subjective sense that all was going well, which in turn resulted in favorable self-assessment and admittance of invested effort. Vice versa, students who did not value doing their math homework tended to experience few positive emotions and many negative emotions (probably because they encountered obstacles during a homework session). This pattern of emotions seemed to be unfavorable for self-assessment and reported effort.

A word of caution is in order here because information on the emotions experienced during the homework session, reported effort, and self-assessment were given at the same time (after finishing the homework session). Hence the relation between these variables could point in the other direction as well, meaning that a favorable assessment of one's performance and admittance of effort produces positive emotions, whereas unfavorable self-assessment leads to negative affect.

WHAT CAN BE LEARNED FROM THESE RESULTS?

Information about students' emotions arising from the mathematics homework experience provides a glimpse into their self-assessment and effort system. That is, students' initial competence and value judgments *interact* with the emotions they experience *during* the homework session. Emotions may amplify, moderate, reduce, or combine with the cognitive information to determine the students' assessment of task outcome and their perception of the effort they have invested.

STUDY 2: TESTING THE MODEL IN OTHER SCHOOL SUBJECTS

In the previous study, we found that students who have an initial sense of competence in mathematics experience many positive and few negative emotions during mathematics homework. We also found that the pattern of positive and negative affect influences their self-assessment of actual homework tasks and, through it, their sense of competence in mathematics several months later.

In the second study that I will report here, I examined whether the emotions experienced during homework sessions in other school subjects have a similar effect as in mathematics. Hypotheses were that students' reported effort, self-assessment, developing competence, and value judgments in the domains of history, native language learning, and foreign language acquisition are affected by the students' emotions in the same way that they are in mathematics. During two 2-week periods in January and May, 357 high school students kept a diary. They completed the online motivation questionnaire before and after homework sessions in different school subjects. Hence we obtained similar information as in Study 1 on the students' perceived competence, task value, self-assessment, and reported effort, as well as their experienced positive and negative emotions.

Structural equation modeling was used to test domain-dependence. We constructed one model for all four subjects. This model was slightly different than the one used in the first study (see Figure 3), namely the students' perceived value, measured in May (second data collection point), was added as an outcome variable to study the (in)direct effects that positive and negative emotions have in the long run on the students' sense of competence and value in a domain of study.

A two-step procedure was used to determine whether the model for the four school subjects had the same underlying structure. First, a series of comparisons was made between the base model (all parameters are fixed) and a series of alternative nested models. Seven hypotheses were formulated concerning the effect of a variable on all the variables to which it was linked (e.g., positive emotions have a domain-specific effect on effort and on self-assessment). Each hypothesis was tested by comparing the χ^2 of the alternative model with the base model, and Bonferroni corrections were applied. This procedure revealed that the models for the four school subjects had a similar underlying structure. Three subject-specific connections emerged, namely links originating in competence, positive affect, and self-assessment. Next, we determined which of the arrows leading from these three variables had produced the subject-specific effects.

This procedure revealed, first, that students' sense of competence is more stable over time for mathematics than for the other three school subjects. Second, positive emotions generated in the homework situation had a much stronger effect on the students' self-assessment of the homework outcomes in mathematics than in the language learning domains, and this effect was not significant in the history domain. Third, students' self-assessment of mathematics homework influenced their sense of competence in the long-term, whereas this effect was not significant in the other school subjects.

Inspection of the resulting model in Figure 3, which had a good fit ($\chi^2 = 123.63$; df $= 79$; RMSEA $= .039$), also reveals that the percentage of variance explained in the students' sense of competence at data collection point 2 is much higher in mathematics than for the other school subjects. Comparison of the results in the mathematics domain with those reported in Study 1 revealed that more variance was explained in reported effort in Study 2 (27% compared to 15%).

WHAT CAN BE LEARNED FROM THIS STUDY?

Again, the first part of Carver and Scheier's (2000) prediction was confirmed. Positive affect signals that one is exceeding one's standards for satisfactory performance and leads to favorable judgments of one's performance (overshooting), particularly in mathematics but also in the language domains. Experiencing negative affect signals that one is falling short of one's standards and decreases self-assessment (undershooting); but it influences self-assessment to a lesser extent than positive affect does. In line with Fredrickson's (2001) theorizing, students who are doing their homework in the presence of positive affect interpret their affective state as a signal that they have enough resources to accept the challenge leading to increased effort, and vice versa.

Similar to what was found in Study 1, positive affect opens extra windows, allowing students to take advantage of the learning opportunity. Experiencing negative emotions during homework sessions does not seem to have a direct negative effect on the learning process itself, since it does not coincide with low effort and low self-assessment. However, it may in the long run affect the

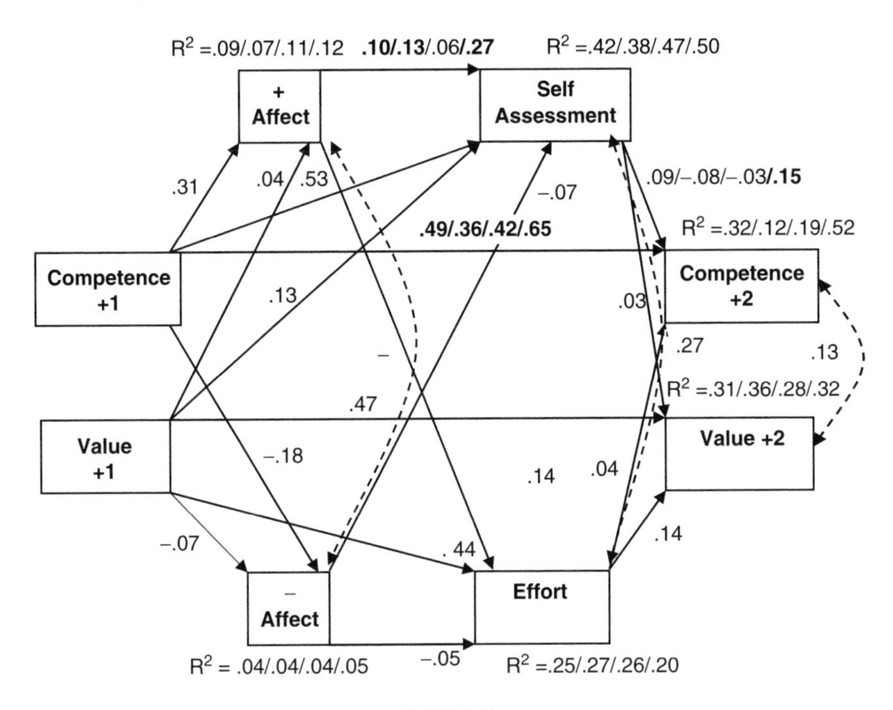

FIGURE 3

Model depicting the relations between the variables for the four school subjects, namely native language learning, foreign language learning, history, and mathematics. R^2 values are printed for all school subjects, parameter values for the paths that differ significantly.

learning process unfavorably, for systematically undershooting one's performance might affect students' sense of competence over time.

Our findings are in line with what other educational researchers found in the classroom. Pekrun, Goetz, Titz, and Perry (2002) showed that emotional experiences are organized in a domain specific way. Goetz et al., (2006) also reported that enjoyment was the most domain-specific emotion, followed by boredom and anxiety. In accordance with their findings, we suggest that the ratio of positive to negative affect that students experience in relation to homework tasks is subject-specific; whether students experience predominantly positive or negative affect in relation to homework depends on the competence and value judgments that they bring to bear on the homework situation.

We consider the students' sense of competence and value expressed before they start on their homework as good indicators of their initial sensitivity to doing homework in a domain of study (see also Fredrickson and Losada, 2005). Students who initially have positive self-images (competence and value) bring information to mind that creates a positive mind-set, which encourages them to explore the task, confirming or falsifying their initial expectations. In other words, students' sensitivity depends on initial conditions and explains why some learning opportunities trigger predominantly positive or negative emotions and are considered as challenging and fun, whereas others are passed by. More classroom research is needed in this important area of investigation. Our studies point to the necessity to validate social-psychological theories in field-based classroom research.

References

Aspinwall, L. G., & Taylor, S. E. (1997). A stitch in time: Self-regulation and proactive coping. *Psychological Bulletin, 121*, 417-436.

Austin, J. T., & Vancouver, J. B. (1996). Goal constructs in psychology: Structure, process, and content. *Psychological Bulletin, 120*, 338-375.

Baumeister, R. F. & Eppes, F. (2005). *The cultural animal, human nature, meaning, and social life*. Florida: Oxford University Press.

Boekaerts, M. (1993). Being concerned with well-being and with learning. *Educational Psychologist, 28*(2), 149-167.

Boekaerts, M. (1996). Personality and the psychology of learning. *European Journal of Personality, 10*(5), 377-404.

Boekaerts, M. (2002a). Meeting challenges in a classroom context. In E. Frydenberg (Ed.), *Beyond coping: Meeting goals, visions and challenges*. (pp. 129-147). Oxford: Oxford University Press.

Boekaerts, M. (2002b). The on-line motivation questionnaire: A self-report instrument to assess students' context sensitivity. In P. R. Pintrich & M. L. Maehr (Eds.), *Advances in motivation and achievement, Vol.12: New directions in measures and methods* (pp. 77-120). New York: JAI.

Boekaerts, M. (2006). Self-regulation with a focus on the self-regulation of motivation and effort. In W. Damon & R. Lerner (Eds.) *Handbook of child psychology, Vol 4, Child psychology in practice* (6th ed.) (pp. 345-377). New York: Wiley.

Boekaerts, M., & Corno, L. (2005). Self-regulated learning in the classroom. *Applied Psychology: An international Review, 54*(2), 199-231.

Boekaerts, M., De Koning, E., & Vedder, P. (2006). Goal-directed behavior and contextual factors in the classroom: An innovative approach to the study of multiple goals. *Educational Psychologist, 41*(1), 33-51.

Boekaerts, M., & Niemivirta, M. (2000). Self-regulated learning: Finding a balance between learning goals and ego-protective goals. In M. Boekaerts, P.R. Pintirch, & M. Zeidner (Eds.), *Handbook of self-regulation* (pp. 417-450). San Diego: Academic Press.

Boekaerts, M., & Röder, I. (1999). Stress, coping, and adjustment in children with a chronic disease: A review of the literature. *Disability and Rehabilitation, 21*(7), 311-337.

Butler, R. (1992). What young people want to know when: Effects of mastery and ability goals on interest in different kinds of social comparison. *Journal of Personality and Social Psychology 62*, 934-943.

Carver, C. S., & Scheier, M. F. (2000). On the structure of behavioral self-regulation. In M. Boekaerts, P. R. Pintrich, & M. Zeidner (Eds.), *Handbook of self-regulation* (pp. 687-726). San Diego: Academic Press.

Carver, C. S. (2003). Pleasure as a sign you can attend to something else: Placing positive feelings within a general model of affect. *Cognition and Emotion, 17*(2), 241-261.

Compas, B. E. (1987). Stress and life events during childhood and adolescence. *Clinical Psychology Review, 7*, 275-302.

Compas, B. E., Malcarne, V. L., & Fondacaro, K. (1988). Coping with stressful events in older children and adolescents. Journal of Consulting and Clinical Psychology, *56*(3), 405-411.
Corno, L. (2004). Work habits and work styles: The psychology of volition in education. *Teacher College Record, 106*, 1669-1694.
Csikzentmihalyi, M. (1990). *Flow: the psychology of optimal experience*. New York: Harper and Row.
Deci, E. L., & Ryan, R. M. (1985). *Intrinsic motivation and self-determination in human behavior*. New York: Plenum Press.
Dweck, C. S. (1986). Motivational processes affecting learning. *American Psychologist, 41*(10), 1040–1048.
Elliot, A. J., & Church, M. A. (2003). A Motivational analysis of defensive pessimism and self-handicapping. *Journal of Personality, 71*(3), 369-396.
Fazio, R. H., Eiser, J. R., & Shook, N. J. (2004). Attitude formation through exploration: Valence asymmetries. *Journal of Personality and Social Psychology, 87*(3), 293-311.
Ford, M. E. (1992). *Motivating humans: Goals, emotions, and personal agency beliefs*. London: Sage.
Forgas, J. P. (1992). Affect, appraisal and action: Towards a multi-process framework. In R. S. Wyer & T. K. Srull (Eds.) *Advances in social cognition*. Hillsdale, NJ: Erlbaum.
Fredrickson, B. L. (2001). The role of positive emotions in positive psychology. *American Psychologist, 56*(3), 218-226.
Fredrickson, B. L., & Losada, M. F. (2005). Positive affect and the complex dynamics of human flourishing. *American Psychologist, 60*(7), 678-686.
Frijda, N. H. (2000). The psychologists' point of view. In M. Lewis & J. M. Haviland-Jones, (Eds.) *Handbook of emotions* (pp. 59-74). New York: The Guilford Press.
Goetz, T,., Pekrun, R., Hall, N., & Haag, L. (2006). Academic emotions from a socio-cognitive perspective: Antecedents and domain-specificity of students' affect in the context of Latin instruction, *British Journal of Educational Psychology, 76*(2), 289-308.
Graham, S., Hudley, C., & Williams, E. (1992). Attributional and emotional determinants of aggression among African-American and Latino young adolescents. *Developmental Psychology, 28*(4), 731-740.
Harter, S. (1986). Cognitive-developmental processes in the integration of concepts about emotions and the self. *Social Cognition, 4*, 119-151.
Krapp, A. (2003). Interest and human development: An educational-psychological perspective. *British Journal of Educational Psychology. Monograph Series II(2) Development and Motivation: Joint Perspectives*, 57-84.
Kuhl, J. (2000). A functional design approach to motivation and self-regulation: The dynamics of personality systems and interactions. In M. Boekaerts, P. R. Pintrich & M. Zeinder (Eds.), *Handbook of self-regulation* (pp.111-163). San Diego: Academic Press.
Lazarus, R. S. (1991). *Emotion and adaptation*. New York: Oxford University Press.
Pekrun, R. (2000). A social-cognitive, control-value theory of achievement. In J. Heckhausen (Ed.), *Motivational psychology of human development: Developing motivation and motivating development* (pp. 143-163). New York: Elsevier.
Pekrun, R., Goetz, T., Titz, W., & Perry, R. P. (2002). Academic emotions in students' self-regulated learning and achievement: A program of qualitative and quantitative research. *Educational Psychologist, 37*(2), 91-105.
Pekrun, R., Frenzel, A., Goetz, T., & Perry, R.P. (2007). Control value theory of achievement emotions: An integrative approach to emotions in education. In P. Schutz & R. Pekrun (Eds.), *Emotion in education* (pp. 9–32). San Diego: Elsevier Inc.
Pekrun, R. (2006). Emotions in Students' Learning and Achievement. Keynote presented at the 26th International Congress of Applied Psychology, Athens, Greece, July 2006.
Pintrich, P. R. (2000). The role of goal-orientation in self-regulated learning. In M. Boekaerts, P. R. Pintrich & M. Zeidner (Eds.), *Handbook of self-regulation* (pp. 452-502). San Diego: Academic Press.

Russell, J. A., & Carroll, J. M. (1999). On the bipolarity of positive and negative affect. *Psychological Bulletin, 125,* 3-30.

Ryan, R. A., & Deci, E. L. (2000). Self-determination theory and the facilitation of intrinsic motivation, social development, and well-being. *American Psychologist, 55,* 68-78.

Seegers, G. & Boekaerts, M. (1993). Task motivation and mathematics achievement in actual task situations. *Learning and Instruction, 3*(2), *133-150.*

Vermeer, H. J., Boekaerts, M., & Seegers, G. (2000). Motivational and gender differences: Sixth-grade students' mathematical problem-solving behavior. *Journal of Educational Psychology, 92*(2), *308-315.*

Wentzel, K. (1998). Social relationships and motivation in middle school: The role of parents, teachers, and peers. *Journal of Educational Psychology, 90*(2), 202-209.

Wolters, C. A., & Rosenthal, H. (2000). The relation between students' motivational beliefs and their use of motivational regulation strategies. *International Journal of Educational Research, 33*(7), 801-820.

CHAPTER

4

Emotion in the Hierarchical Model of Approach-Avoidance Achievement Motivation

ANDREW J. ELLIOT
University of Rochester

REINHARD PEKRUN
University of Munich

Achievement motivation is ubiquitous in educational settings. Administrators, teachers, and students alike desire and strive to attain competence and/or avoid incompetence in their daily activities. Emotion is an integral feature of achievement motivation. It is involved in orienting the individual's competence-relevant concerns, sustaining the individual's competence-relevant interest and effort, and influencing the response of the individual to success and failure. In the following, we overview research we have done on the link between achievement motivation and emotion, using the hierarchical model of approach-avoidance achievement motivation (Elliot, 1997; Elliot & Church, 1997) as a conceptual guide.

The Hierarchical Model of Approach-Avoidance Achievement Motivation

Achievement goals are the centerpiece of the hierarchical model. Achievement goals are conceptualized as cognitive representations of positive or negative competence-relevant possibilities that are used to guide behavior (Elliot & Thrash, 2001). These goals may be organized in terms of two aspects of competence: its definition (i.e., whether an absolute/intrapersonal or normative standard is used to evaluate competence), and its valence (i.e., whether the focus is on a positive possibility or a negative possibility). Crossing these aspects of competence yields four different achievement goals: mastery-approach (focused on a positive absolute or intrapersonal standard), performance-approach (focused on a positive normative standard), mastery-avoidance (focused on a negative absolute/intrapersonal standard), and performance-avoidance (focused on a negative normative standard). This conceptualization is labeled the "2 × 2 achievement goal framework" (Elliot, 1999; Elliot & McGregor, 2001). A subset of this 2 × 2 framework, composed of mastery-approach, performance-approach, and performance-avoidance goals, is labeled the "trichotomous achievement goal framework" (Elliot, 1997; Elliot & Harackiewicz, 1996). This trichotomous framework has been used in most of the research reviewed herein for two reasons: (1) it was introduced into the literature several years before the 2 × 2 framework, and (2) research has shown that mastery-avoidance goals are less commonly pursued than the other three goals in the types of achievement settings and age groups that are typically studied in the achievement motivation literature (see Elliot & McGregor, 2001).

In the hierarchical model, achievement goals are viewed as relatively concrete, situation-specific constructs that emerge from more general motivational energizers such as temperaments and motive dispositions. These underlying forms of motivation serve to orient individuals toward positive or negative possibilities, and goals serve to channel and guide general motivational energy toward precise aims that address the underlying concerns. Thus in achievement settings, general motivational dispositions and more concrete goals work in tandem in the motivational process to energize and direct (respectively) competence-relevant behavior (for a more extensive introduction to the hierarchical model of approach-avoidance achievement motivation, see Elliot & Thrash, 2001; for an introduction to the hierarchical model of approach-avoidance motivation per se, see Elliot, 1999; Elliot, Gable, & Mapes, 2006). Given the central position of achievement goals in the hierarchical model, the achievement goal construct has been at the heart of our empirical work on achievement motivation and emotion.

Research Linking General Affective Dispositions to Achievement Goals

Individuals enter achievement settings with broad, cross-situational, affective propensities. These general affective tendencies are rooted in both biology and socialization, and exert a strong and consistent influence on the way persons are motivated in achievement contexts. Most pertinent to the present discourse, individual differences in general affective dispositions orient persons to the positive and negative possibilities present in the achievement context, and prompt the adoption of particular types of achievement strivings.

We have conducted two lines of research focused on the link between general affective dispositions and achievement goal adoption. The first line of research has focused on temperaments, and the second has focused on competence-relevant motive dispositions. We review this research in the following discussion.

Temperaments

A fundamental way that personality may be conceptualized is in terms of approach and avoidance temperament (Elliot & Thrash, 2002). Approach temperament is viewed as the conceptual core of the extraversion, positive emotionality, and behavioral activation system (BAS) constructs, all of which are integrally grounded in positive affective processes (Depue & Collins, 1999; Lucas, Diener, Grob, Suh, & Shao, 2000; Watson, 2000). Approach temperament is defined as "a general neurobiological sensitivity to positive/desirable (i.e., reward) stimuli (present or imagined) that is accompanied by perceptual vigilance for, affective reactivity to, and a behavioral predisposition toward such stimuli" (Elliot & Thrash, 2002, p. 805). Likewise, avoidance temperament is viewed as the conceptual core of the neuroticism, negative emotionality, and behavioral inhibition system (BIS) constructs, and all of these basic dispositions are integrally grounded in negative affective processes (Gray, 1987; Tellegen, 1985; Watson, Clark, & Tellegen, 1988). Avoidance temperament is defined as "a general neurobiological sensitivity to negative/undesirable (i.e., punishment) stimuli (present or imagined) that is accompanied by perceptual vigilance for, affective reactivity to, and a behavioral predisposition away from such stimuli" (Elliot & Thrash, 2002, p. 805). Approach and avoidance temperament are conceptualized as *temperaments* because they are presumed to possess all of the primary characteristics commonly identified with this category of construct: heritability, presence in early childhood, relative stability across the lifespan, and, importantly, a basic grounding in affect (see Buss & Plomin, 1984). The two temperaments are viewed as separate constructs, not polar ends of a single continuum (see also Cacioppo & Berntson, 1994).

Approach and avoidance temperament evoke immediate propensities in response to encountered or imagined stimuli. The behavior of lower animals is rigidly governed by such biologically-based propensities (Schneirla, 1959); but human functioning is more flexible, in that various forms of self-regulation, such as goals, may be recruited to produce overt behavior beyond immediate propensities. That is, a goal may be directly concordant with its underlying temperament, but if the underlying temperament is not adaptive in a given situation, a different goal may be adopted that overrides the underlying temperament.

Accordingly, in considering the link between approach/avoidance temperaments and achievement goals, we have proposed both valence symmetry and valence override processes (Elliot & Thrash, 2002). Regarding valence symmetry, we posit that approach temperament is a positive predictor of approach goals (both mastery-approach and performance-approach), and avoidance temperament is a positive predictor of avoidance goals (both mastery-avoidance and performance-avoidance). With regard to valence override, we posit that avoidance temperament is also a positive predictor of performance-approach goals, as individuals try to cope with their general avoidance tendencies by striving to approach normative success in achievement situations. Mastery-approach goals are considered a less attractive tool for overcoming general avoidance tendencies, because striving for task mastery and competence development typically entails a protracted process that includes failure experiences. Thus mastery-approach goals are not considered viable candidates for active avoidance (i.e., approaching success to avoid a negative possibility), and are presumed to be unrelated to avoidance temperament.

We have conducted several studies designed to examine the links between various indicators of approach/avoidance temperament and achievement goals. All of the studies have used prospective designs in actual achievement settings, and all have focused on the trichotomous achievement goal framework. In our first study (Elliot & Thrash, 2002, Study 3), we assessed behavioral activation system (BAS) sensitivity and behavioral inhibition system (BIS) sensitivity during the first week of an undergraduate psychology course, and measured students' achievement goals for the course a week later. The results indicated that BAS sensitivity was a positive predictor of mastery-approach goals ($\beta = .26$, $p < .001$); BIS sensitivity was a positive predictor of performance-avoidance goals ($\beta = .20$, $p < .01$); and both BAS sensitivity ($\beta = .17$, $p < .05$) and BIS sensitivity ($\beta = .19$, $p < .01$) were positive predictors of performance-approach goals. Subsequent studies produced this same pattern of results using extraversion/neuroticism, positive/negative emotionality, and approach/avoidance temperament latent variables as predictors of achievement goals (Elliot & Thrash, 2002, Studies 4-6). Thus our empirical work has provided strong, consistent support for each of our tested hypotheses (the mastery-avoidance goal hypothesis has yet to be tested). In addition, other investigators have obtained similar results using concurrent designs and various indicators of approach and avoidance temperament (Day, Radosevich, & Chasteen, 2003; Zweig & Webster, 2004; cf. Tanaka & Yamauchi, 2001).

Motive dispositions

In contrast to temperaments, which are rooted primarily in biological processes, involve general positive and negative affect, and are domain-free, motive dispositions are rooted primarily in socialization processes, involve distinct affective experience, and are domain-specific. Although motive dispositions are domain-specific, they nevertheless represent general tendencies, because within a given domain (e.g., achievement, social), they are presumed to operate as motivational predispositions that individuals carry with them into any and all domain-relevant situations. Within the achievement domain there are two primary motives: the need for achievement (McClelland, Atkinson, Clark, & Lowell, 1953) and fear of failure (Birney, Burdick, & Teevan, 1969). The need for achievement is the dispositional tendency to experience pride upon success, and fear of failure is the dispositional tendency to experience shame upon failure (Atkinson, 1957). *Pride*, in this context, represents both the pleasure of accomplishment and the pleasure of positively evaluating the self (see Elliot, McGregor, & Thrash, 2002; Heckhausen, 1982; cf. Koestner & McClelland, 1990), and *shame* represents a global indictment of the self, not a mere sense of embarrassment (see McGregor & Elliot, 2005).

Persons high in the need for achievement enter achievement settings hoping to experience success and the pride that it affords, and this is presumed to prompt the adoption of self-regulatory forms focused on attaining positive outcomes. Thus we posit that the need for achievement is a positive predictor of mastery-approach and performance-approach goals. In reciprocal fashion, persons high in fear of failure enter achievement situations fearing failure and the shame that it produces, and this is presumed to prompt the adoption of self-regulatory forms focused on avoiding negative outcomes. Thus we posit that fear of failure is a positive predictor of mastery-avoidance and performance-avoidance goals. In addition to this straightforward valence symmetry, valence override processes, like those discussed for temperaments, are also presumed operative with regard to achievement motives. Specifically, persons high in fear of failure may adopt performance-approach goals in an attempt to approach success in order to avoid failure. Thus we posit that fear of failure is a positive predictor of performance-approach goals as well as performance-avoidance goals.

We have conducted several prospective studies on the link between achievement motives and achievement goals. In the first study (Elliot & Church, 1997) we assessed achievement motives via self-report on the first day of an undergraduate psychology course, and assessed students' achievement goals for the course (using the trichotomous framework) three days later. The results indicated that the need for achievement was a positive predictor of mastery-approach goals ($\beta = .22$, $p < .01$), fear of failure was a positive predictor of performance-avoidance goals ($\beta = .45$, $p < .001$), and both the need for achievement ($\beta = .26$, $p < .001$) and fear of failure ($\beta = .41$, $p < .001$) were positive predictors of performance-approach goals (see also Elliot & Thrash, 2004). In subsequent studies we replicated these findings with motive dispositions assessed via projective test (and with latent variables derived from motives assessed via both self-report and projective test; Elliot & McGregor, 1999; Thrash & Elliot, 2002). We also extended these findings to mastery-avoidance goals (Elliot & McGregor, 2001, Study 2), documenting fear of failure as a positive predictor of such goals ($\beta = .28$, $p < .01$; see also Conroy & Elliot, 2004; Conroy, Elliot, & Hofer, 2003). Thus our research has provided robust empirical support for each of our hypotheses. In addition, several other investigators have obtained the same or highly similar results using concurrent designs (Conroy, 2004; Halvari & Kjormo, 1999; Tanaka, 2001; Tanaka, Okuno, & Yamauchi, 2002; Tanaka & Yamauchi, 2000; Vandewalle, 1997; Zusho, Pintrich, & Cortina, 2005).

Before leaving the topic of general affective dispositions, we would like to highlight an important point regarding the issue of the cross-situational nature of the affective dispositions in question. We believe that temperaments and motives represent *predispositions* that individuals bring into specific contexts; we do not believe that temperaments and motives necessarily evoke the same type of goal adoption in every situation. Temperaments and motives may operate differently depending on a host of situation-specific variables (e.g., competence valuation, competence expectancies, prior experience with the task and/or evaluative constraints present in the situation), and a more complete portrait of the role of general affective dispositions in the goal adoption process awaits empirical attention to such complexities.

Research Linking Achievement Goals to Distinct Affect

Achievement goals establish a perceptual-cognitive framework for how individuals construe and interpret achievement settings (Dweck, 1986; Elliot, 1997). Each type of achievement goal establishes a different interpretive framework; therefore it is likely that different achievement goals lead to different emotional experience. These emotional experiences occur throughout the achievement sequence, that is, before, during, and after engagement with the achievement activity (McGregor & Elliot, 2002; Pekrun, Goetz, Titz, & Perry, 2002).

We have conducted three different lines of research focused on the link between achievement goals and distinct affective experience. The first line of research has focused on challenge and threat affect, the second has focused on test anxiety, and the third, and most recent, has focused more broadly on a set of emotions commonly experienced in achievement settings. We review this research in the following, ending with some proposals regarding future research.

Challenge and Threat Affect

Achievement tasks may be appraised by individuals as a challenge or a threat. Challenge appraisals represent perceiving the task as an opportunity for benefit or gain, whereas threat appraisals represent perceiving the task as a possibility for harm or loss (Lazarus & Folkman, 1984). Challenge appraisals are associated with positive anticipatory affect, such as eagerness and hopefulness, whereas threat appraisals are associated with negative anticipatory affect, such as fear and anxiety (Folkman & Lazarus, 1985).

Given that achievement goals dictate the way in which individuals perceive achievement settings, they should influence whether persons experience challenge or threat affect when confronted with an achievement task. Approach goals, much like challenge appraisals, focus on positive possibilities; and avoidance goals, much like threat appraisals, focus on negative possibilities. Thus we propose a valence symmetry in linking achievement goals and challenge/threat affect. Specifically, we posit that mastery-approach and performance-approach goals, with their positive foci on absolute/intrapersonal competence and normative competence, respectively, are positive predictors of challenge affect. Likewise, we posit that mastery-avoidance and performance-avoidance goals, with their negative foci on absolute/intrapersonal incompetence and normative incompetence, respectively, are positive predictors of threat affect. Given that performance-approach goals can be grounded in aversive motivation (e.g., avoidance temperament, fear of failure), as well as appetitive motivation (e.g., approach temperament, the need for achievement), we posit that these goals are also a positive predictor of threat affect.

We have conducted two prospective studies designed to test the link between achievement goals and challenge/threat affect. In the first study (McGregor & Elliot, 2002, Study 1) we assessed undergraduates' achievement goals for a psychology course (using the trichotomous framework) during the second week of the course, and then, three weeks later, measured students' challenge and threat affect regarding an impending course exam. The results indicated that mastery-approach goals were positive predictors of challenge affect ($\beta = .32$, $p < .01$), performance-avoidance goals were positive predictors of threat affect ($\beta = .55$, $p < .01$), and performance-approach goals were positive predictors of challenge affect ($\beta = .19$, $p < .05$); performance-approach goals were unrelated to threat affect. The second study (McGregor & Elliot, 2002, Study 2) replicated the significant findings of the first study, but this time also indicated that performance-approach goals were a positive predictor of threat affect ($\beta = .21$, $p < .05$). Thus this research has provided consistent support for all but one of the tested hypotheses; the proposed positive link between performance-approach goals and threat affect has received inconsistent support (the mastery-avoidance goal hypothesis has yet to be tested).

Test Anxiety

Test anxiety has received more research attention in the achievement motivation literature than any other emotion, and research on achievement goals and emotion is no exception. This oft-studied construct may be defined as "the experience of evaluation apprehension during the examination process" (Spielberger & Vagg, 1995), and the popularity of this construct is likely due to the fact that it is a robust negative predictor of performance attainment (Hembree, 1988). Test anxiety researchers distinguish between trait test anxiety (individual differences in text anxiety across testing situations) and state test anxiety (the experience of test anxiety in a specific situation; Spielberger, 1972); we focus on state test anxiety herein (although see Elliot & McGregor, 1999). It is also common within the test anxiety literature to not only focus on overall test anxiety but also to focus on two primary components of test anxiety: worry and emotionality (Liebert & Morris, 1967; Pekrun, Goetz, Perry, Kramer, & Hochstadt, 2004). Worry represents cognitive reactions, such as self-criticism or concerns about the negative consequences of failure, and has been shown to undermine performance attainment (Sarason, 1972). Emotionality represents physiological and affective reactions, such as accelerated heart rate or nervousness, and has been found to have little impact on performance attainment (Deffenbacher, 1980).

Test anxiety is most pertinent to avoidance goals. Regulating according to performance-avoidance goals is likely to elicit anxiety in testing situations, as individuals anticipate norm-based evaluation and focus on the possibility of failure. Both worry and emotionality are likely to be affected by performance-avoidance regulation. Likewise, focusing on the possibility of a negative outcome in mastery-avoidance regulation is presumed to elicit test anxiety, and both worry and emotionality are likely to be affected here as well. Relations between performance-avoidance goals and the test anxiety variables may be somewhat stronger than those for mastery-avoidance goals and the test anxiety variables, given the added evaluative pressure of the normative (relative to task/intrapersonal) focus of performance-avoidance goals. Mastery-approach goals are presumed to be unrelated to test anxiety, given their appetitive focus and grounding in approach motivation. Similarly, the appetitive focus of performance-approach goals makes it unlikely that these goals will be related to overall test anxiety, but the underlying avoidance motivation that energizes pursuit of these goals may give rise to emotionality in some evaluative situations.

We have conducted several studies on the link between achievement goals and test anxiety. In our first study (Elliot & McGregor, 1999, Study 1) we assessed undergraduates' achievement goals for a psychology course (using the trichotomous framework) during the second week of the course, and assessed the anxiety that they experienced during a course exam three weeks later. The results indicated that performance-avoidance goals were positive predictors of overall test anxiety ($\beta = .43$, $p < .001$), worry ($\beta = .29$, $p < .01$), and emotionality ($\beta = .35$, $p < .001$). Mastery-approach and performance-approach goals were unrelated to all forms of test anxiety. A subsequent study using a short-term longitudinal design (McGregor & Elliot, 2002, Study 3), replicated these findings for overall test anxiety; worry and emotionality were not examined in this work. In another prospective study (Elliot & McGregor, 2001, Study 2), similar in design to Elliot and McGregor's (1999) Study 1 but using the 2 × 2 framework, mastery-avoidance goals were shown to be a marginally significant positive predictor of overall test anxiety (β = .18, p = .059), and a significant positive predictor of both worry (β = .21, $p < .05$) and emotionality (β = .25, $p < .01$). The results for the other achievement goals replicated those of Elliot and McGregor (1999, Study 1). We also conducted an experimental investigation with 13- to 15-year-old children in which achievement goals for a dribbling task were manipulated, and participants reported (on a 1 [low anxiety] to 5 [high anxiety] scale) their overall test anxiety for the task. Participants in the performance-avoidance goal condition reported significantly more test anxiety (M = 3.63) than both mastery-approach (M = 2.89, $p < .001$) and performance-approach (M = 3.12, $p < .01$) participants (who did not differ from each other). Thus our empirical work has provided strong and consistent support for all but one of our hypotheses: the tentative hypothesis that performance-approach goals may be positive predictors of emotionality has only been documented in a single study (Elliot & McGregor, 1999, Study 2). Many other investigators have also obtained results supportive of our hypotheses, using both concurrent (Lopez, 1999; Middleton & Midgley, 1997; Pajares, Britner, & Valiante, 2000; Pajares, 2003; Sideridis, 2003; Skaalvik, 1997; Smith, Duda, Allen, & Hall, 2002; Zusho et al., 2005) and experimental (Cury, Da Fonseca, Rufo, & Sarrazin, 2003) designs.

Emotions commonly experienced in achievement settings

Aside from the aforementioned research on achievement goals and test anxiety, researchers have yet to examine the link between the achievement goals of the trichotomous or 2 × 2 frameworks and discrete emotion (for research focusing on the goals of the trichotomous model and general positive/negative affect see Sideridis, 2003; 2005). Accordingly, we have recently conducted two studies (Pekrun, Elliot, & Maier, in press) designed to link the goals of the trichotomous model to a set of emotions that have been empirically shown to be commonplace in middle school, high school, and college classroom settings: enjoyment, hope, pride, boredom, anger, anxiety, hopelessness, and shame (see Pekrun et al., 2002). These emotions may be organized according to a 2 × 2 taxonomy of achievement emotions (Pekrun, 1992; Pekrun et al., 2002). This taxonomy makes use of two dimensions: object focus and valence. Object focus represents the degree to which an emotion is produced by focusing on the achievement activity per se or on an outcome of activity engagement; outcome-focused emotions may emerge from a prospective or a retrospective outcome focus. Valence represents the degree to which the emotion is positive or negative. Enjoyment is a positive activity-focused emotion, whereas boredom and anger pertaining to achievement activities are negative activity-focused emotions. Hope and pride are positive outcome-focused emotions, with hope having a prospective outcome-focus and pride having a retrospective outcome-focus. Anxiety, hopelessness, and shame are negative outcome-focused emotions, with anxiety and hopelessness having a prospective outcome-focus and shame having a retrospective outcome-focus. It should be noted that these categorizations are offered with respect to a specific achievement context (i.e., the classroom context) and a specific type of emotion assessment (i.e., a broad response to an achievement context, rather than a specific response to a single achievement event). Certain emotions may have different foci within other contexts or other types of assessment (e.g., anger may be categorized as a negative outcome-focused emotion when emotions are assessed immediately following feedback on a specific achievement task).

The achievement goals of the 2 × 2 framework map quite nicely onto the emotions of the 2 × 2 taxonomy. Mastery-based goals focus on the activity itself and the implications of ongoing experience with the activity for intrapersonal development. Mastery-approach goals are presumed to focus on positive activity engagement. Thus we posit that these goals are a positive predictor of enjoyment and perhaps even a negative predictor of boredom and anger. Mastery-avoidance goals are presumed to focus on negative activity engagement. Thus we posit that these goals are a positive predictor of boredom and anger and perhaps a negative predictor of enjoyment. Performance-based goals, on the other hand, focus on normative outcomes in either prospective or retrospective ways. Performance-approach goals are presumed to focus prospective attention on the possibility of attaining positive normative outcomes, and retrospective attention on the positive value of the normative outcome attained. Thus we posit that these goals are a positive predictor of hope and pride. Performance-avoidance goals are presumed to focus prospective attention on the possibility of negative normative outcomes, and retrospective attention on the negative value of the normative outcome attained. Thus we posit that these goals are a positive predictor of anxiety, hopelessness, and shame.

We have conducted two studies designed to examine the link between achievement goals and the emotions in the 2 × 2 taxonomy. Both studies were prospective investigations, one in a German college classroom and the other in a U.S. college classroom. Students' class-relevant achievement goals (using the trichotomous model) were assessed at the beginning of the semester, and their achievement emotions were assessed near the end of the semester. Several variables were controlled for in the two studies, including gender, grade point average, social desirability, approach and avoidance temperament, and competence expectancy. The results across the two studies were largely in accord with our hypotheses, although a few unexpected findings also emerged. A meta-analysis (Rosenthal, 1978) of the two studies yielded the following results: mastery-approach goals were a positive predictor of enjoyment ($Z = 4.79$, $p < .001$), hope ($Z = 4.54$, $p < .001$), and pride ($Z = 4.14$, $p < .001$), and a negative predictor of anger ($Z = -3.38$, $p < .001$) and boredom ($Z = -4.97$, $p < .001$). Performance-approach goals were a positive predictor of pride ($Z = 2.55$, $p < .05$). Performance-avoidance goals were a positive predictor of anxiety ($Z = 3.77$, $p < .001$), hopelessness ($Z = 2.82$, $p < .01$), and shame ($Z = 1.92$, $p \leq .05$). That mastery-approach, rather than performance-approach, goals were positive predictors of hope is likely due to the fact that the items used to assess hope were actually more activity-relevant than outcome-relevant (contrary to the categorization of hope in the 2×2 taxonomy). Likewise, the link between mastery-approach goals and pride was likely due to the fact that the pride items contained some task-relevant material (our hypotheses regarding mastery-avoidance goals have yet to be put to empirical test).

Linking Achievement Goals and Perceived Success/Failure Outcomes to Distinct Emotions

One important topic that has yet to be examined is the link between achievement goals x perceived success/failure outcomes and distinct emotional experience. Although we have no results to report on this issue, we do offer hypotheses as a guideline for future research.

Lazarus (1991) construed emotions as "reactions to the fate of active goals" (p. 92). Goals have different foci, and it is likely that doing well or poorly at achievement goals with different foci produces qualitatively distinct emotions. This proposition is consistent with Mowrer's (1960) "theory of the emotions" (p. 167), in which he highlighted two independent dimensions of events—prospective gains or losses and an increased or decreased probability of occurrence—and linked each of the four possible event combinations (e.g., increased probability of a gain) to distinct affective experience. Appraisal and self-regulation theorists have proffered highly similar conceptualizations (Carver and Scheier, 1998; Higgins, 1997; Ortony, Clore, & Collins, 1988; Roseman, 1984; Stein & Jewett, 1986) in which a motivational state (e.g., seeking reward or avoiding punishment) is presumed to interact with a situational state (e.g., presence or absence) to produce distinct emotions.

In accord with the aforementioned conceptualizations, our contention is that achievement goals interact with perceptions of competence (i.e., doing well, doing poorly) to produce distinct emotions. Each achievement goal/perceived competence combination is posited to evoke a distinct psychological experience that produces a specific emotion. Succeeding at a mastery-approach goal represents doing well at approaching task-based/intrapersonal competence, and we posit that this psychological experience produces joy. Failing at a mastery-approach goal represents doing poorly at approaching task-based/intrapersonal competence, and we posit that this psychological experience produces disappointment. Succeeding at a performance-approach goal represents doing well at approaching normative competence, and we posit that this psychological experience produces pride (specifically, pleasure in positively evaluating the self; not pleasure in accomplishment per se; see Lewis & Sullivan, 2005). Failing at a performance-approach goal represents doing poorly at approaching normative competence, and we posit that this psychological experience produces frustration. Succeeding at a mastery-avoidance goal represents doing well at avoiding task-based/intrapersonal incompetence, and we posit that this psychological experience produces contentment. Failing at a mastery-avoidance goal represents doing poorly at avoiding task-based/intrapersonal incompetence, and we posit that this psychological experience produces embarrassment. Succeeding at a performance-avoidance goal represents doing well at avoiding normative incompetence, and we posit that this psychological experience produces relief. Failing at a performance-avoidance goal represents doing poorly at avoiding normative incompetence, and we posit that this psychological experience produces shame.

Although the aforementioned hypotheses link each goal/perceived competence combination to a distinct emotion, we would be remiss if we did not acknowledge that real world goal pursuit and real world emotional experience are highly complex, and therefore likely to be a bit "messier" in any given situation than our one-to-one pairings might suggest. Indeed, depending on a host of factors (e.g., the extent that future goal pursuit is possible, the importance of the outcome at hand), a given goal can undoubtedly lead to several different emotions, and a given emotion may be linked to different goals. Nevertheless, above this "noise," we do think that certain goal/perceived competence combinations are overall likely to be linked to certain emotions, and we are eager to put our admittedly speculative hypotheses to empirical test.

Summary and Conclusions

In the preceding discussion, we have reviewed research clearly demonstrating that emotion is present in many ways throughout the achievement motivation process. Individuals bring general affective tendencies with them to achievement settings, and these dispositions influence the types of achievement goals that they adopt. In turn, the achievement goals that individuals adopt influence the type of affect that they experience as they anticipate achievement tasks, engage in achievement tasks, and respond to achievement outcomes. Emotion is undoubtedly implicated in the achievement motivation process in other ways as well. For example, the specific emotions that individuals experience in response to achievement outcomes likely have an important impact on their subsequent achievement goal adoption (see Linnenbrink & Pintrich, 2002), as well as their level of continuing commitment to competence itself.

Importantly, the fact that the hierarchical model grounds achievement motivation in deeply engrained personality dispositions does not mean that achievement goal adoption and resultant emotional experience are set in stone. First, although it is true that temperament is quite stable over the lifespan, motive dispositions, although also stable, are likely to remain at least somewhat malleable into adulthood (Elliot, McGregor, & Thrash, 2002; Winter, John, Stewart, Klohnen, & Duncan, 1998). Second, achievement goal adoption is multiply determined; many other factors besides general affective tendencies are involved in goal adoption, including perceived competence (a person factor likely to impact goal valence; Elliot, 1997), implicit theories of ability (a person factor likely to impact goal definition; Dweck, 1999), and numerous properties of the achievement environment (some of which likely impact goal valence [e.g., the harshness of evaluation; Church, Elliot, & Gable, 2001], and others of which likely impact goal definition [e.g., the type of competence evaluation used; Ames, 1992]). Third, emotional experience is also multiply determined; many other factors besides achievement goals influence emotion, including attributions (Weiner, 1985), control beliefs (Pekrun, in press), and instrumental considerations (Husman & Lens, 1999).

Thus both goals and emotions are amenable to change in achievement settings, but such change is undoubtedly constrained to a degree, given the stability of personality. As such we think it is critical to bear in mind both the malleability of student motivation and the limits placed on such malleability by personality when evaluating the promise and efficacy of educational interventions. With many theorists we believe that educators must take great care to structure achievement environments to maximize students' competence, promote their continued interest and investment, and facilitate their personal growth and well-being. Evidence-based motivational interventions designed to foster optimal achievement motivation are of vital importance. However, we also believe that a balanced view of the malleability/stability issue can have the positive effects of relaxing the overly lofty expectations often attached to intervention efforts that inevitably lead to disappointment; of making salient the need to additionally attend to the many other factors beyond school walls (e.g., parental socialization practices, community-based norms, cultural values) that have an impact on how children are motivated when they step into the classroom; and of highlighting the importance of treating each student as a unique individual, motivated by a somewhat different combination of hopes, fears, and foci.

References

Ames, C. (1992). Classrooms: Goals, structures, and student motivation. *Journal of Educational Psychology, 84*, 261-271.

Atkinson, J. W. (1957). Motivational determinants of risk-taking behavior. *Psychological Review, 64*, 359-372.

Birney, R., Burdick, H., & Teevan, R. (1969). *Fear of failure*. New York: Van Norstrand-Reinhold Co.

Buss, D., & Plomin, R. (1984). *Temperament: Early developing personality traits*. Hillsdale, NJ: LEA.

Cacioppo, J., & Berntson, G. (1994). Relationship between attitudes and evaluative space: A critical review with emphasis on the separability of positive and negative substrates. *Psychological Bulletin, 115*, 401-423.

Carver, C., & Scheier, M. (1998). *On the self-regulation of behavior.* New York: Cambridge University Press.

Church, M. A., Elliot, A. J., & Gable, S. L. (2001). Perceptions of classroom environment, achievement goals, and achievement outcomes. *Journal of Educational Psychology, 93,* 43-54.

Conroy, D. E. (2004). The unique psychological meanings of multidimensional fears of failing. *Journal of Sport and Exercise Psychology, 26,* 484-491.

Conroy, D. E., & Elliot, A. J. (2004). Fear of failure and achievement goals in sport: Addressing the issue of the chicken and the egg. *Anxiety, Stress, and Coping: An International Journal, 17,* 271-285.

Conroy, D. E., Elliot, A. J., & Hofer, S. M. (2003). A 2 × 2 achievement goals questionnaire for sport. *Journal of Sport and Exercise Psychology, 25,* 456-476.

Cury, F., Da Fonseca, D., Rufo, M., & Sarrazin, P. (2003). The trichotomous model and investment in learning to prepare for a sport test: A mediational analysis. *British Journal of Educational Psychology, 73,* 529-543.

Day, E. A., Radosevich, D. J., & Chateen, C. S. (2003). Construct- and criterion-related validity of four commonly used goal orientation instruments. *Contemporary Educational Psychology, 28,* 434-464.

Deffenbacher, J. (1980). Worry and emotionality in test anxiety. In I. Sarason (Ed.), *Test anxiety: Theory, research, and applications* (pp. 111-128). Hillsdale, NJ: LEA.

Depue, R., & Collins, P. F. (1999). Neurobiology of the structure of personality: Dopamine, facilitation of incentive motivation, and extraversion. *Behavioral and Brain Sciences, 22,* 491-569.

Dweck, C. S. (1986). Motivational processes affecting learning. *American Psychologist, 41,* 1040-1048.

Dweck, C. (1999). *Self-theories: Their role in motivation, personality, and development.* Philadelphia: Psychology Press.

Elliot, A. J. (1997). Integrating "classic" and "contemporary" approaches to achievement motivation: A hierarchical model of approach and avoidance achievement motivation. In P. Pintrich & M. Maehr (Eds.), *Advances in motivation and achievement* (vol. 10, pp. 143-179). Greenwich, CT: JAI Press.

Elliot, A. J., (1999). Approach and avoidance motivation and achievement goals. *Educational Psychologist, 34,* 149-169.

Elliot, A. J., & Church, M. A. (1997). A hierarchical model of approach and avoidance achievement motivation. *Journal of Personality and Social Psychology, 72,* 218-232.

Elliot, A. J., Gable, S. L., & Mapes, R. R. (2006). Approach and avoidance motivation in the social domain. *Personality and Social Psychology Bulletin, 32,* 378-391.

Elliot, A. J., & Harackiewicz, J. M. (1996). Approach and avoidance achievement goals and intrinsic motivation: A mediational analysis. *Journal of Personality and Social Psychology, 70,* 461-475.

Elliot, A. J., & McGregor, H. A. (1999). Test anxiety and the hierarchical model of approach and avoidance achievement motivation. *Journal of Personality and Social Psychology, 76,* 628-644.

Elliot, A. J., & McGregor, H. A. (2001). A 2 × 2 achievement goal framework. *Journal of Personality and Social Psychology, 80,* 501-519.

Elliot, A. J., McGregor, H. A., & Thrash, T. M. (2002). The need for competence. In E. Deci & R. Ryan (Eds.), *Handbook of self-determination research* (pp. 361-387). Rochester, NY: University of Rochester Press.

Elliot, A. J., & Thrash, T. M. (2001). Achievement goals and the hierarchical model of achievement motivation. *Educational Psychology Review, 12,* 139-156.

Elliot, A. J., & Thrash, T. M. (2002). Approach-avoidance motivation in personality: Approach and avoidance temperaments and goals. *Journal of Personality and Social Psychology, 82,* 804-818.

Elliot, A. J., & Thrash, T. M. (2004). The intergenerational transmission of fear of failure. *Personality and Social Psychology Bulletin, 30,* 957-971.

Folkman, S., & Lazarus, R. (1985). If it changes it must be a process: Study of emotion and coping during three stages of a college examination. *Journal of Personality and Social Psychology, 48,* 150-170.

Gray, J. A. (1987). *The psychology of fear and stress* (2nd ed.). New York: Cambridge University Press.

Halvari, H., & Kjormo, O. (1999). A structural model of achievement motives, performance approach and avoidance goals, and performance among Norwegian Olympic athletes. *Perceptual and Motor Skills, 89,* 997-1022.

Heckhausen, H. (1982). The development of achievement motivation. In W. W. Hartup (Ed.), *Review of child development research* (Vol. 6, pp. 600-668). Chicago: University of Chicago Press.

Hembree, R. (1988). Correlates, causes, effects, and treatment of test anxiety. *Review of Educational Research, 58,* 47-77.

Higgins, E. T. (1997). Beyond pleasure and pain. *American Psychologist. 52,* 1280-1300.

Husman, J., & Lens, W. (1999). The role of the future in student motivation. *Educational Psychologist, 34,* 113-125.

Koestner, R., & McClelland, D. C. (1990). Perspectives on competence motivation. In L. Pervin (Ed.), *Handbook of personality: Theory and research* (pp. 527-548). New York: Guilford Press.

Lazarus, R. (1991). Cognition and motivation in emotion. *American Psychologist, 46,* 352-367.

Lazarus, R., & Folkman, S. (1984). *Stress, appraisal, & coping.* New York: Springer.

Lewis, M., & Sullivan, M. W. (2005). The development of self-conscious emotions. In A. Elliot & C. Dweck (Eds.), *Handbook of competence and motivation* (pp. 185-201). NewYork: Guilford Press.

Liebert, R., & Morris, L. (1967). Cognitive and emotional components of test anxiety: A distinction and some initial data. *Psychological Reports, 20,* 975-978.

Linnebrink, E. A., & Pintrich, P. R. (2002). Achievement goal theory and affect: An asymmetrical bidirectional model. *Educational Psychologist, 37,* 69-78.

Lopez, D. F. (1999). Social cognitive influences on self-regulated learning: The impact of action-control beliefs and academic goals on achievement-related outcomes. *Learning & Individual Differences, 11,* 301-319.

Lucas, R. E., Diener, E., Grob, A., Suh, M. E., & Shao, L. (2000). Cross-cultural evidence for the fundamental features of extraversion. *Journal of Personality and Social Psychology, 79,* 1039-1056.

McClelland, D. C., Atkinson, J. W., Clark, R. A., & Lowell, E. L. (1953). *The achievement motive.* New York: Appleton-Century-Crofts.

McGregor, H. A., & Elliot, A. J. (2002). Achievement goals as predictors of achievement-relevant processes prior to task engagement. *Journal of Educational Psychology, 94,* 381-395.

McGregor, H. A., & Elliot, A. J. (2005). The shame of failure: Examining the link between fear of failure and shame. *Personality and Social Psychology Bulletin, 31,* 218-231.

Middleton, M., & Midgley, C. (1997). Avoiding the demonstration of lack of ability: An under-explored aspect of goal theory. *Journal of Educational Psychology, 89,* 710-718.

Mowrer, O. H. (1960). *Learning theory and behavior.* New York: John Wiley & Sons, Inc.

Ortony, A., Clore, G., & Collins, A. (1988). *The cognitive structure of emotions.* New York: Cambridge University Press.

Pajares, F., Britner, S. L., & Valiante, G. (2000). Relation between achievement goals and self-beliefs of middle school students in writing and science. *Contemporary Educational Psychology, 25,* 406-422.

Pajares, F. (2003). Achievement goal orientations in writing: A developmental perspective. *International Journal of Educational Psychology, 39,* 437-455.

Pekrun, R. (1992). The impact of emotions on learning and achievement: Towards a theory of cognitive/motivational mediators. *Applied Psychology: An International Review, 41,* 359-376.

Pekrun, R. (in press). The control-value theory of achievement emotions: Assumptions, corollaries, and implications for educational research and practice. *Educational Psychology Review.*

Pekrun, R., Elliot, A. J., & Maier, M. A. (in press). Achievement goals and discrete achievement emotions: A theoretical model and prospective test. *Journal of Educational Psychology.*

Pekrun, R., Goetz, T., Perry, R. P., Kramer, K., & Hochstadt, M. (2004). Beyond test anxiety: Development and validation of the Test Emotions Questionnaire (TEQ). *Anxiety, Stress and Coping, 17,* 287-316.

Pekrun, R., Goetz, T., Titz, W., & Perry, R.P. (2002). Academic emotions in students' self-regulated learning and achievement: A program of quantitative and qualitative research. *Educational Psychologist, 37,* 91-106.

Roseman, I. (1984). Cognitive determinants of emotions: A structural theory. In P. Shaver (Ed.), *Review of personality and social psychology* (pp. 11-36). Beverly Hills, CA: Sage.

Rosenthal, R. (1978). Combining the results of independent studies. *Psychological Bulletin, 85,* 185-193.

Sarason, I. (1972). Experimental approaches to test anxiety: Attention and the uses of information. In C. D. Spielberger (Ed.), *Anxiety: Current trends in theory and research* (Vol. 2, pp. 383-403). Washington, DC: Hemisphere.

Schneirla, T. (1959). An evolutionary and developmental theory of biphasic processes underlying approach and withdrawal. In *Nebraska symposium on motivation* (pp. 1-42). Lincoln: University of Nebraska Press.

Sideridis, G. D. (2003). On the origins of helpless behaviour of students with learning disabilities: Avoidance motivation? *International Journal of Educational Research, 39,* 497-517.

Sideridis, G. D. (2005). Goal orientation, academic achievement, and depression: Evidence in favor of a revised goal theory framework. *Journal of Educational Psychology, 97,* 366-375.

Skaalvik, E. M. (1997). Self-enhancing and self-defeating ego orientation: Relations with task and avoidance orientation, achievement, self-perceptions, and anxiety. *Journal of Educational Psychology, 89,* 71-81.

Smith, M., Duda, J., Allen, J., & Hall, H. (2002). Contemporary measures of approach and avoidance goal orientation: Similarities and differences. *British Journal of Educational Psychology, 72,* 155-190.

Spielberger, C. (1972). Anxiety as an emotional state. In C. Spielberger (Ed.), *Anxiety: Current trends in theory and research* (Vol. 1, pp. 23-49). New York: Academic Press.

Spielberger, C. D., & Vagg, P. R. (1995). *Test anxiety: Theory, assessment, and treatment.* Washington, DC: Taylor and Francis.

Stein, N. L., & Jewett, J. L. (1986). A conceptual analysis of the meaning of negative emotions: Implications for a theory of development. In C. Izard (Ed.), *Measuring emotions in infants and children,* (pp. 238-267). Cambridge: Cambridge University Press.

Tanaka, A., Okuno, T., & Yamauchi, H. (2002). Achievement motives, cognitive and social competence, and achievement goals in the classroom. *Perceptual and Motor Skills, 95, 445-458.*

Tanaka, A., & Yamauchi, H. (2000). Causal models of achievement motive, goal orientation, intrinsic interest, and academic achievement in classroom. *The Japanese Journal of Psychology, 71,* 2000.

Tanaka, A., & Yamauchi, H. (2001). A model for achievement motives, goal orientations, intrinsic interest, and academic achievement. *Psychological Reports, 88,* 123-135.

Tellegen, A. (1985). Structures of mood and personality and their relevance to assessing anxiety, with an emphasis on self-report. In A. Tuma & J. Maser (Eds.), *Anxiety and the anxiety disorders.* Hillsdale, NJ: LEA.

Thrash, T. M., & Elliot, A. J. (2002). Implicit and self-attributed achievement motives: Concordance and predictive validity. *Journal of Personality, 70,* 729-755.

Vandewalle, D. (1997). Development and validation of a work domain goal orientation instrument. *Educational and Psychological Measurement, 57,* 995-1015.

Watson, D. (2000). *Mood and temperament.* New York: Guilford Press.

Watson, D., & Clark, L. A., & Tellegen, A. (1988). Development and validation of brief measures of positive and negative affect: The PANAS scales. *Journal of Personality and Social Psychology, 54,* 1063-1070.

Weiner, B. (1985). An attributional theory of achievement motivation and emotion. *Psychological Review, 92,* 548-573.

Winter, D. G., John, O. P., Stewart, A. J., Klohnen, E. C., & Duncan, L. E. (1998). Traits and motives: Toward an integration of two traditions in personality research. *Psychological Review, 105*, 230-250.

Zusho, A., Pintrich, P. R., & Cortina, K. S. (2005). Motives, goals, and adaptive patterns of performance in Asian American and Anglo American students. *Learning and Individual Differences, 15*, 141-158.

Zweig, D., & Webster, J. (2004). What are we measuring? An examination of the relationships between the big-five personality traits, goal orientation, and performance intentions. *Personality and Individual Differences, 36*, 1693-1708.

CHAPTER

5

Examining Emotional Diversity in the Classroom: An Attribution Theorist Considers the Moral Emotions

BERNARD WEINER
University of California Los Angeles (UCLA)

The study of anxiety, and particularly test anxiety, has dominated research on classroom emotions. Having taken many exams as a student, and witnessing scores of others taking exams as a teacher, I have found that test anxiety is an important direction for research, with both theoretical and applied promise.

On the other hand, one could argue that this is a very narrow focus for researchers to take when examining emotions in school settings. To understand what other emotions are left out, we first must consider the perspective from which anxiety is studied. Following this discussion, I identify various properties of emotions to determine what characteristics anxiety does and does not possess. Then I discuss a number of emotions that I believe are prevalent in the classroom and are understudied. I label these "the moral emotions," and examine them from the perspective of an attribution theorist.

ASSUMING EMOTIONS ARE INTRAPSYCHIC AS OPPOSED TO SOCIAL PHENOMENA

Many educational researchers have tended to discuss emotions as intrapsychic phenomena. They are defined as subjective or private experiences having a positive or negative quality. Attesting to the personal quality of emotions are their antecedents and methods for identification and measurement. Among the many antecedents of feeling states are particular thoughts (e.g., "I am going to fail") and hormonal conditions, whereas among their numerous indicators are patterns of actual or reported physiological activity (e.g., "I feel my heart beating faster") and facial characteristics. Emotions (such as anxiety) are thus studied at the level of the individual, and it seems difficult to disagree with that position.

Yet, an argument can be made that emotions are social phenomena. Of course, this does not characterize all emotions (consider fear or anxiety about heights). But love and sadness, for example, two among the most prevalent emotions, typically involve social experiences. The statements "we broke up," "we got back together," and "my heart is broken" reveal that a metaphor for love is the merging of distinct entities into one social unit. On the other hand, sadness often follows the permanent or even temporary loss of another; we are sad when a loved one departs. Love and sadness then can be considered social rather than (or in addition to) personal emotions, with their antecedents and indicators found at the social level (e.g., joining and leaving), outside a particular person. Other emotions also arise in social contexts and, as regulators of behavior, have social consequences. Sympathy promoting giving help and anger increasing aggressive actions are two emotions that play essential roles in social motivation (see Weiner, 1995; 2006).

In sum, the study of anxiety is guided by an intrapsychic rather than a social view of emotions. This limits the range of emotion-related phenomena that might be examined and their methods of study.

CHARACTERISTICS OF EMOTIONS IN (AND OUT OF) THE CLASSROOM

I now examine additional characteristics or properties of emotion to again point out the void when the study of emotions in the classroom is limited to anxiety. Whereas the focus in this chapter is on the classroom, the comments also pertain to the study of affects outside the confines of educational settings.

Positive (Pleasant) vs. Negative (Unpleasant) Feelings

Virtually all emotion theorists agree that emotions have two properties, activation (ranging from high active emotions such as rage to weaker ones such as annoyance) and valence (ranging from positive or pleasant emotions such as love to negative or unpleasant emotions including anxiety). Of course there also are some grey areas when making these distinctions. However, the activation/valence distinctions typically are applicable to describe affects. Anxiety clearly is a negative emotion, with its level of arousal dependent on a variety of conditions. Hence, the main point here is that the focus of anxiety limits the researcher to the study of a negative emotion and excludes positive emotions from consideration.

Emotions Generated by Achievement vs. Affiliative Activities

Most children report that they go to school to be with their friends. It is quite likely that social acceptance and rejection, social activities, and other social concerns generate the majority of emotions in school settings. One need only to observe behavior on the playground during recess as opposed to in the classroom to find differences in emotional prevalence and intensity.

An interesting historical note illustrates this intensity differential among students. When McClelland, Atkinson, and their colleagues were first exploring the usefulness of the Thematic Apperception Test (TAT) as a measurement of motivation (see McClelland, Atkinson, Clark & Lowell, 1953), they reasoned that they must show that the instrument is sensitive to differences in aroused motivation. To arouse achievement strivings, they induced success or failure in an experimental setting. The logic of their approach was to consider the TAT a thermometer, and a good instrument would show a higher "temperature" (TAT achievement imagery) following failure than success. That indeed proved to be the case. They then thought this difference should be replicated in an affiliative setting. To raise the temperature, they had students stand before their peers while these peers indicated their shortcomings (the "failure" or rejection condition). The TAT ratings following this manipulation were to be compared with a condition where peers communicated their positive qualities (the "success" or acceptance condition). However, the experiment had to be called off because the rated individuals were breaking into tears and becoming quite shaken in the rejection condition. This did not occur following task failure in the achievement setting.

As already indicated, the prior research focus has been on test anxiety, to the exclusion of affiliative-related emotions. Of course, there also are some grey areas in this distinction as well. When one feels happy following a problem solution within a group, is this an achievement- or affiliative-related emotion? The answer is uncertain. However, a focus on test anxiety is clearly achievement-related and reflects yet another void or shortcoming by ignoring the social determinants of feeling states.

Self- vs. Other-Directed Emotions

Anxiety is a self-directed emotion. By that I mean the feeling state and the thoughts that generate this emotion relate back to oneself. A great many emotions, including guilt, fear, and happiness, are also self-directed.

On the other hand, many emotions relate to others and the thoughts generating these emotions also concern others. As soon to be discussed, anger and sympathy, for example, tend to be emotions that are other-directed. One is angry at and sympathetic toward other individuals. The thoughts generating anger typically are that someone else has done something aversive that could have been otherwise (e.g., "my roommate did not clean up the kitchen when he was supposed to"). On the other hand, sympathy follows when others are in negative plights that are beyond their personal control (e.g. "I feel sorry for that blind individual"). Another vacuum is thus apparent when research is limited to self-directed anxiety.

Thoughtless vs. Thoughtful Emotions

Some emotions require more cognitive work or more cognitive processing than do others. For example, happiness typically arises when there has been goal attainment. Success at an exam or acceptance in a club tends to produce happiness. Conversely, unhappiness is experienced when goals are not met. In prior work (see Weiner, 1986), I labeled these outcome-dependent, attribution-independent emotions. For example, if one attains the grade of "A" in a class, this is likely to generate happiness regardless of whether this person was one of many who attained this grade (attribution to the ease to the task or the teacher) or was one of the few (attribution to ability or effort). Happiness and unhappiness do not require a great deal of cognitive processing, other than perhaps a comparison between the outcome and one's aspiration level. In a similar manner, fear and anxiety often tend to be automatically elicited in particular stimulus settings. Some information must be processed for these affective experiences, but the need is limited.

On the other hand, many affective experiences require numerous controlled processes among their antecedents. Pride, for example, is not experienced every time that success is attained. The success must also be self-attributed. Self-attribution for success may involve comparison with social norms, recall of prior personal experiences, and so on. In a similar manner, gratitude requires not only that someone benefit us, but also that the benefit was given volitionally (these emotions are examined in greater detail later in this chapter). Anxiety typically is a relatively thoughtless emotion when compared to the antecedent cognitive complexity of feelings such as pride and gratitude, although on occasion it may be associated with more complex processes and not merely be automatically elicited.

In sum, thus far I have conceptualized anxiety in educational settings as a negative, achievement-generated, self-directed, and relatively thoughtless emotion, leaving for study the positive, and/or affiliative-generated, and/or other-directed, and/or more thoughtful emotions.

With this introduction in mind, I now turn to emotions other than anxiety in the classroom. The emotions I want to consider are called *moral emotions*. By that I mean they are associated with such concepts as *ought* and *should*, and they have social norms among their antecedents. Some are positive (pleasant) whereas others are negative (unpleasant); most can have achievement or affiliation as a motivational source; and some are self-directed whereas many are aimed toward other individuals. And all involve a great deal of cognitive work. My goal here is to call attention to some understudied affects in the classroom that deserve attention. I approach this discussion as an attribution theorist, so allow me to first make a small detour to examine a few principles of attribution theory that guide my discussion of emotion.

AN ATTRIBUTIONAL APPROACH TO EMOTIONS

Attribution theorists focus on the perceived causes of events, such as success and failure, and acceptance and rejection. There are, of course, an infinite number of causes of these outcomes. One can succeed at an exam because of high ability, extended effort, good luck, using the correct strategy, cheating, and on and on. In a similar manner, one might be rejected for a date because the other already has a boyfriend, needs to study for an exam, or has the flu. In addition, affiliative rejection may be because the requestor is too tall, unkempt, boring, and on and on. It is evident that these lists of causes can be virtually unlimited.

To understand attributions and attributional processes, a few underlying properties or dimensions of causality have been uncovered. In so doing, this produces a shift from qualitatively distinct to quantitatively different, and it becomes possible to compare and contrast the various causes. Two such properties that are central in the discussion of moral emotions are the locus of causality and the controllability of causality.

The locus of causality contrasts causes that are internal versus external to the person. For example, ability and effort as causes of success are internal to the actor. On the other hand, having a biased teacher, or the desired date already having an appointment, are causes of failure and rejection that reside outside of the actor.

The controllability of causality refers to whether the cause could have been volitionally changed by someone. Failure because of low artistic aptitude might be considered uncontrollable by the actor, whereas failure due to insufficient practice is controllable. In a similar manner, rejection because one is too tall is not controllable by the individual seeking a date, but being unkempt or calling too late would be regarded as subject to volitional change—"it could have been otherwise." Causal controllability can be considered with the actor as the locus ("I did not study hard enough") or from the perspective of an actor's target ("you did not study hard enough").

In the following discussion of moral emotions that are prevalent in the classroom, four emotions relate to uncontrollable causal characteristics of the emotional target, whereas eight concern controllable properties. There are likely to be other emotions in these categories as well, so this is not to be regarded as the definitive list. Rather, the emotions selected were readily identified and are associated with an empirical literature. The four emotions generated by appraisals of uncontrollability are envy, scorn, shame, and sympathy; the eight emotions directed by thoughts about causal control are admiration, anger, gratitude, guilt, indignation, jealousy, regret, and schadenfreude. I believe these feelings are prevalent in and out of the classroom, can be studied, and are worthy of study, yet many have received only scant attention in educational contexts. In addition, virtually all can be aroused in either achievement or affiliation contexts (there might be one or two exceptions to this belief).

Let me briefly introduce these emotions and outline some of their characteristics. In so doing, my goal is not to provide a detailed analysis of any particular feeling state, but rather to point out the wide variety of emotional options that are available for researchers to study. Thus I regard what follows as a "salted peanut" approach, hoping that the taste will produce heightened desires.

The Uncontrollability-Related Moral Emotions

Envy Envy is aroused when a person desires the advantages of another that the individual does not possess. This superiority may lie in material goods, such as a new house or fine car, and be unrelated to thoughts about controllability. However, the desired advantage often is associated with uncontrollable qualities, such as beauty and intelligence (see Feather, 1999; Smith, Parrott, Diener, Hoyle, & Kim, 1999). One does not associate envy with hard work, a controllable quality, because everyone can exert effort. But others typically are unable to become beautiful or intelligent, and individuals with these qualities are targets of envy. In short, envy is in part an ability-linked feeling.

Envy often leads to dislike because the other has what one wants and is unable to obtain. This is not an invariant consequence, since one may envy the high ability of a friend. However, it is a frequent occurrence. We need to study envy in the classroom: when it is experienced, what are its consequences both for the self and the target of envy, and so on. Do children envy their high-performing classmates? Does this lead to social rejection? Might the anticipation of social rejection cause suppression of high ability, resulting in a decrease in actual performance?

Scorn Scorn, like envy, is also a negative emotion directed toward others. Scorn, or contempt, connotes that the other "cannot" (e.g., he or she does not have ability or is perceived as incapable). It therefore involves a downward social comparison. This emotion is said to occur when "one needs to feel stronger, more intelligent, more civilized, or in some way better than another" (Izard, 1977, p. 328), which then elicits disdain. Others might be scorned or held in contempt because of their actions, but these behaviors likely are ascribed to their character, which I suggest drives this emotional reaction. Which children are scorned in the classroom? Does this interfere with their subsequent achievements? Can this emotion be altered?

Shame Shame again is a negative emotion, but it is directed toward the self. Unlike envy and scorn, it has received a great deal of attention from psychologists. Shame involves a belief that the self is uncontrollably flawed, and this deficiency in character has been displayed to others. Being clumsy, unattractive, or having low intelligence evokes shame, which produces behaviors such as helplessness and withdrawal (see Tangney & Fischer, 1995). Shame thus effects the self in the same manner that scorn is likely to effect others. Indeed, if the target of scorn accepts this affect as correct, then he or she should experience shame. Other- and self-directed emotions are thus not independent, but interact in complex ways.

Sympathy When the plight of another is due to a cause that was uncontrollableby the self, we may experience sympathy (or pity). This cause may be external to the individual, such as having a biased or unfair teacher, but often is internal to the distressed person, including lack of ability or some other personal shortcoming. One typically is sympathetic toward those who are less able, such as the physically and mentally handicapped. It has been suggested that if the difference between the experiencer and the target of the emotion is more qualitative than quantitative (e.g., the target is blind or mentally incapable, as opposed to having temporary eye problems or being of marginal intelligence), then the emotional experience is more akin to pity than sympathy (see Weiner, 1986). Here I do not follow-up on this distinction.

A great deal of research has documented that sympathy results in giving help (see reviews in Weiner, 1995, 2006). Hence, it is an extremely important emotion in the context of the school, inasmuch as the reactions of scorn versus sympathy, given the same attribution to lack of ability, result in quite disparate behaviors toward the other: the scorned other is neglected, whereas the arouser of sympathy is aided. Yet, these same feeling are paradoxically aroused by some of the same thoughts.

In sum, a subset of moral emotions (envy, scorn or contempt, shame, and sympathy or pity) is associated with uncontrollable qualities and is ability-linked. These emotions manifestly are not equivalent, and their differences can also be represented in moral terms. One already stated inequivalence between these emotions concerns the behaviors they generate. Sympathy gives rise to prosocial behavior, including helping (going toward), whereas envy and scorn produce antisocial actions (going against), while shame leads to withdrawal (going away from). Furthermore, if others were to judge these emotional experiences (we do praise and criticize others for the emotions they feel), sympathy would be regarded as more moral, correct, suitable, and appropriate, given uncontrollable causality, than are envy, scorn, and shame. That is, emotions, just as behaviors, are considered "right" or "wrong." This contributes to their designation as "moral" emotions.

There is an absence of data to support this position, for pertinent research has not been undertaken. However, the presumptions regarding the perceived fairness of some moral emotions are put forward because individuals failing from an uncontrollable cause do not "deserve to be" the target of negative emotions, either from the self (shame) or from others (scorn). After all, one is not responsible for an uncontrollable cause (e.g., lack of ability), as there is no volitional choice regarding its presence or absence. It also follows that one is not accountable for the effects (e.g., failure) of these causes. In a moral sense, neither shame nor scorn is an "appropriate" emotional reaction to uncontrollable failure; these might be labeled "immoral emotions" because they are undeserved. On the other hand, a prosocial reaction of sympathy is "correct," given an uncontrollable plight.

Envy, in contrast to scorn and shame, is associated with positive rather than negative outcomes of others. Nonetheless, envy also may be considered an antisocial emotion one "ought not" feel. After all, a person should not elicit a negative reaction or be disliked for being smart or beautiful, or for having the benefits these nonvolitional characteristics bring.

The Effort Controllability-Linked Emotions

There are eight listed feelings associated with appraisals related to perceptions of causal control. The eight emotions are:

Admiration Admiration is a positive emotion directed outward. When a positively valued, controllable behavior, such as hard work, results in success, the outcome is perceived as deserved (see Feather, 1991). Deserved success elicits admiration (see Frijda, 1986; Hareli & Weiner, 2000; Ortony, Clore, & Collins, 1988). Admiration, in turn, evokes positive behavioral responses from others, including social acceptance. Admiration is not solely linked to effort ascriptions, for it also is elicited by perceptions of high ability (but perhaps to not as great an extent, which resulted in its classification as an effort-related or controllability-linked emotion).

Anger Anger is a negative emotion directed at an external target, be it an individual, group, or culture. Most emotion theorists agree that anger is generated by a judgment of personal responsibility for a transgression, with the individual experiencing this emotion typically involved in some way in the event (see Averill, 1983). Anger is a value judgment following from the belief that one "could and should have done otherwise." Lack of effort as a cause of failure, which elicits judgments of responsibility, thus arouses anger. However, friends also become angry at one another for failing to live up to a social contract, so anger is not limited to a specific motivational domain.

Gratitude Gratitude, like admiration, is a positive emotion that has an external target. Gratitude connotes a thankful appreciation for received favors (Guralnik, 1971). It is derived from the Latin root *gratia*, meaning "grace" or "graciousness" (see Emmons & McCullough, 2003). Gratitude tends to follow when a personally positive outcome is due to purposive and intentional actions of another. If an individual accidentally or unintentionally benefited another, then gratitude would not be experienced (see Tesser, Gatewood, & Driver, 1968). Gratitude is associated with moving toward others and positively reciprocating, which evens the scales of justice and is experienced when the favor received is valued by the recipient and is costly to the benefactor (see review in Tsang and McCullough, 2004). However, it is a complex emotion with both negative and positive implications for the self, because when gratitude is felt toward another in an achievement setting, the implication is that the other in part caused personal success. This detracts from a self-attribution for success, hence diminishing pride.

Guilt Introducing guilt returns the reader to self-directed, negative emotions. The moral aspect of guilt can be considered in the context of anger, for if communicated anger is "accepted" by the perceived wrongdoer, indicating acknowledgment of personal responsibility, then the recipient of this message will feel guilty (see Graham, 1984; Weiner, 1986). Like anger, guilt is the subject of voluminous writing and speculation. In general, guilt follows volitional acts (or their omission) that violate ethical norms and principles of justice. Hence, guilt has lack of effort rather than lack of aptitude as an antecedent. In opposition to shame, guilt is associated with action rather than character, controllable rather than uncontrollable causality, and the motivated consequences include making amends rather than withdrawal and helplessness (see Tangney & Fisher, 1995; Weiner, 1986).

Indignation For some emotions, particularly indignation, the experiencer of the emotion need not be personally involved in the social transgression (see Dwyer, 2003). For example, we may feel indignant at the bad treatment of B by A, as the case when a teacher is regarded as unfair to a student, even though that student may be a relative stranger. Consistent with the prior discussion, if the bad treatment of B by A was unintended, accidental, etc., then indignation or resentment is considered inappropriate (see Dwyer, 2003). That is, indignation (sometimes used synonymously with resentment) is in part based on the controllability of harm and requires perceived responsibility on the part of the person who does harm. This reaction is difficult to trace to self-interest or personal hedonic gains. Thus the emotion appears to be generated only by moral concerns, with "virtue as its own reward" (see Turillo et al., 2002). Hence, indignation is not readily explained by purely functional perspectives, such as evolutionary theory (although conceptions may be stretched to account for these types of emotions).

Jealousy Jealousy is aroused when one fears being supplanted by another as the recipient of affection from a beloved person. It may occur in a classroom when a student is jealous of the attention a teacher gives to another student. An important determinant of jealousy is why the jealous person believes he or she is no longer the target of affection. "Jealousy [is] more often experienced when a romantic partner was believed to have full control over a transgression, as opposed to being powerless to prevent it" (Bauerle, Amirkhan, & Hupka, 2003, p. 316). Thus jealousy is more likely elicited when the cause of an indiscretion or attention withdrawal is traced to the other rather than to overpowering surrounding circumstances. Jealousy is also more likey aroused when that person intended to act as he or she did (Mikulincer, Bizman, & Aizenberg, 1989). Jealousy is associated with other-blame and judgments of responsibility, revealing that it also has moral components.

Regret Regret has properties similar to guilt in that it is negative and self-directed. Regret is experienced when it is realized an outcome could have been more positive if better choices had been made. Regret has been contrasted with disappointment, which is felt when a decision turns out badly, regardless of the reason for this negative outcome (see van Dijk & Zeelenberg, 2002; van Dijk, Zeelenberg, & van der Pligt, 1999). That is, of these two emotions, only regret has self-agency and personal responsibility among its cognitive antecedents or appraisals (see Frijda, Kuipers, & Ter Schure, 1989). Regret, in a manner similar to guilt, also evokes desires to "kick oneself and to correct one's mistake, and wanting to undo the event and get a second chance" (van Dijk & Zeelenberg, 2002, p. 324). However, unlike guilt, regret results from intrapersonal harm, whereas guilt is more associated with interpersonal harm (see Berndsen, van der Pligt, Doosje, & Manstead, 2004). How often and under what conditions is regret experienced in the classroom? Is it an adaptive or a maladaptive reaction? What classroom behaviors are most regretted, and do they relate to achievement or disciplinary problems?

Schadenfreude Schadenfreude is a positive, self-directed affect. The suffering and misfortune of others at times results in an observer feeling pleasure. This emotional reaction is termed *schadenfreude*, a word literally denoting joy with the damage of another (see Ben-Ze'ev, 1992). Schadenfreude requires a sequence of successes or positive outcomes followed by failures or negative outcomes. Among the antecedents of this emotion are envy, social comparison, and a deserved misfortune. For example, if a student attains a comparatively high rank in class (eliciting social comparison envy) because of cheating (undeserved success), and then fails an exam, fellow students could experience schadenfreude. Often schadenfreude experiences require that the negative event not be too extreme. For example, if the student is expelled from school for cheating on one exam, then joy may not be felt (see Hareli & Weiner, 2002).

In support of this line of reasoning, Feather (1989) conducted a number of investigations documenting reactions to the fall of "tall poppies," or those who stand above others (often politicians, athletes, and persons in the public eye). The more their success is undeserved (e.g., due to luck, inheritance, help) and the more the individual is disliked, the more likely it is that others will feel schadenfreude, given a subsequent failure. Schadenfreude thus also has moral components (see Hareli & Weiner, 2002).

SUMMARY OF THE MORAL EMOTIONS

Recall that I criticized the over concentration on the study of anxiety in part because anxiety is a negative emotion. Yet the moral emotions in the classroom also are primarily negative. However, unlike anxiety, they are more likely to be outer- rather than self-directed. These emotions also are more likely to be linked to beliefs about personal control. Thus the majority of the moral emotions that have been identified are directed towards others, generated by beliefs about abrogation are of responsibility, and negative. That is, the instigating action was controllable, and the transgressor is regarded as responsible for the negative event or outcome.

Why should a disproportionate number of emotions have these characteristics? When considering emotions as social phenomena, it is evident that they serve to regulate social behavior. Anger, indignation, jealousy, and schadenfreude direct others to desist from what they are doing or face punishment. That is, moral emotions often communicate that the person is doing or has done something wrong or bad, and this is not acceptable. Moral emotions are regulators of moral actions directed toward vice-ridden others. This analysis points out yet another shortcoming of a focus on anxiety—anxiety does not appear have a communicative function among its goals, although it may reveal to others the need for help.

SOME CONCLUDING REMARKS

In psychology the research on emotions has been dominated by the study of facial expressions. Today, the main theme seems to be, or soon will be, the neurological substrates of feelings. Whereas these are reasonable pursuits, they do little to illuminate the dynamics of the emotional process in the classroom.

To compound this restriction, many of the leading theories of motivation, and particularly achievement motivation, are devoid of affects. The well-known goal theories and muddy distinctions between them (e.g., mastery versus performance goals, ego versus task focus) do not call upon emotions as motivators of behavior. Neither are emotions implicated in self-efficacy theory or conceptions that revolve around contrasts between intrinsic and extrinsic motivation. Thus one might (controversially) argue that both the study of emotion and the study of motivation are devoid of insights regarding the antecedents and consequences of emotions in educational settings. This greatly detracts from our understanding of "life in the classroom."

I have not suggested here how to study feeling states. Nor have I examined any motivational theories that I regard as giving emotion their just due. I have, however, attempted to provide some guidelines regarding what emotions in the classroom are worthy of pursuit. I also have contrasted these feeling states with the emotion of anxiety, which is of most concern to educational psychologists. This contrast reveals that the study of emotion in achievement contexts needs to be broadened to include its social context and function, to include positive as well as negative emotions, to embrace other- as well as self-directed feelings, and to recognize that emotions in the classroom have social relations as their source.

For this author, which particular theoretical system is being tested or is guiding one's thinking, what particular methodology is being employed, and even which particular emotion is being examined, is not a critical issue. What is important is that some of the emotions suggested here, and many others that were not touched upon, become more central for educational psychologists.

References

Averill, J. R. (1983). Studies on anger and aggression. *American Psychologist, 38,* 1145-1160.

Bauerle, S. Y., Amirkhan, J., & Hupka, R. B. (2002). An attribution theory analysis of romantic jealousy. *Motivation and Emotion, 26,* 297-319.

Ben-Ze'ev, A. (1992). Pleasure in another's misfortune, Iyyan, *The Jerusalem Philosophical Quarterly, 41,* 41-61.

Berndsen, M., van der Pligt, J., Doosie, B., & Manstead, A. S. R. (2004). The determining role of interpersonal and intrapersonal harm. *Cognitian and Emotion, 18,* 55-70.

Dwyer, S. (2003). Moral development and moral responsibility. *The Monist, 86,* 181-199.

Emmons, R. A., & McCullough, M. E. (2003). Counting blessings versus burdens: An experimental investigation of gratitude and subjective well-being in daily life. *Journal of Personality and Social Psychology, 84,* 377-389.

Feather, N. T. (1989). Attitudes toward the high achiever: The fall of the tall poppy. *Australian Journal of Psychology, 41,* 239-267.

Feather, N. T. (1991). Attitudes toward the high achiever: Effects of perceiver's own level of competence. *Journal of Psychology, 43,* 121-124.

Feather, N. T. (1999). *Values, achievement, and justice.* New York: Kluwer Academic.

Frijda. N. H. (1986). *The emotions.* Cambridge, England: Cambridge University Press.

Frijda, N.H., Kuipers, P., & Ter Schure, E. (1989). Relations among emotion, appraisal, and emotional action readiness. *Journal of Personality and Social Psychology, 57,* 212-228.

Graham, S. (1984). Communicated sympathy and anger to black and white children: The cognitive (attributional) consequences of affective cues. *Journal of Personality and Social Psychology, 47,* 40-54.

Guralnik, D. B. (1971). *Webster's new world dictionary.* Nashville, TN: Southwestern Co.

Hareli, S., & Weiner, B. (2000). Accounts for success as determinants of perceived arrogance and modesty. *Motivation and Emotion, 24,* 215-236.

Hareli, S., & Weiner, B. (2002). Dislike and envy as antecedents of pleasure at another's misfortune. *Motivation and Emotion, 26,* 257-277.

Izard, C. E. (1977). *Human emotions.* New York: Plenum.

McClelland, D. C., Atkinson, J. W., Clark, R. A., & Lowell, E. L. (1953). *The achievement motive.* New York: Appleton-Century-Crofts.

Mikulincer, M., Bizman, A., & Aizenberg, R. (1989). An attributional analysis of social-comparison jealousy. *Motivation and Emotion, 13,* 235-258.

Ortony, A., Clore, G. L, & Collins, A. (1988). *The cognitive structure of emotions.* Cambridge, England: Cambridge University Press.

Smith, R., Parrott, W., Diener, E., Hoyle, R., & Kim, S. H. (1999). Dispositional envy. *Personality and Social Psychology Bulletin, 25,* 1007-1020.

Tangney, J. P., & Fischer, K. W. (1995). *Self-conscious emotions.* New York: Guilford.

Tesser, A., Gatewood, R., & Driver, M. (1968). Some determinants of gratitude. *Journal of Personality and Social Psychology, 9,* 233-236.

Tsang, J., & McCullough, M. E. (2004). Annotated bibliography of research on gratitude. In R. A. Emmons & M. E. McCullough (Eds.), *The psychology of gratitude* (pp. 291-341). New York: Oxford University Press.

Turillo, C. J., Folger, R., Lavelle, J. J., Umphress, E. E., & Gee, J. O. (2002). Is virtue its own reward? Self-sacrificial decisions for the sake of fairness. *Organizational Behavior and Human Decision Processes, 89,* 839-865.

van Dijk, W. W., & Zeelenberg, M. (2002). Investigating the appraisal patterns of regret and disappointment. *Motivation and Emotion, 26,* 321-331.

van Dijk, W. W., Zeelenberg, M., & van der Pligt, J. (1999). Not having what you want versus having what you don't want: The impact of negative outcome on the experience of disappointment and related emotions. *Cognition and Emotion, 13,* 129-148.

Weiner, B. (1986). *An attributional theory of motivation and emotion*. New York: Springer-Verlag.
Weiner, B. (1995). *Judgments of responsibility: A foundation for a theory of social conduct*. New York: Guilford.
Weiner, B. (2006). *Social motivation, justice, and the moral emotions*. Mahway, NJ: Erlbaum.

CHAPTER

6

A Macro Cultural-Psychological Theory of Emotions

CARL RATNER
Institute for Cultural Research and Education

Educators typically emphasize conveying information and facts; rarely have they articulated or modeled the full learning process replete with emotions of confusion, fear, sorrow, apathy, anger, jealousy, pride, and enthusiasm. Because emotions are integral to such educational practices as learning, persuasion, concentrating, and cooperating on projects, it is vital to understand and address them. Understanding emotions requires comprehending both their specific, distinctive qualities (e.g., palpable visceral qualities) and their general psychological features that they share with other psychological phenomena. We may imagine a model that looks like a funnel. At the top are general aspects of psychological phenomena that they have in common. As the funnel narrows, we find "emotions," which are specific psychological phenomena. At the bottom of the funnel stand specific emotions in specific situations (e.g., classrooms). Each lower level of the funnel incorporates upper levels. Consequently, comprehending and addressing specific emotions (in education) requires understanding emotions in general, and psychological phenomena in general.

This chapter shall elucidate an explanation of emotions as psychological phenomena that has relevance to the cultural practice of education. This theory is called "macro cultural psychology." It explains how emotions, and all psychological phenomena, are rooted in macro cultural factors, such as social institutions, artifacts, and cultural concepts. Emotions have cultural origins, characteristics, and functions. Macro cultural psychology has the potential to illuminate the cultural features of emotions that are not only cultivated by the cultural institution of the school (system), but also by other macro cultural factors (e.g., consumerism, entertainment, social class, and cultural ideologies about the origins of intelligence in ethnic groups and individuals), and brought into academic activities inside and outside the school (e.g., homework).

MACRO CULTURAL FACTORS ARE THE BASIS OF GENERAL FEATURES OF PSYCHOLOGICAL PHENOMENA

Vygotsky and his colleagues pioneered macro cultural psychological theory under the name of cultural-historical psychology (cf. Ratner, 2006 for historical and contemporary developments). The theory strives to *explain* the cultural basis, character, and function of emotions, and psychological phenomena in general as grounded in macro cultural factors. We can deduce this explanation from Darwinian theory:

1) Culture is a unique adaptive mechanism that humans use to enhance their survival and fulfillment. Culture is the greatest adaptive mechanism because it coordinates and objectifies the strengths of many individuals to enhance the capability of each. A collective is more powerful, supportive, knowledgeable, stimulating, and enriching than separate individuals who may physically coexist together.
2) Cultural behavior is a distinctive behavior that requires and selects for special behavioral (and biological) mechanisms to direct it. (Behavioral mechanisms, such as instincts, that guide noncultural behavior will be deselected by culture.)
3) Human psychological phenomena are the mechanisms that enable cultural behavior. Psychological phenomena must have special attributes (and biological underpinnings) that are capable of creating, maintaining, and reforming culturally distinctive behavior on which our survival and fulfillment depend.
4) Human culture fundamentally consists of macro factors, such as social institutions (government, corporations, educational systems, family structure, religious organizations), artifacts (technology, art, buildings, clothing), and cultural concepts (time, sex, children, privacy, private property, a fetus).
5) Therefore psychological phenomena must have features that are capable of generating and sustaining macro cultural factors.

The general and specific features of psychological phenomena are based on, geared toward, and congruent with the general and specific features of macro cultural factors.[i] Consequently, understanding the nature of macro factors provides indispensable insight into the nature of psychology.

[i] Culture and psychology develop together phylogenetically and ontogenetically. Conversely, biological programs for behavior, a lack of culture, and a lack of psychology are equally

Macro cultural factors are social (institutional), material (artifacts), and conceptual formations. They are vast, complex, planned, coordinated, administered, objectified, and enduring. They are humanly constructed through struggles among competing groups. They are political in the sense that they are contested and controlled by vested interests. And they are modifiable through conscious, collective action at the macro level. Macro cultural *factors* are the *environment* that exercised selective pressure for the formation of psychological phenomena that have features capable of sustaining *macro cultural factors*. Macro cultural factors selected for psychology in several ways.

On a basic level, cultural construction fosters psychological phenomena by restraining behavior. In order to coordinate behavior, everyone has to restrain their action and consult with others before acting. Social coordination requires separating action from impulses. We do not directly act to obtain food when we feel hunger. Instead, we coordinate a social effort for collectively obtaining food. This separation of behavior from impulse enables the organism to form a symbolic image, or idea, of the object before acting (Greenspan & Shanker, 2004, pp. 36–37; Ratner, 1991, chap. 1; Vygotsky, 1978, p. 26, 35, 40, 49–51). This is the origin of consciousness, the mind, and psychology. They occupy a "space" that is created between impulse and action by social restraint. The impulse loses its power to determine action that it has in the case of noncultural organisms, such as animals and infants. This power is acceded to consciousness, the mind, and psychological phenomena.

Cultural coordination is the impetus for communicating information across individuals, so that each individual is expanded/extended to include the information/knowledge of many. As Vygotsky said: "Social interaction based on rational understanding, on the intentional transmission of experience and thought, requires some system of means. Human speech, a system that emerged with the need to interact socially in the labor process, has always been and will always be the prototype of this kind of means" (Vygotsky, 1987, p. 48).

inseparable. In both sets, any one element depends on the others in its set. It cannot coexist with elements from the other set.

Animals that do not have developed emotions cannot construct and maintain large-scale, coordinated, planned, reformable macro cultural factors; conversely, animals that lack culture cannot develop emotions (Ratner, 1989, 1991, chap 1). Instinctual behavior precludes both culture and emotions (and all psychological phenomena). Instinctual behavior is an automatic, involuntary, unconscious, stereotypical, fixed, invariant (throughout the species), immediate response to a stimulus. There is no mental space, mind, consciousness, or psychology to mediate action. Furthermore, instinctual behavior is essentially individual behavior, not cultural. Each organism is internally (genetically) programmed to act in response to a stimulus. Any apparent coordination of behavior is the result of different individual programs. Thus a queen bee is biologically determined to act in one way, whereas worker bees are programmed to act in another way. This integrated division of labor is the product of each individual acting out its own program. It is not the result of individuals collectively coordinating their behaviors. Animal "society" is based upon individual action that is not enhanced by others; nor can it be improved by its members.

Social communication, in turn, is the impetus for symbolic representation and the entire domain of the mental, the mind, the psychological. Social communication requires that each individual encode and store his or her particular experience in symbolic forms inside his or her head, and then recall and communicate these symbols to others at another time and place. All the others must develop the capacity to decode the symbols to comprehend what they refer to.

The vastness of macro cultural factors is also a main impetus for abstract concepts. Macro cultural factors cannot be known or managed by sensory impressions because they are too vast. One cannot see or hear a government, a war, a university, a transportation system, democracy, or the French language, *en toto*. The entirety of vast macro cultural factors can only be known conceptually. I propose that people develop abstract concepts to manage vast, complex macro cultural factors. The more complex the objects one deals with, the more abstract the concept one needs to develop.

Macro cultural factors operate according to abstract rules that require (select for) abstract concepts. "Hand in homework on time:" "You go to the third tree on the left and I will go to the second big rock on the right to trap the animal that was here last night." "If the forward is double-teamed, pass the basketball to the point guard." "Stay three car lengths behind the truck in front of you." These are examples of abstract cultural rules that require abstract cognitive competencies. Vygotsky explained the relation of communication and abstract concepts as follows: "To communicate an experience or some other content of consciousness to another person, it must be related to a class or group of phenomena. This requires *generalization. Social interaction presupposes generalization and the development of verbal meaning*; generalization becomes possible only with the development of social interaction" (Vygotsky, 1987, p. 48).[ii]

Macro cultural factors additionally select for the capacity to learn and remember cultural norms (van Schalk, 2004), and to merge self/behavior/consciousness with those of others in group action, joint intentionality, collective agency, and collective rationality (Pettit & Schweikard, 2006). In addition, the ability to understand others' intentions (theory of mind, social referencing, social learning, identification) is important for coordinating and predicting behavior, as well as learning and teaching information.

Importantly, the more coherent, stable, and extended over time and space that macro cultural factors are, the more advanced psychological functions

[ii] Vygotsky also emphasized the reciprocal importance of psychology for culture: "social interaction mediated by anything other than speech or another sign system is extremely primitive and limited. Indeed, strictly speaking, social interaction through the kinds of expressive movements utilized by nonhuman animals should not be called social interaction. It would be more accurate to refer to it as *contamination*. The frightened goose, sighting danger and rousing the flock with its cry, does not so much communicate to the flock what it has seen as contaminate the flock with its fear" (Vygotsky, 1987, p. 48).

must be to enable them. Transient, informal, small cultural factors necessitate simpler psychological functions.

Advanced psychological phenomena require a corresponding anatomy (Ratner, 2004b). They require a neocortex. In addition, cultural learning requires that infants are born immature and acquire skills through extended parental nurturing rather than being equipped with them at birth. The neocortex and neoteny are thus ultimately selected for by macro cultural factors. In addition, they depend upon macro factors to provide the resources for nurturing them. The nutrition and time necessary to "feed" the neocortex and neoteny are only supplied by collective labor that raises output beyond what individuals can produce on their own.

This macro model of psychological phenomena applies to emotions. Emotions have all the foregoing properties of psychological phenomena that have been selected for by macro cultural factors and are functional for them:

(a) Emotions animate and sustain cultural behavior. Their passion animates and sustains long-range, persistent behavior that is necessary for forming and sustaining complex macro cultural factors that extend over time and space and encompass millions of individuals—for example a nation's government. Emotions are shaped by, and socially learned and nuanced from, experience with macro factors. Socially organized and shared emotions are cultural factors in their own right (Harre, 1986; Hochschild, 1978). Saturated with the cultural content of macro cultural factors, emotions can direct and respond to macro cultural factors (cf., Ratner, 1991, pp. 156–157, 214–217; Ratner, 1997, pp. 104–105; Ratner, 2000, pp. 22–24).

(b) Human emotions include consciousness of macro cultural factors. Emotions include love for one's country, anger at injustice, dejection about political trends, resentment of a rival country's technical superiority, and admiration for a form of government. Such emotions would not be possible unless they were informed by consciousness of abstract phenomena. Culturally-conscious emotions enable people to develop and respond to social institutions, artifacts, and cultural concepts.

For instance, when students are afraid to hand in homework late, the students' fear is based upon an understanding of the system's (abstract) rule that requires timely homework and punishes tardiness. The fear may motivate action that conforms to and sustains this rule. Students' emotions must be based on a conscious understanding of abstract rules if the rules are to be maintained. If emotions were only sensitive to physical colors and odors, they would not relate to complex macro cultural factors, such as school rules, and they would not contribute to constructing and maintaining them. The macro cultural factors that enhance our survival and fulfillment would then be rendered impossible.

(c) Conscious emotions are consciously known to the individual. We not only become angry, we know that we are angry. Human emotions are conceptualized, or intellectualized, emotions, as Vygotsky said (cf., Ratner, 2004a). Reflecting on our emotions enables us to analyze them, evaluate them, and alter them. This is vital for delaying individual behavior so that stronger, supportive, richer collective cultural behavior can be coordinated. It is also vital for animating new macro cultural factors that can improve on problematical ones.

Construing emotions as subjective processes that form and maintain macro cultural factors eliminates the aura of mystery and irrationality that surrounds them. Emotions are not irrational, uncontrollable, unfathomable, animalistic, unpredictable phenomena that have a life of their own, localized in a special, separate, primitive, nonconscious part of the brain, and overwhelm cognition, behavior, and social life (Ratner, 1989, 2000, 2006 pp. 256–259). On the contrary, emotions *facilitate* social life, thinking, and deliberate behavior (as Aristotle and Vygotsky observed), and are adapted to macro cultural factors. We love culturally-valued things (e.g., country, democracy, consumer goods, body types), and our love may motivate us to preserve them. Our jealousy maintains a cultural value of competing against others in school, work, and consumption. Fear of punishment/failure motivates socially acceptable behavior, such as studying for tests. Excitement motivates us to finish reading the book and solving math problems. Hatred of injustice motivates us to work for democracy.

Because emotions are distinctive aspects of culture and psychology, they can serve as an entrée to culture and psychology. Emotions can be cultivated and aroused to stimulate social behavior. Teachers frighten students to encourage them to study; football coaches stimulate feelings of loyalty in fans to encourage them to support the team.

Emotions arose phylogenetically as adults struggled to construct social institutions, artifacts, and cultural concepts. Rudimentary emotions and rudimentary cultural behavior reciprocally drove each other forward to more sophisticated forms. Emotions did not evolve as a means of individual expression or from infants' "natural tendency" to express themselves and communicate with caretakers or from interpersonal interactions about personal matters.[iii]

The conceptual nature of emotions originated in the need to engage with vast, complex, macro cultural factors such as government, educational institutions, war, economic inflation, racism, or democracy. We could not

[iii] Nor did emotions (and all psychological phenomena) evolve via genetic mutation of individual organisms. They were socially constructed by groups of people in their struggle to form macro cultural factors as adaptive mechanisms. The individualistic model of individual genetic mutation that evolutionary psychologists (and Darwin) espouse does not apply to changes in human behavioral competencies (Ratner, 2006, pp. 201–209).

emotionally respond to them on the basis of simple sensory properties such as a noise or color because they are far more vast than these attributes. This conceptual character of emotions was then extended to all stimuli. For example, we become afraid of an animal in the woods because we recognize it to be a bear that we believe to be dangerous. We do not simply become afraid because of its size or gestures. If we didn't believe it to be dangerous, or if we had a gun with which we could kill it if necessary, we would not fear the bear. Our emotion depends on abstract, conceptual cultural knowledge about things ("bears are dangerous," "this gun will kill the bear"), which is required by cultural life.

The emotions we employ in face-to-face interactions similarly originate at the macro level (Hochschild, 1978). Anger and guilt are based upon ethical and legal values. If a student injures someone in her physical proximity and she caused the injury, she feels guilty. If she did not cause the injury she may feel sad and compassionate for the victim; however, she will not feel guilty. The reason is that guilt is instigated by personal responsibility for a misdeed. If we are not responsible for the misdeed, we do not feel guilty over it. We must (implicitly) know the cultural concept of personal responsibility to feel guilt. Personal responsibility is also the conceptual basis of anger. If Jill injures John by mistake, John has "no right to get angry" (as his teacher will say) because it was a mistake. But if Jill deliberately injures him, he legitimately becomes incensed. The reason is that anger is triggered by the ethical and legal principle that deliberate, willful injury is wrong. Western legal principle distinguishes between willful and accidental injury and dispenses very different punishments for them. This legal distinction is the basis of anger. People must know this cultural concept to become angry.[iv]

Personal expression and communication are derivative functions of macro cultural emotions. The latter are capable of explaining the former because broader, more complex phenomena can explain smaller, simpler ones. The converse is not possible. Personal emotions that convey information about

[iv] The conceptual character of emotions that originate in macro cultural factors is qualitatively different from animal emotions, which have no such origin. Love for one's country, informed by a social consciousness, is incomparable to the attachment a cat feels for her master. Even the human appreciation of a delicious meal is incomparable to an animal's contentment after eating. The human satisfaction is permeated with thoughts, memories, social feelings toward one's companions, and those that are not present. Culinary satisfaction is also permeated by a conscious reflection on the palatability of the food and the quality of the taste. Animals are simply satiated by the full stomach.

Infants' emotional outbursts are also qualitatively different from adult emotions. In Vygotsky's terms, the former is a natural, spontaneous, unconscious, nonsocial, biologically programmed reaction. Adult emotions are informed by consciousness and culture as we have explained above. The fact that animals and infants cannot respond emotionally to abstract things, such as democracy or Michelangelo's *David*, testifies to the fundamental difference between animal and human emotionality.

individuals in face-to-face interactions do not have the scope to generate emotions that are necessary to initiate and sustain and reform broad macro cultural factors.

MACRO CULTURAL FACTORS ARE THE BASIS OF EMOTIONS' SPECIFIC CONTENT

Just as the general properties of macro cultural factors are the basis of general features of psychological phenomena and emotions, so the culturally-specific form and content of macro cultural factors structure the specific form and content of psychological phenomena. Emotions' content, form of expression, intensity, interrelation (or organization), and socialization must be congruent with and vary with particular social institutions, cultural concepts, and artifacts (cf. Ratner, 1991, pp. 76–83; 2000; 2006, pp. 105–108). Romantic love, maternal love, and children's emotions illustrate this point.

Romantic Love

Romantic love in Western countries, and increasingly the world, has a historically unique quality. It is a delirious, impulsive, sensuous, fun-loving, emotion that is elicited by idiosyncratic personality traits in one's partner. Romantic love is divorced from real-life concerns. It is a magical, irrational feeling that overpowers analytical thinking. Modern romantic love is qualitatively different from puritanical love during the colonial period in America, which was a restrained, spiritual, rational affection based upon a partner's moral qualities (cf. Stearns & Knapp, 1993).

The two forms of love embody different institutions, artifacts, and cultural concepts. Puritan love embodied the frugal, hard-working, serious, patriarchal, communitarian features of the petty bourgeois family economy, and Christianity. The family was the economic unit, so familial relations and personal feelings were part of work activity. This is why personal attraction was based upon socioeconomic norms of work: dutiful responsibility, thoughtfulness, self-control, and realism.

Modern romantic love (as opposed to aristocratic, medieval love) was developed during the 18th and 19th centuries in Western Europe and the United States by a different class—the bourgeoisie—as part of living with different macro cultural factors. The macro cultural factors that the bourgeoisie developed, to which romantic love was adapted, included the exclusion of personal relations from commoditized work relations, and their confinement in a private domain. With love situated in a realm of personal relations segregated from serious, calculating, disciplined, impersonal, routine public concerns, it took on the opposite qualities of escapism, irrationality, playfulness, and personalism.

Romantic love is additionally made giddy and delirious by the fact that it is desperately sought and extremely hard to find. When it is encountered, it is cause for exhilaration. Modern capitalist society makes it difficult to establish an intimate relationship with another person. The breakdown of community and the depersonalization of public life are contributing factors. The individualistic ideology of capitalism is another. It makes people attracted to idiosyncratic qualities of individuals, not to their common, social attributes. Encountering idiosyncratic traits in someone that happen to match those of oneself is improbable. When it does happen and romantic love is kindled, a feeling of exhilaration and giddiness ensues.

The sensuousness of romantic love is generated by the sensationalism, hedonism, and materialism that pervade middle-class personal life in the form of consumerism. With personal life so occupied by these issues, romantic love is infected by the heavy emphasis on sensual gratification.

The irrational, hedonistic antisocial qualities of romantic love are cultural phenomena that are shaped by macro cultural factors. These qualities are not due to a natural autonomy, animalism, and irrationality of emotions. They did not characterize Puritan love. In addition, the irrational, delirious quality of romantic love is based on cultural interpretive schemas; it is not devoid of cognition. People *look for* "the right, unique person" because that is the normative ideal; they *reject* the ordinary person; they *know* intimacy is rare and special; they *hope for* it; they *think about* how exhilarating it will be; they are *primed* to experience delirious love. It is not a natural experience.

Romantic love (implicitly) embodied elements of cultural concepts, social institutions, and artifacts of a particular class of people (cf. Ratner, 2000, pp. 12–16; 2006, pp. 105–106). Romantic love also reciprocally animated the bourgeois family structure and personal relations. It promoted a middle-class family structure based upon personal choice, individualism, and separation from the public domain. Romantic love motivated people to regard themselves and others as unique individuals and to become attracted to each other based on idiosyncratic personality attributes. Romantic love also reinforced the distinction between personal life and social life that capitalist work relations initiated. The irrational, magical, impulsive, playful quality of romantic love could only occur outside serious social concerns and public activities. Work, education, health care, and government could never incorporate personal concerns and love because the latter were inimical to serious, public activity. The escapist quality of love thus legitimated the depersonalization of public life. A different cultural form of love that was elicited by humane public treatment of people, and by one's social beliefs and contributions, would motivate people to humanize public activities to find love in that domain instead of in private relations exclusively.

Romantic love demonstrates that the form and content of emotions is culturally organized. Whether emotions are controlled or impulsively expressed, whether they are strong or weak, serious or playful, rational or irrational is a function of macro cultural factors.

Maternal Love

Certain emotional experiences may be developed by social leaders to advance their own control of and profit from macro cultural factors. Military leaders stir feelings of patriotism toward fellow citizens and hatred for the enemy to recruit soldiers for war. Religious authorities cultivate feelings of blind devotion to the faith to recruit religious disciples.

Sociologist Daniel Cook (2004) explains how new emotional experiences in parents and children were cultivated in the 1920s and 1930s by clothing manufacturers and marketers. They hoped emotions would induce them to consume quantities of expensive clothing. This strategy was spelled out in the trade journal, *Infants' Department*, in its inaugural issue in 1927: "If mothers bought for their babies only what was absolutely required, a few yards of diaper cloth, a knitted undergarment or two, and a few dresses would be the limit of their purchases. But the maternal instinct that desires everything that will contribute to the comfort and welfare of the baby is enlisted on the side of the merchant who knows how to create desire and inspire confidence" (Cook, 2004, p. 58).

Clothing merchants cultivated a distinctive new form of mother love that was manifested in continuously seeking out every imagined desire the child had and indulging them through consumer products. Merchants wrote massive quantities of articles and advertisements in trade journals and popular magazines, expressing the following psychological themes that inculcated this maternal love (and children's emotions to be discussed momentarily):

- Mothers should express love for their children effusively.
- Children have an insatiable need for love.
- Children's needs are great and must be satisfied quickly.
- Children like new stimulation; they are dissatisfied (bored) with stable, familiar conditions.
- Children are entitled to things.
- The good life is defined as having more material possessions.
- Children's appearance is very important to their success and happiness.
- Children know themselves and are capable of making choices to make themselves happy.
- Parents do not really understand children's needs and development and should not interfere.
- Children are impulsive, hedonistic, and egotistical.
- Children identify themselves with material objects.
- Children are concerned with peers' opinions.
- Children are unique individuals who require special conditions to thrive. Ordinary things will not bring out the uniqueness of each individual.

- Children pass through distinctive stages of development in quick succession.
- Children want to grow up quickly.
- Mothers are insecure about rapidly changing social norms and how to best bring up their children.

Each one of these emotions or needs had a commercial function involving the consumption of clothes and other commodities. Mothers' love was not only to be satisfied in new ways through clothing and consumer products. The very quality of emotions, needs, and self-concept was transformed. Mothers' love was intensified and extended. It was now to be manifested continuously and effusively so that children would be constantly aware of it and never in doubt about it. Additionally, mothers' love took the form of anxiously seeking out and indulging every desire the child might have.

Children's Emotions

The foregoing list of emotional themes cultivated a new emotionality in children as well as mothers. In contrast to the restrained, rational, diffident emotionality of Puritan children, bourgeois children's emotions were cultivated to be intense, impulsive, insistent, irresistible, egocentric, unquestioned, overtly expressed, hedonistic, and immediately gratified through consumer products. Marketers thus cultivated the generation gap. They unleashed children's desires from parental control so that children could demand more products. Children's independence and individuality had a commercial motive. "Markets and market mechanisms are inseparable from the historical process of elevating the child to more inclusive levels of personhood" (Cook, 2004 p. 68). As the advertising director of *Child Life* magazine said in 1938, "An important factor in the growth and development of the juvenile market is the trend toward stimulating greater self-expression in children" (Cook, 2004, p. 77).

Clothing merchants cultivated an additional psychological phenomenon to instigate consumption of children's clothing. They promoted the idea that biopsychological development occurs in distinctive, sequential stages of short duration, which children want to traverse rapidly. Delayed development in one stage was claimed to be emotionally frustrating and psychologically damaging. Stage psychology had the economic purpose of enabling merchants to market distinctive clothing (toys, games, and other products) to each stage (just as creating distinctive psychological disorders creates a market for new medications). Each psychological stage was converted into a market for new products. Children were said to need distinctive clothes that were appropriate to their momentary psychological stage. Outmoded clothing would retard psychological development to the next stage because peers and teachers—and the child herself—would treat the child according to the younger styles that she wore. The more stages, and the more rapidly that children traversed them, the more new, distinctive clothes (and products) could be sold.

Whereas clothing merchants said they were simply designing clothes to meet the natural developmental stages and needs of the child, they actually cultivated the child's stages and needs to meet the economic demand for profit. For example, clothing designed for toddlers was the mechanism for instantiating the notion of toddler as a psychosocial stage of life: "In 1936 the 'toddler' *as a commercial persona or construct* began to take shape. . . . The term 'toddler' began to be used with great frequency as a size range and as a merchandising category, and soon after, as an age-stage designation" (Cook, 2004, p. 86, emphasis added). "Commercial interests and concerns coalesce and interact to essentially institutionalize a new category of person and new phase of the life course" (Cook, 2004, p. 85). Department stores also segregated age-graded products into separate departments (Cook, 2004, pp. 115–116), thereby physically objectifying and promoting the notion of distinct developmental stages.

The ontogenetic category, toddler, that parents (and psychologists) regard as natural, originated as a commercial category invented by businessmen who objectified and promulgated it through clothing products (Cook, 2004, pp. 18, 19).[v] Psychology became a commodity that served money; things and money did not serve to express psychology. Capitalism harnessed psychology to generate more capital (Money - Psychology - Money). Psychology did not utilize monetary things to express and develop itself (Psychology - Money - Psychology). Clothing did not simply *express* psychological stages, clothing *defined* the stages; psychological stages conformed to the distinctions that were displayed in clothing styles (Cook, 2004, p. 97). This case study exemplifies an important way that psychology is formed at the macro level.

PRACTICAL APPLICATIONS OF THE MACRO CULTURAL THEORY OF EMOTIONS TO EDUCATION

Macro cultural psychology emphasizes that the cultural character of emotions must be considered and altered to resolve interpersonal and social problems. The commercial qualities of emotions (and needs, motives, attention,

[v] Clothing merchants did not create age distinctions on their own, although they articulated these ideas in ways that specifically induced consumerism. Other macro cultural factors contributed as well. Schools had already divided students into age-graded classes (Cook, 2004, p. 98). In addition, child psychologists such as Hall and Freud, who had articulated the notion of developmental stages and the notion that child psychology was different from adult psychology and could not be understood by ordinary parents, had articulated some of the psychological ideas of the merchants.

concentration, memory, and reasoning) that we have just discussed (cf. Dawson, 2003) impede serious learning. Under the pressure of consumerism and its domination of entertainment, news, sports, products, and advertisements, many students have become habituated to attend to appearances; seek continuous sensual pleasure and material satisfaction; seek novel and more intense forms of sensory stimulation; have an attention span that is limited to transitory sensational images (for every two minutes of TV programming there is one minute of commercials); engage in uncritical reasoning that accepts superficial, preposterous images and associations; are titillated by trivia and have low motivation for intellectual activity; and desire success for minimal effort.[vi]

Teachers must pointedly address students' cultural psychology and transform it to one that is conducive to serious study. Teachers must point out to students how their psychology has been adversely affected by consumerism and other cultural influences. They should systematically remediate each psychological function (attention, emotions, reasoning, memory, sensationalism) to make it conducive to serious learning. This requires helping students alter their activities outside school—to choose different forms of entertainment, interests, and peer groups—as books, such as *Beyond the Classroom*, urge. It involves working with parents to support these changes. Teachers will fail to educate students if they treat them as having neutral needs and emotions that can be readily oriented toward serious learning by simple encouragement to "pay attention," "study hard," and "notice how interesting the material is." Such an approach fails to consider the cultural basis, character, and function of psychological phenomena that are involved in academic work.

Macro cultural psychology is applicable to resolving psychological problems on the group and individual levels. In dealing with a classroom, or school we deal with the largest common denominator of cultural psychology that encompasses the most individuals. If research, or logic, indicates that consumerism has a pervasive influence on academic habits, then if we address

[vi] The pedagogy of American schools recapitulates these anti-academic psychological tendencies. Stigler & Perry (1988) found that American classrooms are more fragmented and incoherent than Japanese classes (p. 46). American teachers shift among topics far more rapidly than Japanese teachers. Seventy-five percent of all five-minute instructional segments [of a math lesson in fifth grade] in Japan focused on only one problem, compared to only 17% of the segments in Chicago (p. 47). Japanese teachers frequently devote an entire 40-minute math lesson to one or two problems. This never happened in American classes. Additionally, American teachers rarely explain relationships among different math topics and problems. They concentrate on individual problems discretely. "More time is spent making sure students have a blue crayon than to conveying the purpose of the three segments on measurement" (p. 50). Similarly, "in American first grade classrooms, a total of 21% of all segments contain transitions or irrelevant interruptions [such as handing out materials, checking on crayons, chatting] compared to 7% in Japan" (p. 46). Additionally, teachers lead student academic activities far less in the United States than in Asia: "No one was leading the students' [mathematical] activity 9% of the time in Taiwan, 26% of the time in Japan, and 51% of the time in the U.S." (p. 37).

and counter this influence in a class or school, we are likely to help a large number of individuals, even though some individuals have a different cultural psychology because they were exposed to contradictory cultural factors, such as art.[vii] Dealing with the largest common denominator is like immunizing a risky population against a common disease, even though not every individual is likely to contract the disease.

A macro cultural approach to psychological problems on the individual level involves tracing an individual's problem to his or her particular cultural experiences. Poor academic habits, or debilitating emotional reactions, are not only due to consumerism. They may spring from any number of cultural influences to which one is exposed.[viii] In addition, cultural models engender psychological problems that one employs as coping mechanisms to handle cultural conditions and treatments. Individuals draw on cultural models, such as body image, sexual behavior, violent behavior, self-blame, masculine and feminine stereotypes, spiritual notions of body and soul, and guns as ways of coping with cultural experiences (Ratner, 2006, pp. 170–174). Debilitating cultural coping mechanisms that an individual employs can be identified and remediated by informed macro cultural psychologists.

A cultural psychological perspective on emotional problems not only emphasizes their cultural content, but their conscious psychological character as well. Recall that cultural psychological phenomena are conscious because that is essential for their ability to react to and animate vast, complex macro cultural factors.

Beck (1988) has demonstrated the cognitive underpinnings of emotions through analyses of individual cases. Extreme anger or depression is generated by interpretations and assumptions about another's behavior. A student becomes enraged with a classmate because he interprets her action as insulting. The "emotional problem" is not a problem with emotions, per se; it is not due to defects in endemic mechanisms, such as faulty neurotransmitters and neural connections in certain areas of the brain that distort emotions. The term "emotional problem" is thus a misnomer. It creates the impression that there is something internal to the operation of emotions themselves that generates a localized deficit within emotions. However, a problem that has emotional manifestations, such as angry outbursts, is the expression of a more general psychological problem that involves attention, perception, memory, self-concept, and a vast cultural setting that includes social treatment, physical infrastructure, incidence of crime, job opportunities, and cultural values.

[vii] A society does not necessarily require every individual to adopt a given behavior, such as consumerism, as long as the majority of people adopt it, and as long as the non-normative behavior does not directly threaten the norm.

[viii] Individual problems may also have idiosyncratic sources in unique family interactions (Laing, 1964). However, these are not the province of cultural psychology (cf. Ratner, 2002, p. 93 for discussion of idiosyncratic and cultural factors in individual psychology).

CONCLUSION

Macro cultural psychology is a Copernican shift in our understanding of emotions and psychology in general. Whereas mainstream psychology explains culture in terms of the individual, adults in terms of childhood experiences, the human in terms of animal processes, the large in terms of the small, the complex in terms of the simple, and the extrinsic (culture) in terms of the internal (mind, biology), macro cultural psychology explains the small, the simple, the individual, the child, and the internal in terms of stimulation and organization by the large, the complex, the adult, and the extrinsic (culture).[ix] Of course individuals, psyches, and biologies are the active agents that form macro cultural factors. However, (1) they do so collectively, not as separate individuals, and (2) the macro cultural factors they form then outrun or transcend individuals and actually constrain their behavior. For instance, individuals form a school; however, the rules, budgets, and physical infrastructure they create then become structures that require that individuals maintain them. Individuals adjust their behavior to maintain the structure. Social structures are not reified; however, they do structure behavior.

Macro cultural psychology takes facts that are traditionally overlooked or regarded as marginal—for example, cultural variations in emotions and culturally-oriented emotions, such as love for one's country and delight over viewing Michelangelo's sculpture *David*—and construes them as prototypes of human emotionality. Rather than being extensions of simpler, natural, universal "basic emotions," such as fear and love, *these cultural emotions are the basic form of human emotions*. Emotions that are invoked on the interpersonal level are extensions of macro features of emotions, not vice versa. Consequently, micro level emotions should be identified with cultural terms. We should speak of bourgeois romantic love, or Puritanical love, or capitalist maternal love, rather than love, in general. We should speak of Buddhist sadness and Western sadness, rather than sadness in general. Speaking of emotions in general, abstract terms creates the impression that emotions are universal and natural and should be studied from the perspective of natural science. This is a false impression that is contradicted by the cultural basis, character, and function of emotions and all psychological phenomena.

[ix] A formulation of this is the extended mind hypothesis in cognitive psychology. It emphasizes that cognitive processes depend upon artifacts, and are extended in artifacts (exograms), beyond the individual mind and body. (cf. Sutton, 2005; Clark, 2006; Clark & Chalmers,1998; Carruthers, 1996).

References

Beck, A. (1988). *Love is never enough*. New York: Harper

Carruthers, P. (1996). *Language, thought, and consciousness*. Cambridge, England: Cambridge University Press.

Clark, A., & Chalmers, D. (1998). The extended mind. *Analysis, 58*, 7-19.

Clark, A. (2006). Material symbols. *Philosophical psychology, 19*, 291-307.

Cook, D. (2004). *The commodification of childhood: The children's clothing industry and the rise of the child consumer*. Durham, NC: Duke University Press.

Dawson, M. (2003). *The consumer trap: Big business marketing in American life*. Urbana: University of Illinois Press.

Greenspan, S., & Shanker, S. (2004). The first idea: How symbols, language, and intelligence evolved from our primate ancestors to modern humans. Cambridge, MA: Da Capo Press.

Harre, R. (1986). *The social construction of emotions*. Oxford: Blackwell

Hochschild, A. (1978). Emotion work, feeling rules, and social structure. *American Journal of Sociology, 85*, 551-575.

Laing, R.D. (1964). *Sanity, madness, and the family*. New York: Basic Books.

Pettit, P., & Schweikard, D. (2006). Joint action and group agents. *Philosophy of the Social Sciences, 36*, 18-39.

Ratner, C. (1989). A social constructionist critique of naturalistic theories of emotion. *Journal of Mind and Behavior, 10*, 211-230.

Ratner, C. (1991). *Vygotsky's sociohistorical psychology and its contemporary applications*. New York: Plenum.

Ratner,. C. (1997). *Cultural psychology and qualitative methodology: Theoretical & empirical considerations*. New York: Plenum.

Ratner, C. (2000). A cultural-psychological analysis of emotions. *Culture and Psychology, 6*, 5-39.

Ratner, C. (2002). *Cultural psychology: Theory and method*. New York: Plenum.

Ratner, C. (2004a). Vygotsky's conception of child psychology. In R. Rieber (Ed.), *The essential Vygotsky* (pp. 401-413). New York: Plenum.

Ratner, C. (2004b). Genes and psychology in the news. *New Ideas in Psychology, 22*, 29-47.

Ratner, C. (2006). *Cultural psychology: A perspective on psychological functioning and social reform*. Mahwah, NJ: Erlbaum.

Stearns, P., & Knapp, M. (1993). Men and romantic love: Pinpointing a 20th century change. *Journal of Social History, 27*, 769-793.

Stigler, J. & Perry, M. (1988). Mathematics learning in Japanese, Chinese, and American classrooms. In G. Saxe & M. Gearhart (Eds.), *Children's mathematics* (pp. 27-53). San Francisco: Jossey-Bass.

Sutton, J. (2005). Exograms and interdisciplinarity: History, the extended mind, and the civilizing process. In R. Menary (Ed.), *The extended mind*. London: Ashgate.

Van Schalk, C. (2004). *Among orangutans: Red apes and the rise of human culture* Cambridge, MA: Harvard University Press.

Vygotsky, L.S. (1978). *Mind in society*. Cambridge, MA: Harvard University Press.

Vygotsky, L.S. (1987). *Collected works, vol. 1*. New York: Plenum.

PART III

Students' Emotions in Educational Contexts

CHAPTER

7

The Role of Affect in Student Learning: A Multi-Dimensional Approach to Considering the Interaction of Affect, Motivation, and Engagement

ELIZABETH A. LINNENBRINK
Duke University

With the growing interest in the role of emotions in education, it is important to consider both the theoretical background for current work and existing empirical findings. In the current chapter I focus on the development of a dynamic model of motivation, affect, and engagement. The intent of this chapter is to highlight the theoretical basis of the proposed model and to discuss the empirical support for this model, based on a series of experimental and correlational studies conducted in laboratory and classroom settings. I begin by briefly discussing the theoretical model of affect that I use to frame this work. I then turn to the integration of affect and motivation, followed by a discussion of affect and engagement in schooling. Finally, I consider whether affect mediates the relation between motivation and engagement, and make suggestions for future research and educational practice based on these findings.

A MULTI-DIMENSIONAL MODEL OF AFFECT

Following Rosenberg (1998), affect can be thought of in terms of states and traits. Trait-like affect reflects a general way of responding to the world, which varies by person but is relatively stable over time. In contrast, state-like affect reflects a response to the changing environment that is based on the situation rather than on personality differences or personal tendencies of responding. Affective states consist of mood states and emotion states. Thus while moods and emotions can be differentiated in terms of intensity and duration (Rosenberg, 1998; Schwarz, 1990; Schwarz & Clore, 1996), this distinction is not made in the current chapter, as the focus is on the more general category of affective states.

Affective states can be further defined based on the underlying dimensions of valence (pleasantness)[1] and activation (e.g., Russell & Feldman Barrett, 1999; Tellegen, Watson, & Clark, 1999; Thayer, 1986). In particular, Feldman Barrett and Russell (1998) argue that the intersection of *activation*, which refers to arousal, mobilization, and energy, and *pleasantness*, which refers to valence or hedonic tone, can be used to describe an array of core affects resulting in a circumplex model of affect (see Figure 1). In this model the crossing of these dimensions is also important, such that activated pleasantness (e.g., excitement) is distinct from deactivated pleasantness (e.g., relaxation). Each affective state within the circumplex can also vary in terms of intensity. In educational settings the distinction between activation and valence may be especially important. For instance, activated unpleasant affect may lead to more intense engagement than deactivated unpleasant affect. Happiness (pleasant, neutral activation) may also lead to different patterns of learning and engagement than excitement (activated pleasant).

Circumplex models of affect differ from other current perspectives on affect in education. For instance, the control-value theory of emotions in education (Pekrun, Goetz, Titz, & Perry, 2002) focuses specifically on emotions and does not consider more general mood states. In addition, rather than differentiating emotional states based on valence and activation, it categorizes emotions based on appraisal processes and highlights the dimensions of object focus and valence (Pekrun, in press). Other research on students' affect in educational settings focuses on specific affective states, such as test anxiety (Hill & Wigfield, 1984), interest as an affective state (Ainley, in press), and shame (Turner, Husman, & Schallert, 2002) rather than general models of affect. In contrast to these other approaches, the circumplex model presents a more comprehensive model that can be used to integrate a variety of affective states, including moods and emotions, in educational settings.

[1] In prior work, I have referred to the valence dimension as positive/negative.

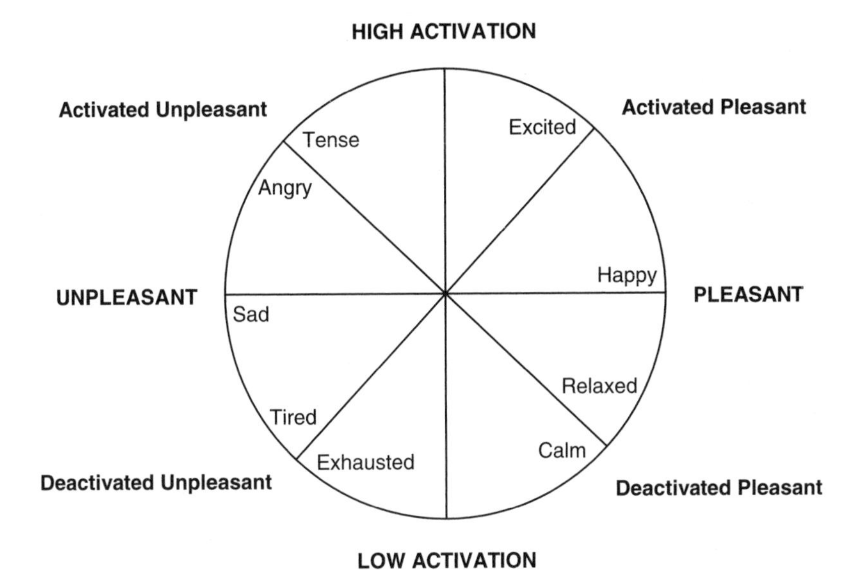

FIGURE 1

Affective circumplex. Figure adapted with permission from Feldman Barrett and Russell (1998), Independence and bipolarity in the structure of current affect, *Journal of Personality and Social Psychology, 74*(4), 967–984, published by the American Psychological Association.

INTEGRATING AFFECT, MOTIVATION, AND ENGAGEMENT

Using a circumplex model of affect, my colleagues and I have developed a general model that integrates research on achievement goal theory, affective states, and school engagement (Linnenbrink & Pintrich, 2002a, 2003, 2004). This integration across the three major areas is rather complex; therefore I begin by considering how to integrate research on motivation and affect. Next I turn to the research on affect and engagement. Finally I consider a larger model that integrates all three components.

Motivation and Affect

For the purposes of this chapter, I use achievement goal theory as the theoretical lens through which to consider motivation. Achievement goal orientations are useful in predicting a variety of school-related outcomes including affect and engagement (Dweck & Leggett, 1988), and thus fit well into the proposed integrative model. According to achievement goal theory, goal orientations provide a framework for interpreting and reacting to events (Dweck & Leggett, 1988). There are thought to be two primary goal orientations that provide the reasons why students engage in achievement behavior:

a mastery goal orientation, where the focus is developing one's competence, and a performance goal orientation, where the focus is demonstrating one's competence. These goal orientations are thought to emerge and develop in response to one's schooling experiences; as such the context has an important influence on the goal orientations that students endorse in any particular setting. Recently, both Elliot (1999) and Pintrich (2000) have suggested that goal theory be expanded to include approach and avoidance dimensions. However, the empirical studies discussed in this chapter focus on approach goal orientations only; therefore avoidance goal orientations are not discussed.

To integrate affect into achievement goal theory, Paul Pintrich and I (Linnenbrink & Pintrich, 2002a) developed an asymmetrical, bidirectional model of achievement goals and affect. The proposed model suggests there is a reciprocal but somewhat asymmetric relation between affect and goal orientations. Our hypotheses regarding achievement goals as predictors of affect are based largely on Carver and Scheier's (1990) control-process model of self-regulation as well as research on achievement goal orientations. The control-process model suggests that affect may vary based on whether one is approaching or avoiding a particular goal and one's rate of progress towards (or away from) those goals. Approach goals are generally associated with elation (when one is approaching the goal at a standard or close to standard rate) or sadness (when one is not approaching the goal at a standard or close to standard rate). Accordingly, a student with a mastery-approach or performance-approach goal orientation would be expected to experience elation if he was successfully approaching his goal and sadness if he was not.

However, we might expect some differentiation in the frequency of these types of affect depending on the type of goal orientation. It seems plausible that a person may be more successful in approaching a mastery goal, because the standard of progress is set relative to the self (improvement or learning), making it more likely that one would experience a pleasant affective state. Even with insufficient progress, a student with a mastery-approach goal would not be expected to feel any strong unpleasant affect, as the lack of progress should signal that one is not trying hard enough and should not reflect negatively on one's view of oneself. In this sense, the attribution for lack of progress would likely be towards lack of effort rather than lack of ability, and would thus help to protect a negative evaluation of the self. In fact, individuals with mastery goals often view difficult situations as challenging and may take pleasure in their attempts to master a difficult task, even when success is not readily apparent.

In contrast, a student who endorses a performance-approach goal orientation will likely evaluate her performance relative to her peers. This normative standard may make a student with a performance-approach goal more likely to feel as if she is not making sufficient progress (as only a limited number of people can outperform their peers), resulting in sadness. Furthermore, given the emphasis on demonstrating competence, anxiety is expected, since one is putting one's sense of self on the line. However, when a student is making sufficient progress in his orientation to demonstrate his competence, he should experience elation and pride.

The Linnenbrink and Pintrich (2002a) model also proposes that affective states, specifically moods, can influence students' goal adoption. This is thought to occur through perceptions of the classroom goal structure and through a direct influence on personal goal adoption. I focus here on the proposed link between affect and personal goal adoption. Experiencing pleasant affect may make adopting an approach rather than avoidance goals more likely, but may not distinguish between mastery and performance. That is, students who experience pleasant affect may perceive that they have the resources available to approach a certain outcome, making it more likely that they focus on approaching a goal for understanding or a goal for demonstrating their competence. In contrast, students who experience unpleasant affect may feel that they do not have the resources to approach a particular goal; instead, these students may focus on trying to avoid unwanted outcomes. Unpleasant affect may also signal to a person that there is a threat (Schwarz, 1990), making it more likely that the person focuses on avoiding the threatening situation or preventing a potential threat from occurring.

My colleagues and I have found some support for this model in a series of correlational studies. We assessed the goal orientations that upper elementary, middle school, and college students endorsed for specific tasks (e.g., series of mathematics problems, reading a passage on Newtonian physics) and examined the affective states that emerged during these tasks. Across all of the studies, mastery-approach goal orientations were positively related to pleasant affect and negatively related to unpleasant affect (Linnenbrink, 2005; Linnenbrink, Hruda, Haydel, Star, & Maehr, 1999; Linnenbrink & Pintrich, 2002b, 2003; Linnenbrink, Ryan, & Pintrich, 1999). In most of these studies, we were not able to differentiate between the various levels of activation, as the pleasant and unpleasant measures included both activating and deactivating indicators. However, a few studies included scales that allowed for more fine-tuned analyses. For instance, we have some evidence that mastery goal orientations are associated with an increase in activated pleasant affect (e.g., excited) and a decrease in activated unpleasant affect (e.g., anxious) (Linnenbrink, Hruda et al., 1999; Linnenbrink & Pintrich, 2002b, Study 2; Linnenbrink, Ryan et al., 1999). Using three bipolar indicators of affective states, we also found that middle school students who endorsed mastery-approach goals while solving number sequences reported feeling more happy than sad (valence), more calm than tense (deactivating pleasant-activating unpleasant), and more excited than tired (activating pleasant-deactivating unpleasant) (Linnenbrink & Pintrich, 2003).

The pattern of findings was not as clear for performance-approach goal orientations. In general, performance-approach goals were either unrelated (Linnenbrink & Pintrich, 2003) or positively related to pleasant affect (Linnenbrink, 2005; Linnenbrink & Pintrich, 2002b). This pattern is not surprising given that performance-approach goals may lead to pleasant affect, especially if students perceive that they are making sufficient progress towards those goals. For unpleasant affect, the findings ranged from no relation (Linnenbrink, 2005; Linnenbrink & Pintrich, 2002b, 2003, Study 1) to a positive relation (Linnenbrink, Hruda et al., 1999; Linnenbrink & Pintrich, 2003, Study 2; Linnenbrink, Ryan et al., 1999). When we differentiated based on activation and valence, we found that students who reported endorsing performance-approach goals reported higher levels of activated unpleasant affect (Linnenbrink, Hruda et al., 1999; Linnenbrink, Ryan et al., 1999). However, we did not find a significant relation when we used a bipolar measure of activating unpleasant affect (e.g., tense to calm) or when we used a measure that assessed neutral to high unpleasant affect (Linnenbrink & Pintrich, 2002b, Study 2, 2003, Study 1). A study examining unpleasant affect in small groups further suggested the comparison group (students within one's small group or the whole class) may change students' affective experience, such that students who evaluated their competence relative to others in their group experienced unpleasant affect (high and low activation), but those who judged their competence relative to the whole class did not report experiencing more unpleasant affect (Linnenbrink & Pintrich, 2003, Study 2). This pattern of findings for performance-approach goals is generally in line with the proposed model; the prevalence of in-significant and mixed findings may suggest that there are moderators that have not been considered or may simply reflect variability in the degree to which students are successfully approaching a goal to demonstrate competence.

Two experimental studies (Linnenbrink & Pintrich, 2001) conducted with college students complement the correlational studies and help to tease apart the potential bidirectional relation between affective states and goal orientations. In the first study, students were induced into one of the three mood states (pleasant, unpleasant, neutral) using a Velten mood induction and asked to work on a series of reasoning problems; students' self-reported goal orientations were assessed in relation to the reasoning problems to see if students endorsed different goal orientations depending on their affective state. Participants in the unpleasant mood condition were less likely than those in the pleasant or neutral mood condition to endorse mastery-approach goals during the reasoning problems; however, the mood condition was not related to the endorsement of performance-approach goals. Although we had hypothesized that pleasant affective states would be associated with the endorsement of both mastery and performance goals, the findings suggest that mastery goal adoption may be more easily influenced by affective states than performance-approach goal adoption.

In the second study we experimentally manipulated achievement goal orientations (mastery-approach, performance-approach) and examined the resultant affective state that emerged while students solved reasoning problems. The goal condition effects were significant for females but not for males. Female students in the performance-approach condition reported more activated, unpleasant affect than females in the mastery condition. There were no significant differences in activated, pleasant affect based on goal condition. Again, the failure to find any difference for pleasant affect may be because performance-approach goals can be associated with pleasant affect or may be unrelated; thus in future studies it may be important to determine whether there are other factors, such as perceived success, that moderate this relation.

Affect and Engagement

A second critical piece for developing a model that integrates affect, motivation, and engagement is the link between affect and engagement. Here, I focus on both behavioral and cognitive engagement. Behavioral engagement refers to effort and persistence and is distinct from cognitive engagement in that the emphasis is on the amount or quantity of engagement rather than the quality of thought or type of engagement (Fredricks, Blumenfeld, & Paris, 2004; Pintrich, 2000). In contrast, cognitive engagement refers to the *quality* of one's thinking in terms of cognitive strategies (e.g., elaboration, rehearsal), metacognitive strategy use, and self-regulated learning. Also included in this discussion of cognitive engagement are students' actual learning or achievement and their general cognitive functioning (e.g., working memory).

Behavioral Engagement and Affect A number of social psychological theories are relevant to understanding how affect might influence behavioral engagement. For instance, Schwarz and colleagues' (Schwarz, 1990; Schwarz & Clore, 1996) affect-as-information model suggests that when a person is in an unpleasant mood, he is motivated to respond to and pay attention to the details in the situation. When a person is in a pleasant mood, she is not motivated to attend to the situation and will therefore use strategies requiring less effort, such as schemas, to interpret and react to the situation. Similarly, Carver and Scheier's (1990) control-process model suggests that pleasant affect signals that one is progressing at a sufficient rate towards one's goals. As such, an individual might "ease-off" her goal pursuit, especially if multiple goals are being pursued, allowing her to turn her limited resources to pursuing a different goal. In contrast, sadness (for approach goals) and anxiety (for avoidance goals) signals that one is not making sufficient progress towards one's goals; accordingly, unpleasant affect may lead to prolonged and perhaps intensified engagement.

This view that pleasant affect reduces engagement is not supported by all theoretical perspectives. For instance, Bless' (2000) mood-and-general-knowledge theory suggests that the reliance on general schemas under a pleasant mood occurs because these schemas are typically useful in benign situations, but does not signal a lack of motivation; thus the individual may use those "saved" resources for processing other aspects of the situation. Fredrickson's (2001) view on pleasant affect also suggests that it helps to broaden one's thought-action repertoire and build resources. In this sense, pleasant affect should not lead to disengagement, but may instead lead to engagement in other ways.

In summary the majority of social psychological theories suggest that unpleasant affect is beneficial in terms of behavioral engagement, but are less clear regarding pleasant affect. These perspectives, however, do not consider the interplay between valence and activation. For instance, one might expect that activated unpleasant affect, such as anxiety, should lead to some type of prolonged engagement, as anxiety should enhance vigilance. However, deactivated unpleasant affect, such as feeling tired or exhausted, should lead to decreased levels of engagement. Indeed, Baumeister (2000) suggests that ego depletion, which is characterized in terms of psychic exhaustion, lowers one's ability to self-regulate and may therefore be associated with decreased engagement. For pleasant affect, the majority of the findings focus on either pleasantly valenced affect (and ignore activation) or activated pleasant affect (e.g., Fredrickson, 2001).

The empirical evidence regarding the relation of affect to behavioral engagement in academic settings is complex, with different patterns emerging for different types of affect. For instance, Pekrun et al. (2002) found that pleasant, activating emotions such as enjoyment were associated with effort. In contrast, both activating (anger, anxiety) and deactivating (boredom) unpleasant emotions were associated with lower levels of effort, although the relation was much stronger for boredom than anger and anxiety.

In a series of correlational studies, my colleagues and I have investigated how affective states relate to students' persistence and effort while working on science and math activities. Using scales that assessed pleasant and unpleasant affective states, we found that pleasant affect was positively correlated with behavioral engagement, whereas unpleasant affect was negatively correlated with behavioral engagement (Linnenbrink & Pintrich, 2003, Study 2). When we considered both valence and activation, we found that both activating and deactivating unpleasant affective states were associated with lower levels of behavioral engagement for both children and young adults, but that activating and deactivating pleasant affect was unrelated to engagement (Linnenbrink, 2004; Linnenbrink, Kelly, & Kempler, 2005). Using bipolar measures, we found that high levels of excitement were associated with increased engagement, whereas valence (sad-happy) was unrelated (Linnenbrink et al., 2005, Study 2; Linnenbrink & Pintrich, 2003, Study 1). The results were mixed for calm-tense; one study showed no significant relation and another indicated that feeling more tense than calm was associated with disengagement.

In general, these findings suggest that pleasant affect does not undermine behavioral engagement, and may even enhance it, especially when it is activated pleasant affect. Unpleasant affect, however, seems to undermine behavioral engagement, regardless of activation level. These findings are not in line with theories arguing that unpleasant affect plays a regulatory role and instead suggest that both activating and deactivating unpleasant states undermine engagement. This highlights the difficulty with applying laboratory studies to real world settings, such as classrooms. Indeed, the structure of the classroom may be critical in understanding the relation between affect and behavioral engagement. For example, in several of our studies, engagement was assessed within a small group context. Thus other factors, such as how well one got along with other group members, might have played a significant role in shaping students' affect. Furthermore, it is possible that the direction of the relation of affect and behavioral engagement is reversed, such that students may have first disengaged and then felt tired or frustrated, rather than the affect predicting the disengagement.

Cognitive engagement and affect With respect to the quality of students' engagement (e.g., cognitive processing, cognitive strategy), social psychology theories are again potentially useful. These theories, which focus on how cognitive processing influences affect as well as how affect influences cognitive processing, are rather wide-ranging (for reviews, see Dalgleish & Power, 1999; Forgas, 2000). I focus here on how affect influences the way in which information is processed and the way in which one approaches a particular situation (e.g., Bless, 2000; Fredrickson, 1998; Schwarz, 1990). Whereas these models differ in important ways, they generally suggest that pleasant affect leads to heuristic processing, including the use of scripts to more easily process information. The model also suggests that unpleasant affect leads to more systematic, analytic processing, in which the individual must focus on the details of the environment (Bless, 2000; Schwarz & Clore, 1996). Fredrickson (1998) also suggests that positive, activating emotions, such as joy, help to broaden possible thought-action repertoires, resulting in the pursuit of novel, creative, unscripted thoughts and actions. In academic settings we might expect that metacognitive strategies and elaboration would be activated under pleasant affective states, but rehearsal and other detail focused strategies would be activated under unpleasant affective states. With respect to actual performance, the relation of affect to learning might depend on the task demands, such that pleasant affect is beneficial if broad, heuristic processing is needed, and unpleasant affect is beneficial when more detail-focused, analytical processing is necessary.

An alternative perspective focuses on differences in cognitive capacity. For example, the resource allocation model suggests that one's cognitive capacity is limited when one is in either a depressed or happy mood state (Ellis, Seibert,

& Varner, 1995; Ellis & Ashbrook, 1988). In essence, being in a pleasant or unpleasant mood results in task-irrelevant processing that "clutters" working memory, making it more difficult to attend to and process the task at hand. The detrimental effects of affect on cognitive processing are expected for complex tasks that require high levels of cognitive processing; simple tasks that do not require extensive use of working memory should not be affected by one's current mood state. This same argument can be applied to activation as well as valence. Specifically, Revelle and Loftus (1990) suggest that arousal (activation) facilitates working memory functioning for low-load tasks and hinders processing for high-load tasks. In this way, pleasant and unpleasant affective states and high arousal states should undermine cognitive processing on complex tasks. In academic settings this perspective suggests that neutral valence states and low activation states are beneficial for complex tasks. Furthermore, the use of complex cognitive strategies would likely be inhibited by the experience of pleasant or unpleasant affect, since the "cluttering" of working memory would interfere with these strategies that require a substantial role for the central executive processor.

Prior research examining the relation between cognitive processing and affect in academic settings found that pleasant activating emotions (hope, enjoyment) were associated with flexible modes of thinking, a tendency to elaborate or draw connections among ideas, and a tendency to engage in metacognitive or self-regulatory strategies (Pekrun et al., 2002). Unpleasant emotions (anger, anxiety, boredom) were negatively related to elaboration and were associated with external regulation rather than self-regulation. In our own work, we focused on elaborative and metacognitive strategies rather than rehearsal.

For unpleasant affect, we would expect that elaborative and metacognitive strategy use would be undermined, following either the mood-and-general knowledge structures model (Bless, 2000) or the resource allocation model (Ellis & Ashbrook, 1988). Across a variety of correlational studies conducted with children, adolescents, and young adults, we found that general measures of unpleasant affect, activating unpleasant affect, deactivating unpleasant affect, and bipolar indicators of activating unpleasant affect (calm-tense) were unrelated to elaborative and metacognitive strategy use in the domains of mathematics and science (Linnenbrink, 2004; Linnenbrink et al., 2005; Linnenbrink & Pintrich, 2002b, 2003). This pattern of findings is not consistent with the resource allocation view, but does fit with the notion that unpleasant affective states are associated with detailed rather than heuristic processing. Under this perspective, it is possible that negative affective states are unrelated to the use of heuristic strategies, such as elaboration or metacognition, rather than negatively related.

For pleasant affect, the mood-and-general-knowledge-structures model (Bless, 2000) would predict a positive relation, whereas the resource allocation model (Ellis & Ashbrook, 1988) would predict a negative relation. Our findings regarding pleasant affect were inconsistent. Pleasant affect was either positively related to metacognitive and elaborated strategy use or unrelated (Linnenbrink, 2004; Linnenbrink et al., 2005; Linnenbrink & Pintrich, 2002b, 2003). Unfortunately, using different indicators of affect did not help to clarify the picture, as the findings were mixed regardless of whether general scales of pleasant affect (deactivating and activating pleasant states), indicators of activating pleasant affect, or bipolar measures were used. Indeed, the only consistent findings within a particular type of measure were for the tired-excited bipolar measure in which we found that both upper elementary and middle school students who reported feeling more excited than tired were more likely to engage in metacognitive strategy use (Linnenbrink et al., 2005, Study 2; Linnenbrink & Pintrich, 2003, Study 1). In general, these findings provide more support to the link between pleasant affect and the activation of heuristic strategies than the suggestion that pleasant affect "clutters" working memory functioning and thus interferes with complex cognitive strategies. But additional studies are needed to clarify this link.

Finally, with respect to cognitive engagement, one can also consider how affective states relate to actual performance. Overwhelmingly, across studies investigating conceptual change in science understanding and performance on mathematics tasks, we have found little support for a direct relation between affect and learning, regardless of how we assessed affect (Linnenbrink, 2004; Linnenbrink et al., 2005; Linnenbrink & Pintrich, 2001, 2002b, 2003). For instance, general measures of pleasant affect were consistently unrelated to students' learning during a mathematics unit on statistics and graphing and to the change in conceptual understanding in science across four studies. Activating pleasant affect and bipolar indicators of sad-happy were also not related to students' learning during mathematics (three studies).

With respect to unpleasant affect, there seemed to be more evidence that unpleasant affect could undermine learning. For instance, using general measures of unpleasant affect, unpleasant affect was negatively related to students' learning in science and math, although this finding was no longer significant in one study once pretest knowledge was included as a control (Linnenbrink & Pintrich, 2002b, Study 1, 2003, Study 2). When activating and deactivating unpleasant affect was assessed, two studies showed no relation to students' conceptual change in science understanding (Linnenbrink, 2004, Studies 1 and 2), but two other studies suggested that activated unpleasant affect was associated with decreased working memory functioning (Linnenbrink, Ryan et al., 1999) and lower levels of conceptual change (Linnenbrink & Pintrich, 2002b, Study 2). One experimental study actually revealed a positive relation between activated unpleasant affect and students' reasoning on an analytical reasoning task (Linnenbrink & Pintrich, 2001). Finally, three studies using bipolar measures found no significant relation between learning and the extent to which students felt more tense than calm (Linnenbrink, 2004, Study 2; Linnenbrink et al., 2005, Study 2;

Linnenbrink & Pintrich, 2003, Study 1). Thus, overall, there is some evidence that unpleasant affect may interfere with learning, but the pattern is not consistent across studies.

DOES AFFECT MEDIATE THE RELATION BETWEEN ACHIEVEMENT GOALS AND ENGAGEMENT?

Thus far I have considered the relation between affect and achievement goal orientations as well as affect and engagement. In summarizing these findings (see Figure 2), I collapse across activation levels, as there are too few studies that actually distinguished between valence and activation. However, it will be critical for future research to make this distinction between valence and activation. Overall, the studies that my colleagues and I have conducted suggest that mastery-approach goal orientations are associated with higher levels of pleasant affect and lower levels of unpleasant affect. The findings for performance-approach goal orientations are rather mixed, with performance-approach goals either unrelated or positively related to both pleasant and unpleasant affect. With respect to engagement, we found that pleasant affect does not undermine behavioral engagement and may even enhance it, especially when it is activated pleasant affect. Unpleasant affect, however, seems to undermine behavioral engagement, regardless of activation level. The relation for cognitive engagement is more complex; however, we generally

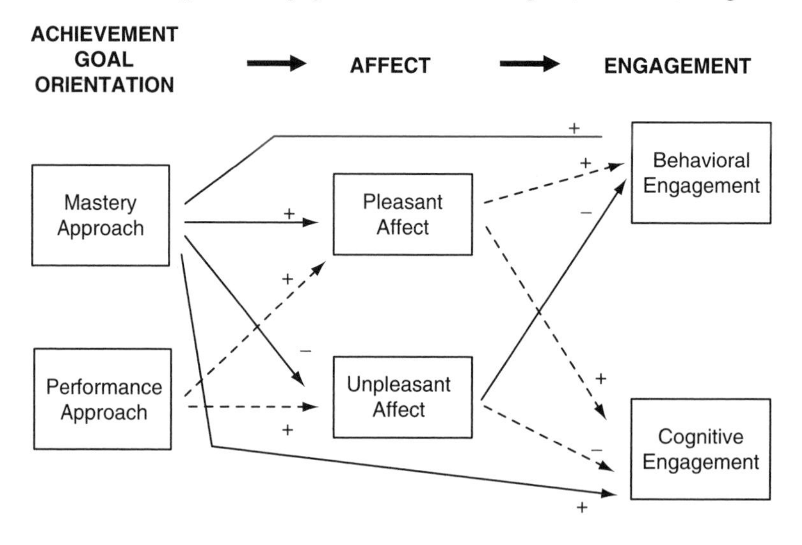

FIGURE 2

Proposed mediational model linking motivation, affect, and engagement. Black lines indicate consistent findings; dashed lines indicate general patterns based on less consistent findings.

found no relation between unpleasant affect and cognitive engagement, including elaborative or metacognitive strategy use as well as learning. Of note, however, were a few studies suggesting that unpleasant affect undermined learning and working memory functioning. With respect to pleasant affect, the results were mixed, but generally suggested pleasant affect was either positively related to metacognitive and elaborated strategy use or unrelated; there does not, however, appear to be a direct relation between pleasant affect and learning.

Based on these studies, it certainly seems plausible that pleasant and unpleasant affect might mediate the relation between achievement goal orientations and behavioral and cognitive engagement (see Figure 2). Following Baron and Kenny (1986), four conditions are necessary for mediation: (1) the predictor variable must relate significantly to the mediator, (2) the mediating variable must relate significantly to the dependent variable, (3) the predictor variable must relate significantly to the dependent variable, and (4) the relation between the predictor variable and the dependent variable must be significantly reduced when the mediating variable is included in the regression equation. Thus far, I have provided evidence for the first two conditions linking achievement goals to affect (1) and affect to engagement (2). With respect to the third condition, mastery-approach goals are generally associated with higher levels of behavioral and cognitive engagement, although the findings are less consistent when learning/achievement is the outcome (for a review, see Pintrich, 2000). The findings for performance-approach goals are quite mixed, making it difficult to make clear predictions regarding the proposed model (for reviews, see Linnenbrink, 2002; Pintrich, 2000).

With respect to the fourth condition, I begin by considering the findings regarding mastery-approach goal orientations. With my colleagues (Linnenbrink & Pintrich, 2002b, 2003; Linnenbrink, Ryan et al., 1999), I have conducted several studies directly testing whether the relation between the mastery-approach and engagement decreases when pleasant or unpleasant affective states are included in the model. In two of these studies (Linnenbrink & Pintrich, 2002b, Study 2; Linnenbrink, Ryan et al., 1999), we found evidence that unpleasant affect partially mediated the relation between mastery-approach goals and learning. In both studies, undergraduates who strongly endorsed mastery-approach goals experienced less unpleasant affect while working on either a working memory task or a science reading, and this unpleasant affect was associated with lower levels of performance on these tasks. Furthermore, entering unpleasant affect into the model reduced the strength of the positive relation between mastery-approach goals and learning/performance. We found no evidence, however, that pleasant affect mediated the relation between mastery-approach goals and learning, behavioral engagement, or cognitive engagement (Linnenbrink & Pintrich, 2002b, 2003).

Thus there is some support for the idea that a reduction in unpleasant affect for mastery-oriented students may help to explain higher levels of learning, but affect does not help to explain the patterns of behavioral engagement and cognitive strategy use that are typically associated with mastery-approach goals. In some studies (Linnenbrink & Pintrich, 2003) the relation between affect and engagement became nonsignificant when achievement goal orientations were included in the model, which might even suggest that mastery-approach goals mediate the relation between affect and engagement, rather than vice versa.

With respect to performance-approach goal orientations, it is more difficult to test for mediation, as the findings relating performance-approach goals to affect and engagement are less consistent. In most of the studies described previously the first three criteria for mediation were not met so we could not test for mediation for either pleasant or unpleasant affect (Linnenbrink & Pintrich, 2002b, 2003). In one study, however, we found that performance-approach goals were associated with enhanced working memory functioning for males, but only after controlling for unpleasant affect (Linnenbrink, Ryan et al., 1999). This suggests that rather than mediating a relation between performance-approach goals and working memory, unpleasant affect may undermine any benefits that performance-approach goals might have. As such, variation in unpleasant affect among students with performance-approach goals may help to explain the somewhat inconsistent findings linking performance-approach goal orientations and cognitive performance more generally. This should certainly be more closely examined in future studies.

FUTURE DIRECTIONS

The somewhat inconsistent pattern of findings relating achievement goal orientations, affect, and engagement suggests that there is a strong need for continued research in this area. Accordingly, I suggest several avenues for future research that may help to clarify these findings and thus help us move forward in the difficult task of integrating affect into models of motivation and engagement. First, it may be useful to move away from self-report measures of affect and instead use physiological indicators, such as heart-rate and galvanic skin response, or observational measures, such as facial action coding (FACS) (Linnenbrink & Tyson, 2006; Pekrun, in press). By using alternative methods for assessment, we may more accurately assess affect and thus may be able to create a clearer picture regarding the role of affect in educational settings. Physiological and observational coding of affect also allows for real-time assessment of affect. Thus one can get a better sense of affect as it emerges while working on academic tasks and can better assess the dynamic interplay between affect and engagement.

Relatedly, many of the studies reported in this chapter assessed achievement goal orientations, affect, and engagement after the completion of the task. This single time point assessment makes it difficult to ascertain the directionality of the relation and leaves one open to the possibility that the findings have more to do with students' affective response to their performance. Thus future research would benefit from assessing motivation, affect, and engagement at multiple time points to better understand the causal ordering.

Second, the time frame of study should be considered. Most of the studies reviewed here assessed affect in relation to performance on a single academic task, lasting 30 to 60 minutes. If we are to truly understand affect as it emerges in school contexts, it would be useful to vary the time frame of the study. Detailed analyses of smaller time segments such as 1-minute intervals might provide greater insight into the link between affect and cognitive processing. Longer intervals, such as days, weeks, or even months, would provide insight into how personal achievement goal orientations relate to emotional well-being over time, and how consistent affective states in educational settings relate to long-term engagement or disengagement.

Third, the type of affective state being assessed needs to be more carefully considered. At the beginning of the paper, I suggested that affective states included both moods and emotions, but did not make that distinction in the studies reviewed. By differentiating between moods and emotions, we may gain a better understanding of the differences in motivation and engagement that emerge when we compare intense, emotional reactions to general, pervasive mood states. Furthermore, the utility of the affective circumplex as a model for understanding affect in educational settings should be further explored. Based on the studies reviewed in this chapter, we see some evidence that the differentiation between both activation and valence is important. Yet, we have not established that this is more useful than considering discrete emotions or using other theoretical models to consider affect, such as Pekrun's control-value theory (Pekrun, in press; Pekrun et al., 2002). Thus future research should directly compare the utility of affective circumplex assessments versus other theoretical models.

Fourth, moderators need to be more carefully considered. For example, it may be that certain types of affect are beneficial for certain types of engagement. Thus, we need to more carefully consider the type of task on which students are being asked to engage. Regulation of affect may also alter the relation among motivation, affect, and engagement. Consideration of moderators may be especially important in understanding the mixed findings linking performance-approach goal orientations to both affect and engagement. For instance, one's perceived success or failure may also alter one's emotional response while completing an academic task. The consideration of these issues should help to improve research in this area and thus increase our ability to develop both theoretical and practical suggestions for the role of affect in educational settings.

CONCLUSION

With respect to the proposed mediational model, there is some evidence that the reduction in unpleasant affect associated with mastery-approach goals may help to explain why mastery-approach goals are beneficial for learning. However, this appears to be the only relation where affect might serve as a mediator. There does, however, appear to be preliminary support for the notion that achievement goals predict students' affect, with mastery-approach goal orientations relating to increased pleasant affect and decreased unpleasant affect, and performance-approach goal orientations increasing activated unpleasant affect and showing mixed relations to pleasant affect. Furthermore, there is evidence that the social psychological models regarding affect and engagement do not cleanly map onto engagement in educational settings. Behavioral engagement appears to be undermined by unpleasant affect rather than enhanced; and there is some evidence that activating pleasant affect is associated with both behavioral and cognitive engagement.

This suggests that rather than developing a mediational model, it may be useful to considered a triarchic model of reciprocal causation (similar to Bandura, 1991), in which there are reciprocal relations among achievement goal orientations and affect, affect and engagement, and achievement goal orientations and engagement. A triarchic model such as this also gives equal weight to motivational and affective outcomes, rather than giving engagement and learning precedence. This would help to further highlight that both motivation and emotional well-being are critical outcomes of schooling and thus should not be ignored (Brophy, 1999; Roeser, Eccles, & Strobel, 1998).

Based on the empirical evidence presented in this chapter, a few preliminary suggestions can be made to educators; however, one must keep in mind that the findings thus far are somewhat inconsistent. In general, it seems that creating classrooms that promote the adoption of mastery-approach goal orientations should benefit emotional well-being by enhancing pleasant affect and decreasing unpleasant affect. Furthermore, there is some evidence that neutral to pleasant affective states are more beneficial to students' behavioral and cognitive engagement and should thus be promoted. Future research, however, will need to more carefully examine whether unpleasant affect may serve a useful role, especially in situations where rehearsal or detailed processing is more beneficial than heuristic processing.

Author's Note: I thank Diana F. Tyson for her helpful feedback and suggestions on this manuscript.

References

Ainley, M. (in press). Connecting with achievement activities: Motivation, affect and cognition in interest processes. *Educational Psychology Review*.

Bandura, A. (1991). Social cognitive theory of self-regulation. *Organizational Behavior and Human Decision Processes, 50*, 248-287.

Baron, R. M., & Kenny, D. A. (1986). The moderator-mediator variable distinction in social psychological research: Conceptual, strategic, and statistical considerations. *Journal of Personality and Social Psychology, 51*, 1173-1182.

Baumeister, R. F. (2000). Ego depletion and the self's executive function. In A. Tesser, R. B. Felson, & J. M. Suls (Eds.), *Psychological perspectives on self and identity* (pp. 9-33). Washington, DC: American Psychological Association.

Bless, H. (2000). The interplay of affect and cognition: The mediating role of general knowledge structures. In J. P. Forgas (Ed.), *Feeling and thinking: The role of affect in social cognition* (pp. 201-222). Paris: Cambridge University Press.

Brophy, J. (1999). Toward a model of the value aspects of motivation in education: Developing appreciation for particular learning domains and activities. *Educational Psychologist, 34*, 75-85.

Carver, C., & Scheier, M. (1990). Origins and functions of positive and negative affect: A control-process view. *Psychological Review, 97*, 19-35.

Dalgleish, T., & Power, M. J. (1999). Cognition and emotion: Future directions. In T. Dalgleish & M. Power (Eds.), *Handbook of cognition and emotion* (pp. 799-805). Chichester, England: John Wiley & Sons Ltd.

Dweck, C., & Leggett, E. (1988). A social-cognitive approach to motivation and personality. *Psychological Review, 95*, 256-273.

Elliot, A. J. (1999). Approach and avoidance motivation and achievement goals. *Educational Psychologist, 34*, 169-189.

Ellis, H., Seibert, P., & Varner, L. (1995). Emotion and memory: Effects of mood states on immediate and unexpected delayed recall. *Journal of Social Behavior and Personality, 10*, 349-362.

Ellis, H. C., & Ashbrook, P. W. (1988). Resource allocation model of the effects of depressed mood states on memory. In K. Fiedler & J. Forgas (Eds.), *Affect, cognition, and social behavior: New evidence and integrative attempts* (pp. 25-43). Toronto: C. J. Hogrefe.

Feldman Barrett, L., & Russell, J. A. (1998). Independence and bipolarity in the structure of current affect. *Journal of Personality & Social Psychology, 74*, 967-984.

Forgas, J. P. (2000). *Feeling and thinking: The role of affect in social cognition*. Paris: Cambridge University Press.

Fredricks, J. A., Blumenfeld, P. C., & Paris, A. H. (2004). School engagement: Potential of the concept, state of the evidence. *Review of Educational Research, 74*, 59-109.

Fredrickson, B. L. (1998). What good are positive emotions? *Review of General Psychology, 2*, 300-319.

Fredrickson, B. L. (2001). The role of positive emotions in positive psychology: The broaden-and-build theory of positive emotions. *American Psychologist, 56*, 218-226.

Hill, K. T., & Wigfield, A. (1984). Test anxiety: A major educational problem and what can be done about it. *Elementary School Journal, 85*, 105-126.

Linnenbrink, E. A. (2002). *The dilemma of performance goals: Promoting students' motivation and learning in cooperative groups*. Unpublished dissertation, University of Michigan, Ann Arbor, MI.

Linnenbrink, E. A. (2004, April). *Considering positive and negative affect in learning from science texts*. Paper presented at the annual meeting of the American Educational Research Association, San Diego, CA.

Linnenbrink, E. A. (2005). The dilemma of performance-approach goals: The use of multiple goal contexts to promote students' motivation and learning. *Journal of Educational Psychology, 97*, 197-213.

Linnenbrink, E. A., Hruda, L. Z., Haydel, A., Star, J., & Maehr, M. L. (1999, April). *Student motivation and cooperative groups: Using achievement goal theory to investigate students' socio-emotional and cognitive outcomes*. Paper presented at the annual meeting of the American Educational Research Association, Montreal, Canada.

Linnenbrink, E. A., Kelly, K. L., & Kempler, T. K. (2005, August). *Affect during small group instruction: Implications for students' engagement and learning*. Paper presented at the biennial meeting of European Association for Research on Learning and Instruction, Nicosia, Cyprus.

Linnenbrink, E. A., & Pintrich, P. R. (2001, April). *The relation between achievement goals and affect: Moods, emotions, and directionality*. Paper presented at the annual meeting of the American Educational Research Association, Seattle, WA.

Linnenbrink, E. A., & Pintrich, P. R. (2002a). Achievement goal theory and affect: An asymmetrical bidirectional model. *Educational Psychologist, 37*, 69-78.

Linnenbrink, E. A., & Pintrich, P. R. (2002b). The role of motivational beliefs in conceptual change. In M. Limon & L. Mason (Eds.), *Reconsidering conceptual change: Issues in theory and practice* (pp. 115-135). Dordrecht, The Netherlands: Kluwer Academic Publishers.

Linnenbrink, E. A., & Pintrich, P. R. (2003, April). *Motivation, affect, and cognitive processing: What role does affect play?* Paper presented at the annual meeting of the American Educational Research Association, Chicago, IL.

Linnenbrink, E. A., & Pintrich, P. R. (2004). Role of affect in cognitive processing in academic contexts. In D. Dai & R. Sternberg (Eds.), *Motivation, emotion, and cognition: Integrative perspectives on intellectual functioning and development* (pp. 57-87). Mahwah, NJ: Lawrence Erlbaum.

Linnenbrink, E. A., Ryan, A. M., & Pintrich, P. R. (1999). The role of goals and affect in working memory functioning. *Learning & Individual Differences, 11*(2), 213-230.

Linnenbrink, E. A., & Tyson, D. (2006, April). *Measuring student affect in educational contexts*. Paper presented at the annual meeting of the American Educational Research Association, San Francisco, CA.

Pekrun, R., Goetz, T., Titz, W., & Perry, R. P. (2002). Academic emotions in students' self-regulated learning and achievement: A program of qualitative and quantitative research. *Educational Psychologist, 37*, 91-106.

Pekrun, R. (in press). The control-value theory of academic emotions: Assumptions, corollaries, and implications for educational research and practice. *Educational Psychology Review*.

Pintrich, P. R. (2000). The role of goal orientation in self-regulated learning. In M. Boekarts, P. R. Pintrich & M. Zeidner (Eds.), *Handbook of self-regulation: Theory, research and applications* (pp. 451-502). San Diego: Academic Press.

Revelle, W., & Loftus, D. (1990). Individual differences and arousal: Implications for study of mood and memory. *Cognition and Emotion*, 4, 209-237.

Roeser, R. W., Eccles, J. S., & Strobel, K. R. (1998). Linking the study of schooling and mental health: Selected issues and empirical illustrations at the level of the individual. *Educational Psychologist, 33*, 153-176.

Rosenberg, E. L. (1998). Levels of analysis and the organization of affect. *Review of General Psychology, 2*, 247-270.

Russell, J. A., & Feldman Barrett, L. (1999). Core affect, prototypical emotional episodes, and other things called emotion: Dissecting the elephant. *Journal of Personality and Social Psychology, 76*(5), 805-819.

Schwarz, N. (1990). Feelings as information: Informational and motivational functions of affective states. In E. T. Higgins & R. M. Sorrentino (Eds.), *Handbook of motivation and cognition: Foundations of social behavior* (Vol. 2, pp. 527-561). New York: Guilford Press.

Schwarz, N., & Clore, G. L. (1996). Feelings and phenomenal experiences. In E. T. Higgins & A. W. Kruglanski (Eds.), *Social psychology: Handbook of basic principles* (pp. 433-465). New York: Guilford Press.

Tellegen, A., Watson, D., & Clark, L. A. (1999). On the dimensional and hierarchical structure of affect. *Psychological Science, 10*, 297-309.

Thayer, R. E. (1986). Activation-deactivation adjective checklist: Current overview and structural analysis. *Psychological Reports, 58*, 607-614.

Turner, J. E., Husman, J., & Schallert, D. L. (2002). The importance of students' goals in their emotional experience of academic failure: Investigating the precursors and consequences of shame. *Educational Psychologist, 37*, 79-90.

CHAPTER

8

A Dynamical Systems Perspective Regarding Students' Learning Processes: Shame Reactions and Emergent Self-Organizations

JEANNINE E. TURNER
Florida State University

RALPH M. WAUGH
University of Texas at Austin

Receiving performance feedback regarding academic tasks can be an emotional time for students (Folkman & Lazarus, 1985; Turner & Schallert, 2001). It can, however, initiate mindful consideration of students' plans and of their self-efficacy for strategic goal attainment. When students receive confirmation from or grapple with their own cognitive-emotional evaluations, they can initiate interrelated cognitive, emotional, motivational, and behavioral feedback processes that will transform their specific patterns and tendencies, thus influencing simultaneous cognitive, emotional, motivational, behavioral, and psychophysiological patterns concerning their next performance feedback experience. Academic evaluations represent both important culminations and important initiations. Hence, when viewed as an ongoing progression, each student's learning processes demonstrate a dynamic story. Through investigating students' stories, the complexity of learning unfolds.

In this chapter we discuss aspects of dynamical systems theories (e.g., Kelso, 1995; Kugler & Turvey, 1987; Mascolo, Harkins & Harakal, 2000; Waugh, 2002, 2003) that may deepen our understanding and guide our investigations of students' multidimensional, emergent academic processes. Dynamical systems theories (DST) have been used to articulate new conceptualizations of complex processes, including, transformations, developments, and emergence of novelty in multiple fields of natural and behavioral sciences. Lewis and Granic (2000) have explained:

> At the heart of this perspective is the idea that natural systems behave very differently than the simple, idealized systems so well described by Newtonian mechanics. Systems in nature are characterized by interactions among many components whether molecules in a fluid, cells in a body or brain, organisms in an ecosystem, or individuals in a society. This complexity is not just a detail. The interactions among the elements of complex systems are reciprocal, with constituents influencing each other simultaneously, and they recur over time, as systems continue to evolve or perpetuate their own stability. . . . The most important and dramatic result of this kind of system dynamics is the emergence of novel forms at higher levels of organization. (p. 1–2)

Dynamical systems theories (DST) can offer a fresh perspective for understanding the complex processes involved within students' academic experiences. In terms of learning, we find that this framework explains novel, individualistic student experiences as well as the establishment and maintenance of higher-scale recurrent patterns of similar students' behaviors (e.g., students who are "mastery-oriented," "underachieving," "self-handicapping," "avoidant"). By integrating micro-and macro-processes, DST sheds light on moment-to-moment processes, recurring cyclical processes (such as a semester or school year), and overarching trait-like orientations (such as mastery-orientation or performance-orientation). Therefore, DST explains both individual emergent fluctuations within each student's learning "event" (Turner & Husman, 2001; Turner & Waugh, 2003) and multidimensional profiles (Shell & Husman, submitted for publication) of students' patterns.

This chapter attempts to deepen our understanding of the recurring complex, mutually influential multivariate interactions and feedback loops that are embedded in the context of students' goal-strivings and academic learning. To explore these issues, we will describe the framework of dynamical systems theories and then use findings from our research to illustrate these principles. In particular, we will focus on students' cognitive, emotional, and motivational dynamics as they experience the emotion of shame and then proceed through a semester. Through an understanding of students' experiences and recovery from in-the-moment shame, within the context of their short-term and long-term goal strivings, the complex interactions among students' cognitions, emotions, motivations, behaviors, and psychophysiology can be more fully appreciated.

DYNAMICAL SYSTEMS APPROACH TO UNDERSTANDING STUDENTS' LEARNING, MOTIVATION, EMOTIONS, AND SELF-REGULATION PROCESSES

Despite our best efforts to control and predict learning-related processes and outcomes, each student's learning experience is a unique tapestry of cognitions (ongoing goals, perceptions, interpretations, understandings, etc.), emotions, behaviors, and psychophysiology. No matter how we slice academic experiences into segments of learning (i.e., academic "events" or "episodes") for research purposes, students' learning remains highly individualistic. DST provides a useful framework to understand these processes.

By its very nature, a dynamical system is *intrinsically dynamic*. This means that a dynamical system may evince changes even in the absence of external influences (i.e., internal causation; Waugh, 2002, 2003). This feature helps to explain how individuals can be the initiators of their own thoughts, feelings, and behaviors in the midst of external stimuli and within their own goal-striving behaviors. This occurs because of the interconnected, *bidirectional*, and *multidirectional feedback* processes and mutual influences among internal, *system elements* (variables) and *subsystems*. Such subsystems include the perceptual system, the cognitive system, and the emotional system (Mascolo et al., 2000; Op 't Eynde & Turner, in press; Waugh, 2002, 2003).

Each individual system is highly complex and contains nested, integrated subsystems that are interconnected via bidirectional feedback loops. These bidirectional influences of feedback loops between and among various subsystems activate the transformation-creation of continually self-organizing and changing structures. In terms of academic motivation, internal subsystems, such as ongoing perceptions, appraisals, and emotions, can influence the choices students make at any given moment regarding engaging in academic activities. Indeed, dynamical systems are marked by *sensitive dependence upon initial conditions*. Hence, a small change in a particular system variable can induce very large, *qualitatively different* system behaviors (i.e., discontinuities, bifurcations, nonlinear interdependencies explained by ordinary difference equations). When critical thresholds are exceeded with respect to one or more system variables, sensitive dependence on initial conditions can lead to discontinuity(ies) and large qualitative changes in system behaviors (Waugh, 2002, 2003).

In dynamical systems, an important element is that of *periodicity*, or *cyclicity*, which produces changes in *initial conditions*. A new cycle (a new learning "event") may be initiated by students' shifts in attentional foci towards learning about a particular concept. Students' initial conditions (e.g., affective-cognitive, motivational, and physiological states) influence their learning processes during that cycle (e.g., students' cognitive, emotional, and motivational interpretations), thereby initiating moment-to-moment fluctuations and dynamical changes in system elements (variables) and subsystems. The culmination of a cycle creates new values for the initial conditions, states, characteristics, and changes in system elements and subsystems. An entire learning "event" or "episode" may involve one or more cycles (Waugh, 2002, 2003).

The dynamical system's sensitive dependence upon ever-changing initial system conditions produces continuously and iteratively cascading affects on an extraordinarily complex, ongoing, multifaceted array of each individual's system components. For example, all students enter a learning event with their own initial conditions, such as their unique history of academic success and failure, reasons for engaging in the learning activity, future goals, motivational orientation(s), personal values, beliefs, and future goals. Additionally, students bring to the learning event their own unique levels of prior knowledge, academic abilities, and learning skills (e.g., note taking, reading, study skills, and emotion regulation). These initial conditions may change throughout the learning event, with some changes in initial conditions producing significant impact, whereas other changes may not (Waugh, 2002, 2003).

Sensitive dependence upon initial system conditions also highlights the power of specific internal or external informational cues, such as contextual elements, students' perceptions of control issues (e.g., self-efficacy), and students' consideration of personal values (e.g., future goals) (Pekrun, 2000; in press) for instigating changes within one or more of the system components and invoking *self-organization*. Self-organization occurs from the dynamic interplay among system elements and subsystems (Kugler & Turvey, 1987; Turvey, 1990). These interactive system elements and subsystems exert dynamic influences upon a person's "state" at any particular moment during any specific event through continual self-organization (Waugh, 2002, 2003).

The continual process of self-organizing transmutation is characteristic of dynamical, open, living systems. Within academic settings and events, each student may be thought of as a self-organizing system that acts and reacts to both external and internal informational signals (Kelso, 1995). These processes may explain the unique, individual facets of students' learning-related cognitions, emotions, motivations, and behaviors.

Performance Feedback Cycles and Dynamical Systems Principles

One overarching cycle for students' self-organizing processes include their preparations for academic assessments, performance on assessment tasks, and the consequent performance feedback (usually in the form of "scores" or "grades"). Receiving performance feedback is a naturally occurring opportunity for qualitative shifts, or "phase transitions" (Kelso, 1995), in students' learning-related cognitions, emotions, motivations, and behaviors. Within students' ongoing academic strivings, feedback from academic evaluations provides a powerful impetus for students' confirmation or reassessment of their goal-related values, commitments, strivings, and conflicts. When looking across the semester, students' unique intraindividual patterns of motivation, emotions, behavior, and achievement can be detected (Turner & Waugh, 2003). These patterns are consistent with a dynamical systems perspective.

The cycle of being exposed to new information, working with the new information, and demonstrating that one has integrated and mastered the new information provides an exemplary illustration of the dynamic systems' principle of periodicity. Consistent with DST, the period that begins with being exposed to new material and culminating with performance feedback contains numerous microcycles. Students' participation in several class sessions and in several self-initiated study sessions can provide successive knowledge-building opportunities to work with, internalize, and integrate class information. Additionally, students' learning-directed participation can initiate self-organization of cognitions, emotions, and behaviors. Within each microcycle lies the potential for abrupt changes in students' motivation. For example, if a student has difficulty initiating study sessions, if he becomes overwhelmed and frustrated with trying to understand a poorly-written or complicated textbook, if he has difficulty understanding the information in his class notes, or if he is interrupted by a telephone call informing him that his father has experienced a heart attack, these conditions could immediately and significantly alter the student's path that leads to another performance outcome.

The hallmark of spontaneous self-organization is emergent patterns of higher-order properties. Our research consistently supports a theoretical framework in which students' control-related and value-related cognitions, emotions, and behaviors form higher-order organizational properties. This is consistent with Pekrun's control-value theory (Pekrun, 2000; in press) of academic emotions. Because his theory encompasses so many aspects of our conception of students' cognitive, emotional, and motivational initial conditions, we briefly describe its importance in the next section, knowing that readers may gain a deeper understanding of his theory within this book (see Chapters 1, 2, 4, 18).

CONTROL AND VALUE: HIGHER-ORDER PROPERTIES OF COGNITIONS, EMOTIONS, AND MOTIVATIONS

Pekrun (2000; in press) has proposed that students' control-related and value-related appraisals of academic tasks and outcomes are the main sources of students' academic motivation and emotions. Control-value theory also describes reciprocal processes among cognitions, emotions, and motivations that influence students' academic processes, which fit well with complex systems theories.

Perceptions and emotions of control and value Students' perceptions about their ability to control aspects of their learning include their beliefs about themselves as learners (e.g., academic self-concept; Harter, Bresnick, Bouchey, & Whitesell, 1997), their attributions about the causes of successes and failures (e.g., self-generated or other-generated, Weiner, 1985), and their perceptions and beliefs about their ability to regulate actions and attain desired outcomes (e.g., self-efficacy; Bandura, 1997). Research has shown that students who believe they are competent and capable are more likely to demonstrate motivated behaviors, even after they have experienced prior difficulties or received feedback indicating they are not doing well (e.g., Rosenthal & Zimmerman, 1978; Schunk, 1990).

The companion and interconnected concept to control-related beliefs is that of value-related beliefs. Students rarely engage in activities that have absolutely no value-related component(s), such as valuing learning outcomes, valuing grade outcomes, or valuing outcomes of another's approval. Within academic settings, a primary focus of the value component has been on students' goals and reasons for engaging in academic tasks. These goals can be as vague as wanting to be a good student, or they can be as specific as obtaining a grade of 95 on an exam. Goals can be represented as future end points, for example, wanting to work in an environment where rockets are designed. As end points, goals can contain sequential plans of actions that contain many subgoals and related actions (Oatley, 1992). Raynor (1981) described an individual's perceived connections between a current task and that of a future goal as having *instrumentality* value. If obtaining a goal is on a contingent path to a future, superordinate goal, the present goal is perceived as having a higher instrumental value than a goal or action that is not on a contingent path (Husman & Lens, 1999; Raynor, 1981).

Students' appraisals of personal control and personal value influence their motivation and emotions for engaging in academic tasks. For example, depending on students' sets of control and value perceptions (i.e., initial conditions), they may feel excitement in learning something new, hope that outcomes will be successful, and pride in accomplishments. The experience of distressing emotions, such as fear, shame, or despair, may occur with different control and value appraisals.

One fundamental premise of appraisal theories (e.g., Frijda, 1986; Lazarus, 1991) that links emotions and cognitions is that one responds emotionally to important contextual perceptions. If a situation holds little or no personal value, a person will not attune to it or will easily dismiss it, with little or no emotional connection. The importance and value that emotions represent to an individual regarding a cognition-emotion interpretation tie directly into his or her values and identity. This is particularly true for the classification of emotions that are self-referential and self-conscious, such as the emotion of shame. Most current emotion theories agree that shame is triggered by perceptions of failure to meet important internalized standards, rules, or goals (e.g., Lewis, 1993). In the moment of shame, a person feels that he or she has failed in a context of personal importance (high value) and that there is nothing one can do about it (low control). This type of powerful event reflects intrinsically dynamic processes that influence formulation of subsequent motivational goals, expectations of self-efficacy, and effort-regulation processes. Therefore perceptions of failure and shame influence significantly the students' trajectories of thoughts, emotions, psychophysiology, and behaviors for the next episode of exam preparation and feedback. In the next section, we illustrate how, with the instantiation and recovery from shame reactions, students' intraindividual dynamics influence their unfolding achievement-appraisals, emotional experiences, and self-regulation processes regarding personal goals, values, and outcome expectancies, thus creating their personal path.

DYNAMICAL SYSTEMS THEORY AND THE EXAMPLE OF ACADEMIC SHAME

Typically, shame is viewed as one of the most distressing and disruptive unpleasant emotions. This section illustrates the intrapersonal dynamics of precursors, concomitants, and consequences of experiencing shame reactions in academic settings. These examples come from research that was focused on understanding college students' motivation, emotions, and self-regulation in courses that were part of students' goal-oriented paths (e.g., Turner, Husman, & Schallert, 2002; Turner & Schallert, 2001; Turner & Waugh, 2003). Our research has investigated students' initial conditions gathered at the beginning of a semester (e.g., goal orientations, future goals, academic self-concepts, global self-esteem, self-efficacy for obtaining desired outcomes, ability to regulate negative moods), self-perceptions and emotions after receiving exam feedback (e.g., perceptions that exam scores represented "failures," state self-esteem, and various concomitant emotions), and ongoing motivations, effort regulation, study time, use of study strategies, and exam outcomes. Research designs included quantitative measures and linear analysis (e.g., Turner, 1998; Turner, Schallert, Wicker, & Waugh, 1998; Turner & Schallert, 2001) as well as qualitative interviews and analysis (e.g., Turner & McCann, 2000; Turner & Husman, 2003). In this chapter, we focus on the analysis of qualitative interviews that provided rich content and descriptions of students' shame-experiencing cognitions, physiological sensations, emotional feelings, motivational tendencies, and subsequent actions.

Initial Conditions of Shame-Related Controls and Values

Consistent findings across several studies of students' experiences of academic shame (e.g., Turner, Husman, & Schallert, 2002) showed that students who had shame reactions to exam feedback indicated they had not received the grade they wanted, the grade they wanted was important to them, and they perceived the grade as representing failure—even if the grade was not an objective failure (e.g., answering 85% of the exam items correctly). Congruent with DST, these aspects represent some of the initial conditions in the instantiation of feeling shame.

Our research has consistently demonstrated that the underside of having high academic self-concepts and high expectations of success is that perceptions of academic failure can be particularly devastating for these students. Contrary to theoretical perspectives that posit having high dispositional self-esteem protects individuals from experiencing shame (Brown & Dutton, 1995; Dutton & Brown, 1997), our findings have consistently demonstrated that having high self-esteem, high perceptions of efficacy, and a mastery-focused orientation will *not* protect students from feelings of in-the-moment shame when they perceive they have failed an important academic task (Turner & Schallert, 2001; Turner & Whohlblatt, 1999), although, these characteristics may assist in motivational resiliency following shame reactions (Turner & Schallert, 2001). When students hold values for academic success and they expect academic success, these specific initial conditions can trigger feelings of shame when they experience a personal failure.

Our interviews with shame-experiencing students have revealed the dissonance many students experienced between their perceived failures (i.e., lower than expected exam grade) and experiences of past successes and expectations for continued accomplishments. For example, one upper-division psychology student who had experienced shame upon receiving a "C" on an exam stated, "I've done really well in psychology and my GPA in psychology is high. . . . I've never made a 'C' in a psychology class. I made a 'B' in [Introductory Psychology] but that was two years ago." Another shame-experiencing student articulated the feeling of being a failure relative to her history of success, "Well, I thought that I was basically a failure in the class. I had just failed to do anywhere near what I'm used to doing." These statements support the notion that, at the heart of in-the-moment shame experience is the exposure to oneself of one's failure to live up to ideals and standards (Lewis, 1993). Placing high value on academic accomplishments and attributing personal control for outcomes can be initial conditions that instantiates perceptions of failure and feelings of shame.

Dynamical Instantiations of Shame Experiences and Concomitant Processes

The experience of in-the-moment shame demonstrates the DST principles of interrelated components systems and subsystems, nonlinear qualitative shifts, and emergent self-organization. Indeed, the emotion of shame is an interesting and appropriate emotion to explore with respect to illustrating dynamical systems principles. To begin, humans seem to be "hardwired" to experience this emotion, yet it cannot occur until certain cognitive developmental milestones have been met (Izard, Ackerman, Schoff, & Fine, 2000; Tompkins, 1970; 1987). Important for the activation of shame, an individual must have cognitive structures in place for self-representations, appraisal processes, and internalized values (i.e., standards, rules, or goals; Lewis, 1993). With those elements in place, preconscious processes that initiate self-organization among the physiological, cognitive, affective, and behavioral systems may trigger the experience of shame.

Integrative Emergence of Shame According to Lewis (2000) and Lewis & Granic (2000), emotions begin with "coarse processing" (p. 43) of attention shifts; for example, a global cognitive-affective perception that an external stimulus may be important. During global processing, cognitive perceptions and interpretations, along with integration of physiological components, result in a holistic "gist" of the situation and its potential impact. Once the preconscious gist is in progress, recursive feedback processes among the constituent elements, subsystems, and higher-order components influence more and more refined evaluative processes (microcycles) that culminate in the emergence of coordinated cognitive awareness, appraisals, physiological responses, and behavioral action tendencies. For example, early preconscious appraisals may highlight an exam score as a potential threat to one's identity, ignite physiological arousal processes, and initiate blushing. These initial processes instigate further coordinative processing, resulting in an emergent appraisal-emotion amalgam, or *emotional interpretation* (p. 43, Lewis, 2000; Lewis & Granic, 2000) that is experienced as shame. This spontaneously coordinated emergence eliminates the question of which process comes first, physiological arousal or cognitive perceptions? From a DST point of view, all systems are simultaneously and iteratively involved. As described by Lewis (2000), "appraisal processes can be reconceptualized as emergent order in the cognitive system corresponding to, but not preceding, emotion" (p. 43). The emergence of shame illustrates coordinative instantiation of emotional interpretations.

Nonlinear, Concomitant Instantiations of Feeling Shame

Shame has been described as being a shocking emotion that occurs suddenly and unexpectedly (Goldberg, 1991; Lewis, 1971; Lynde, 1958). Similar to other "intense" emotions, such as rage or anguish, the instantiation of shame does not occur as a linear progression from mild forms (i.e., embarrassment) to extreme forms (i.e., humiliation). Shame emerges as a discontinuous, nonlinear "boom"—a leap from preconscious threat to heightened physiological arousal, self-consciousness, and cognitive confusion. Shame is experienced as an emotional implosion. This aspect is consistent with DST that explains how small changes in one system variable can create spontaneously large changes in one or more interdependent variables. Our research supports this description.

In terms of quantitative measures, our research suggests that feeling shame is uniquely correlated with concomitant feelings of shock—*not* surprise (Turner & Waugh, 2001). Additionally, interviews with students who experienced shame highlighted the simultaneous integration of students' feelings of emotional shock, cognitive confusion, and physiological arousal (Turner & Husman, 2003; Turner, Husman, & Schallert, 2002; Turner & McCann, 2000). For example, one shame-experiencing student was able to articulate cognitive disbelief along with physiological feelings when she explained, "I initially felt that this can*not* possibly be.... It felt like my heart sank to my stomach and I broke out in a prickly kind of feeling.... [I felt] very prickly all over. I couldn't believe it." When asked what he remembered about initially seeing his exam score, another student described escalating psychological and physiological arousal and discomfort; "I didn't feel good at all. I felt cold, and insecure, and I just felt really bad. I felt really hollow inside—and then—you start to get butterflies in your stomach ... [then] you've got alligators biting at you from inside. I felt all that." Other students' descriptions resonated with feelings of shock integrated with cognitive disorientation. For example one student explained that her initial reaction felt "like the bottom fell out from my universe.... [I kept thinking] 'this is not happening; this is not happening.' I had felt initially that this cannot possibly—I could not possibly have gotten a 'D.' There must be something very wrong here."

Interference with Cognitive Processing

The intensity of shame interferes with cognitive processing, because, "...shame disrupts the natural functioning of the self" (p. 7, Kaufman, 1989). According to DST, the disruption of cognition, accompanied with feelings of high arousal and emotional distress, occurs because of the integrative bidirectional feedback loops within and between constituent elements, subsystems, and higher-order components. With shame, the shocking emotional-cognitive interpretation floods the physiological system with overwhelming arousal and floods the cognitive system with confusion.

Coinciding with this description, many of our interviewed students mentioned the difficulty they experienced concentrating on the class lecture after they had received their exam grades and experienced shame. Students either described recurring thoughts or disruptive thoughts while they were taking notes on the subsequent lecture. When one student was asked if she remembered the instructor lecturing, she responded, "I don't remember the lecture; not at all. I know [the professor] did lecture, but I don't remember [the content]." When asked if she remembered thinking or feeling anything during the lecture she replied, "Not the lecture, but more . . . What am I going to do? What do I have to do? I cannot fail this class. [I was thinking all this] while he was talking. I have lecture notes . . . but I don't remember." Another student described her cognitive conflict, explaining, "I was trying to force myself to pay attention a little bit more because, after I looked over the test, I realized a lot of the stuff I was missing was from lecture[s]. . . . So I remember trying to pay attention a little bit more. But at the same time, I felt I wasn't really connected to it because of my disappointment. I was kind of thinking how I could do better instead of paying attention to do better." We believe these students were highlighting DST principles of the integrative, simultaneous, interconnected nature of constituent elements, subsystems, and higher-order components.

Self-Reflection Following Shame: Initiating and Preparing for the Next Cycle of Learning

Shame has been described as being a powerful cognitive "interrupt" signal (Lewis, 1991; 1993). The shocking intensity of shame seems to provide a signal to an individual that one's course of action is not working and seizes a person's immediate and total attention (a sudden qualitative shift in system behavior). The result is that of abruptly halting current activities (Lewis, 1991; 1993) and demanding an appraisal with respect to the violation of expectations (Tangney & Fischer, 1995). In this sense, the awful feelings that students experience with shame can be interpreted anew as a powerful, riveting, cognitive-affective refocus of attention towards self-relevant information.

Our evidence from student interviews suggests the experience of shame does, indeed, engage affective-cognitive evaluation processes. This finding is consistent with Pekrun's (1992) proposal that unpleasant emotions in academic settings may evoke complex thought processes while supporting the DST perspective of emergence through changing initial conditions. After the physiological and cognitive intensity of shame had decreased, interviewed students expressed ways in which they were forced to assess, evaluate, and weigh their personal abilities, circumstances, and options relative to their learning-related motivations and behaviors. Consistent with DST, the outcomes of students' assessments initiated emergent trajectories of subsequent cognitions, behaviors, and learning (Turner & Schallert, 2001; Turner, Husman,

& Schallert, 2002; Turner & Husman, 2003). Students' self-reflection and subsequent outcomes highlight the individualistic nature of students' emergent learning paths described in the next section.

Considering Personal Context

Consistent with DST, each individual's context affects his or her initial conditions and emergent states. In order to initiate the next cycle of cognitions, emotions, motivations, and psychophysiology that followed shame reactions and self-reflection processes, students considered their next actions relative to their personal contexts. For example, in thinking through courses of actions, several shame-experiencing students explained that their employment hours and/or coursework load interfered with the amount of time and energy they could devote to this particular class. One student explained,

> The problem that I'm having this semester is that I've got a really heavy workload with my classes. I didn't plan too well going into [the semester]. This is my fifth year here and, with a double major, I'm kind of throwing in classes at the last minute, just to get my two degrees done. I ended up having to take three labs and three writing components at the same time this semester. So, [this class], not being a lab or writing component, tends to get pushed on the back burner and not really looked at until two days before the exam.... It's really hard to make one class a priority when you've got five or six classes.

On the other hand, during their self-reflections and self-assessments, some shame-experiencing students considered their important future goals that dominated their focus. For several of the students, their future goals contained powerful incentives for persistent or increased motivation. One student explained:

> I'm trying to get into the pharmacy school and I've been put on a provisional acceptance. If I can make all 'B's, I'm in. I'm under a lot of stress. My family owns a chain of pharmacies and in order for me to get into the pharmacy business, I *have* to become a pharmacist. I'm under a huge amount of pressure....and it all rides on whether I can make a 'B' in this course....I guess what I need to say is that it has life-long repercussions.

Simons and her colleagues have argued that when students see an academic task as part of a larger goal path they are less likely to see failure as nonrecoverable, thus lessening the negative impact of failure (Simons, Dewitte, & Lens, 2000). Our findings have shown that when students have future goals for which the current course outcome is instrumentally connected, they are more likely to take actions that support their long-term goal-striving processes after they have experienced a perceived failure.

Interest and Intrinsic Motivation

Tompkins (1987) proposed that the experience of shame inhibited interest, enjoyment, or both; while Kaufman (1980) has suggested that shame disables learning because of concomitant feelings of being overwhelmed, feeling confused, and being out of control. Our findings have supported these concomitant experiences. However, what happens *after* students experience shame? For example, how might feeling shame affect a person's subsequent intrinsic motivation (i.e., interest and value of course information) for learning? Pekrun (1992) proposed that unpleasant emotions could interfere with the dynamics of intrinsic motivation in two ways: (1) unpleasant emotions could reduce task enjoyment because unpleasant and pleasant emotions may be incompatible; and (2) unpleasant emotions may induce negative intrinsic motivation (Pekrun, 1992) whereby students avoid an academic task because it is associated with unpleasant emotions. In this case, students who experience unpleasant emotions may develop an aversion to the academic task associated with those unpleasant emotions. Our research demonstrated three outcomes with respect to "resetting" initial conditions of intrinsic motivation: (1) students remained intrinsically motivated and increased their motivated behavior; (2) students remained interested in the course content but did not increase motivated behavior; or (3) students' positive intrinsic motivation was supplanted by negative intrinsic motivation.

With respect to maintaining intrinsic motivation, it does not seem coincidental that students who emerged from self-assessment processes with increased motivational energy also tended to maintain intrinsic motivation. Other students indicated that, following assessment processes, they remained *interested* in the class material but they did not increase their motivation. One student explained, "... [The class] was interesting. I wasn't going to get out of it what I wanted—which was a good grade—but I was in it for the 'long haul.' " When asked how she had approached the class after receiving her first exam grade she replied, "If anything—this is awful to say—I probably slacked off a bit." Thus, even though this student felt the class was interesting, she did not engage additional effort towards learning the class material or preparing for exams. Although students may find intrinsic value in the information, perceptions of lowered control may inhibit students' energy for studying. This finding is compatible with studies that have demonstrated that interest alone is not enough to induce motivated behavior (e.g., Reed & Schallert, 1993; Reed, Hagen, Wicker & Schallert, 1996).

A final example from student interviews was from a student who would fit within the third classification of outcomes, that of feeling negative intrinsic motivation, (i.e., an intrinsic distaste to going to class after feeling shame). She explained, "So going to class and sitting there has just really been kind of a burden ... You know, I go to class because I have to go to class. I've got to take the notes, and I've got to study, and I've got to get the best grade that

I can get for the class. It's not really a pleasure kind of thing like it was at the beginning when I first started the class." For some students, the self-assessment process resulted in lowered levels of personal control and lower intrinsic valuing of course information. Additionally, some students may have reacted more strongly by developing negative intrinsic motivation or an aversion to the class. So, while all students may initially enter the course with initial conditions of interest, self-control, hope, and optimism, for some students the experience of shame reset those variables through the initiation of self-organizing processes and motivational trajectories.

Emergent Self-Organization and Trajectories

Students' self-reflections and self-assessments resulted in decisions—not necessarily conscious decisions—that initiated a path of subsequent cognitions, emotions, motivations, and behaviors. Following students' self-assessments, they instigated ongoing decision-commitments to engage, maintain, or withdraw motivation for learning.

Self-Organization

Student interviews suggested that, when their future-goal commitment was high and the course outcome was instrumental in obtaining the goal, decisions to invigorate goal strivings was instantaneous. These students emerged from their self-assessment with "clarity of focus." They were committed to clear important future goals for which the course information and/or grade had instrumental connections. As one student explained, "Well, . . . [because of] the long-term repercussions of succeeding, . . . at this point, I don't really have a choice. It's either I succeed or not." These students viewed their exam grade as intolerable, and they had a sense of urgency to "do things differently." One student recalled, "I thought I knew enough to at least make a "B" and when I took the test I thought I'd done really well—and I did poorly. I thought, 'Golly, I'm going to have to do a whole lot more than what I did.' " Although some students quickly committed increased motivation, the perceived failure and shame feelings necessitated ongoing self-regulation processes as students encountered motivational roadblocks. Not only did these students need to adjust their study strategies, they also needed motivational boosts, that is, the use of volitional strategies.

Volitional Strategies

Students who emerged from their self-reflection and self-assessment with increased motivation used both negative (avoidance-based) and positive (approach-based) thoughts to keep themselves on track (i.e., volitional strategies). However, they placed more emphasis on positive thoughts than negative thoughts. For example, when asked if she ever thought about the first exam outcome while studying for the second exam, one student replied, "That 61 was right there on my mind," but she also told herself, "You have got to get an 'A.' " She chose to focus her thoughts and energy on getting the 'A.' Another student put up a written message that she would see first thing each morning that read, "Failure is not an option" as a daily motivational support. When asked about his motivation, one student stressed his positive thoughts of future goals:

> My motivation is not just making an 'A' in this course, or making a 'B' or whatever. My motivation for doing well in this class is to succeed, to retire, and die around my grandkids in a nice house.... It goes way beyond the end of the semester. And it doesn't feel like my motivation was more driven than just [that]—this course is just a small, small, very small part of the big picture and so my motivation is long-term more than just trying to succeed in this class. This is... one of the first stumbling blocks, I suppose, to what the long-term picture is.

In addition to using volitional thoughts and actions to maintain motivation, students also used volitional thoughts to neutralize negative cognitive and emotional doubts. To enhance feelings of self-efficacy, they used statements such as, "I know I can do it. I'm not dumb," and "I know I have the intelligence to succeed." Interestingly, our research with quantitative measures and analysis also demonstrated that students who were able to increase motivated energy gave themselves higher scores with respect to their certainty of their academic abilities on the first day of class (Turner & Schallert, 2001).

On the other hand, students who emerged from self-assessments with lower motivation revealed the difficulty they experienced with wavering motivational energy, as one student said:

> I was really determined to do a lot [of studying] [and] determined to use different methods in studying for the next test. I thought about different things that I was going to do and [tried to] do them from the very beginning.... Like I wanted to meet with [friends] every week so I [could] know... that my notes are in accordance with everyone else's. So that was one thing. And then I thought, 'if I review every week in addition to that, then keep on-task with the readings and make outlines as we're going through the weeks, then it will be a lot easier,' and so, I tried to do that. It didn't work too successfully, though.

These students demonstrated a minimum of volitional self-regulation. They experienced difficulty in garnering the discipline and effort required to initiate their studying or to sustain their momentum during studying. Consequently, the likelihood of becoming cognitively engaged in the material was hindered. Thus the outcomes of students' self-assessment processes instigated changes in their motivational trajectory and achievement path.

Trajectories

Students who increased their motivation made conscious decisions to increase the number of learning strategies they used and the types of strategies they used. Here, students' illustrated the DST principle of emergent adaptation. They emerged from their self-reflection and self-assessment with renewed energy, volitional effort regulation strategies for keeping themselves on track, and concrete plans to support their academic goal-striving processes. Other students emerged from their self-assessments with perceptions of low control (i.e., low self-efficacy) and lack of volitional self-regulation. For example, one student emphasized that she couldn't envision doing more than what she had done previously. When asked if she had approached her studying differently following the first exam she replied, "If anything, this is awful to say, I probably slacked off... I did everything in my power in the first test... What else could I do?"

Other students talked about the conflict between knowing they needed to study and not being able to take action. One student described her conflict as, "My thoughts and feelings were completely different from my actions, because I kept telling myself, 'you have to do this. You have to study.' But I didn't listen to myself." When they did study, these students tended to use passive, surface-level strategies for studying and learning. They did not seem to have a wide repertoire of study skills from which to draw. When questioned about how she prepared for the second exam, one student recalled, "I think I read the chapters a little bit more beforehand... I made my flash cards and I went through the notes—I always make flash cards and highlight my notes." Another student recalled, "I just continued doing what I was doing earlier... I just read and I recopied my notes after every class."

Closing the Cycle, Beginning the Next

Although some students had used volitional strategies to project their intentions to succeed, and increased their study time and study strategies, they expressed having feelings of anxiety upon beginning the second exam. However, as they completed the exam, they gained confidence because they *knew* the answers. One student recalled, "I didn't feel very confident... on the second exam... That frightened me because I didn't know that I knew [the answers]. I was afraid... almost a little panicky. Once I saw the questions, [I thought] 'Yes! I've got this. I know I can do it.' " Obtaining higher exam scores consequently brought feelings of relief and confirmation of efficacy, as exemplified by one student who said, "When I made a 'B' on the second one... I felt a whole lot more relieved because I knew I could do it."

In contrast, students whose context did not allow increased energy for this course, or who did not use volitional self-regulation to support their motivation and actions for increasing their scores, had different experiences. Most likely, because of the decrease in study time and use of learning strategies, these students did not obtain higher grades on their second exams; they either received a similar grade or received a lower grade.

The outcomes of the second exam initiated another cycle of self-reflection and self-assessment. For students who experienced positive results, they did not need to change their strategies. As these students approached the final exam, they continued to use the strategies they had used successfully for the second exam. One student explained, "I feel better because I know that the way that I studied seemed to have worked, so I'm more confident." For students who had not obtained increased scores, their feelings about their end-of-semester final exam and final course grade were characterized by feelings of resignation and further lack of motivation. One student explained, "I guess in a way, [I felt] resignation since I didn't get the grade [on the second exam] that I wanted to. Because I didn't get the grade that I needed, then I guess that's why . . . as time went on, I kind of lost motivation." As these students approached the final exam, they continued to use the strategies they had used unsuccessfully for the first and second exams. Interestingly, one student described her motivation for the final exam as, "I always have some hope. Even now, I'm hoping for a miracle. But I know I can't get what I want so that's more of a driving force."

CONCLUSION AND DISCUSSION

Several facets of DST have been proposed for understanding students' learning and goal striving (e.g., Op't Eynde & Turner, in-press). To illustrate students' dynamical cognitive-emotional-motivational processes throughout a semester, we used examples of instantiation, concomitants, and consequences of students' feeling academic shame to illustrate dynamical systems theories. Our research supports a dynamical systems approach, suggesting that students' complex "initial conditions" are ongoing sources of emergent influence throughout cycles of academic endeavors. Students may experience perceived failure and experience shame, initiating self-reflection, self-evaluation, and trajectories of cognition, emotions, motivations, and behaviors. For some students, this process will start a trajectory of increased motivation and academic success; for others it will initiate a trajectory of lowered goals and decreased motivation.

Still, one might ask, "Even though perceptions of failure and feelings of shame might lower some students' motivation and learning, could they be "resilient" within a larger perspective? Might their "failure" experiences initiate a reevaluation of what is personally important to them, perhaps changing their academic focus, and thus pursuing different goals for which they are successful?" Within a broader context, perhaps one's "failure" initiates a significant shift in a trajectory towards different values, goals, and ultimate success. Dynamic systems theories provide theoretical support to explain this scenario. Our recent research (Turner, Simmons, & Goodin, 2005) has also illustrated that, throughout the lifespan, failures or unexpected events may occur that invoke "mental turbulence" (p. 267, Vallacher & Kaufman, 1996), creating instability followed by emergent self-organization and changes in life paths. Throughout life experiences, mental turbulence presents an opening for emergent reorganization. Vallacher and Kaufman (1996) explain:

> . . . instances of instability represent more than noise or breakdown in [a] system. To the contrary, far from being unavoidable at best or dysfunctional at worst, instability plays a critical role in the functioning of many different kinds of systems. Simply put, the fluctuation among different states and patterns characterizing an unstable system provides the raw material for subsequent self-organization in the system. (p. 265)

Because the focus of our research revolves around what happens during ongoing academic processes over time, we hope to renew interest in the importance of time as a variable. The focus of dynamical system theories considers for analysis "slices" of time—from microscopic sets of moments to macroscopic observations, such as a semester or academic year(s). A primary focus of using dynamical systems theory to frame academic research is to address the reality that learning begins and ends during "instants" of interconnected time and episodes, and through interconnected constituent elements, subsystems, and higher-order components. Similar to all theoretical perspectives highlighted in this volume, at the heart of students' academic processes, we believe that emotions play a substantial role. As academic research moves forward, focusing on multiple interrelated time-sequences, the complexity and dynamics of students' intraindividual, cognitive-affective processes will be revealed.

References

Bandura, A. (1997). *Self-efficacy: The exercise of control.* New York: Freeman.

Brown, J. D. & Dutton, K. A. (1995). The thrill of victory, the complexity of defeat: Self-esteem and people's emotional reactions to success and failure. *Journal of Personality and Social Psychology, 68*, 712-722.

Dutton, K. A., & Brown, J. D. (1997). Global self-esteem and specific self-views as determinants of people's reactions to success and failure. *Journal of Personality and Social Psychology, 73*(1), 139-148.

Folkman, S. & Lazarus, R. (1985). If it changes it must be a process: Study of emotion and coping during three stages of a college examination. *Journal of Personality and Social Psychology, 48*(1), 150-170.

Frijda, N. (1986). *The emotions.* Cambridge: Cambridge University Press.

Goldberg, C. (1991). *Understanding shame.* Northvale, NJ: Jason Aronson, Inc.

Harter, S., Bresnick, S., Bouchey, A. A., & Whitesell, N. R. (1997). The development of multiple role-related selves during adolescence. *Development and Psychopathology, 9*, 835-853.

Husman, J., & Lens, W. (1999). The role of the future in student motivation. *Educational Psychologist, 34*, 113-126.

Izard, C. E, Ackerman, B. P., Schoff, K. M., Fine, S. E. (2000). Self-organization of discrete emotions, emotion patterns, and emotion-cognition relations. In M. D. Lewis & I. Granic

(Eds.), *Emotion, development, and self-organization: Dynamic systems approaches to emotional development* (pp. 15-36). Cambridge, UK: Cambridge University Press.

Kaufman, G. (1980). *Shame: The power of caring*. Rochester, VT: Schenkman Books.

Kaufman, G. (1989). *The psychology of shame: Theory and treatment of shame-based syndromes*. New York: Springer.

Kelso, J. A. (1995). *Dynamic patterns: The self-organization of brain and behavior*. Cambridge, MA: MIT Press.

Kugler, P. N., & Turvey, M. T. (1987). *Information, natural law and the self-assembly of rhythmic movement*. Hillsdale, NJ: Lawrence Erlbaum Associates.

Lazarus, R. S. (1991). *Emotion and adaptation*. New York: Oxford University Press.

Lewis, H. B. (1971). Shame and guilt in neurosis. *The Psychoanalytic Review, 58*, 419-438.

Lewis, M. (1991). *Shame: The exposed self*. New York: The Free Press.

Lewis, M. (1993). Self-conscious emotions: Embarrassment, pride, shame, and guilt. In M. Lewis & J. Haviland (Eds.) *Handbook of emotions*. New York: Guilford Press.

Lewis, M. (2000). Emotional self-organization at three time scales. In M.D. Lewis & I. Granic (Eds.), *Emotion, development, and self-organization: Dynamic systems approaches to emotional development* (pp. 125-152). Cambridge, UK: Cambridge University Press.

Lewis, M. D. & Granic, I. (1999) Self-organization of cognition-emotion interactions. In T. Dalgleish & M. J. Power, (Eds.) *Handbook of cognition and emotion*. (pp. 683-701). New York: Wiley.

Lewis, M. D. & Granic, I. (2000). Introduction: A new approach to the study of emotional development. *Emotion, development, and self-organization: Dynamic systems approaches to emotional development* (pp. 1-15). Cambridge, UK: Cambridge University Press.

Lynde, H. (1958). *On shame and the search for identity*. New York: Harcourt, Brace & Co.

Mascolo, M.F., Harkins, D., & Harakal, T. (2000). The dynamic construction of emotions: Varieties of anger. In M.D. Lewis & I. Granic (Eds.), *Emotion, development, and self-organization: Dynamic systems approaches to emotional development* (pp. 125-152). Cambridge, UK: Cambridge University Press.

Op 't Eynde, P. & Turner, J. E. (in press). Focusing on the complexity of emotion-motivation issues in academic learning: A dynamical component systems approach. *Educational Psychology Review*.

Oatley, K. (1992). *Best laid schemes: The psychology of emotions*. Cambridge UK: Cambridge University Press.

Pekrun, R. (1992). The impact of emotions on learning and achievement: Towards a theory of cognitive/motivational mediators. *Applied Psychology: An International Review, 41*(4), 359-376.

Pekrun, R. (2000). A social cognitive, control-value theory of achievement emotions. In J. Heckhausen (Ed.), *Motivational psychology of human development* (pp. 143-163). Oxford, England: Elsevier.

Pekrun, R. (in press). The control-value theory of emotions: Assumptions, corollaries, and implications for educational research. *Educational Psychology Review*.

Raynor, J.O. (1981). Future orientation and achievement motivation: Toward a theory of personality functioning and change. In G. Ydewalle & W. Lens (Eds.), *Cognition in human motivation and learning* (pp. 199-231). Hillsdale, NJ: Erlbaum.

Reed, J. L. & Schallert, D. L. (1993). The nature of involvement in academic discourse tasks. *Journal of Educational Psychology, 85*, 253-266.

Reed, J. L., Hagen, A. S., Wicker, F. W., & Schallert, D. L. (1996). Involvement as a temporal dynamic: Affective factors in studying for exams. *Journal of Educational Psychology, 88*(1), 101-109.

Rosenthal, T. L., & Zimmerman, B. J. (1978). *Social learning and cognition*. New York: Academic Press.

Schunk, D. H. (1990). Goal setting and self-efficacy during self-regulated learning. *Educational Psychologist, 25*(1), 71-86.

Shell, D. & Husman, J. (submitted for publication). Motivation, affect, and strategic self-regulation in the college classroom: A multidimensional phenomena.

Simons, J., Dewitte, S., & Lens, W. (2000). Wanting to have versus wanting to be: The effect of perceived instrumentality on goal orientation. *British Journal of Psychology, 91*, 335-351.

Tangney, J. & Fischer, K. W. (1995). *Self-conscious emotions: The psychology of shame, guilt, embarrassment, and pride*. New York: Guilford Press.

Tompkins, S. (1970). Affect as the primary motivational system. In M. B. Arnold (Ed.) *Feelings and emotions: The Loyola symposium* (pp. 101-110). New York: Academic Press.

Tompkins, S. (1987). Shame. In D. Nathanson (Ed.) *The many faces of shame* (pp. 133-161). New York: Guilford Press.

Turner, J. E. (1998). *An investigation of shame reactions, motivation, and achievement in a difficult college course*. Unpublished dissertation. The University of Texas at Austin.

Turner, J. E. & Husman, J. (2001, April). Exploring predictors and consequences of negative affect clusters. In R. Pekrun & P. Schutz (Co-chairs) *Goals, affect, and motivation in students and teachers*. Symposium conducted at the American Educational Research Association National Conference, Seattle, WA.

Turner, J. E. & Husman, J. (2003, August). *Resiliency from shame reactions: The importance of students' future goals*. In R. Pekrun & P. Op't Eynde (Co-chairs), *Emotions in students' self-regulated learning and achievement*. Symposium conducted at the European Association of Research on Learning and Instruction Conference, Padova, Italy.

Turner, J. E., Husman, J., & Schallert, D. L. (2002). The importance of students' goals in their emotional experience of academic failure: Investigating the precursors and consequences of shame. [Special Issue: Emotions in Education]. *Educational Psychologist, 37*(2), 79-89.

Turner, J. E. & McCann, E. J. (2000, April). *The importance of student goals and academic context: Investigating the consequences of experiencing shame upon students' subsequent motivational behavior, volitional strategy use, and academic achievement*. In J. E. Turner (Chair), *Exploring the interlinkages: Motivation, self-regulation, emotion, and academic achievement*. Symposium conducted at the American Educational Research Association National Conference, New Orleans, LA.

Turner, J. E. & Schallert, D. L. (2001). Expectancy-value relationships of shame reactions and shame resiliency. *Journal of Educational Psychology, 98*(2), 320-329.

Turner, J. E., Schallert, D. L., Wicker, F. W., & Waugh, R. M. (1998, April). *Capturing a shame reaction to exam feedback and investigating antecedent student characteristics and consequential behavior*. Paper presented at the American Educational Research Association National Conference, San Diego, CA.

Turner, J. E., Simmons, C. E., Goodin, J. B. (2005, August). Investigating the role of emotions in negotiating multiple goal-strivings. Paper presented at the European Association of Research on Learning and Instruction Conference, Cypress.

Turner, J. E. & Waugh, R. M. (2001, August). *Feelings of shame: Capturing the emotion and investigating concomitant experiences*. Poster presented at the American Psychological Association National Conference, San Francisco, CA.

Turner, J. E. & Waugh, R. M. (2003, April). An ideographic investigation of students' learning processes throughout a semester from an emergent, dynamical systems perspective. In P. Schutz & J. E. Turner (Co-chairs) *The dynamic interplay of students' emotions, motivation, and self-regulation within classroom contexts*. Symposium conducted at the American Educational Research Association National Conference, Chicago, IL.

Turner, J. E. & Wohlblatt, K. A. (1999, April). *The importance of self-relevant context for students' experiences of negative feelings of self worth*. In R. Pekrun & J. E. Turner (Co-chairs) *The role of emotions in students' learning and achievement*. Symposium conducted at the American Educational Research Association National Conference, Montreal, Canada.

Turvey, M. T. (1990). Coordination. *American Psychologist, 45*, 938-953.

Vallacher, R. R. & Kaufman, J. (1996). Dynamics of action identification: Volatility and structure in the mental representation of behavior. In P. M. Gollwitzer & J. A. Bargh, (Eds*). The psychology of action: Linking cognition and motivation to behavior* (pp. 260-282). New York: Guilford Press.

Waugh, R. M. (2002). *A grounded theory investigation of dyadic interactional harmony and discord: Development of a nonlinear dynamical systems theory and process-model*. Unpublished dissertation.

Waugh, R. M. (2003, May). *Development of a dyadic, interactional, nonliner dynamical systems theory and process-model*. Paper presented to the Annual Association for Psychological Science, Atlanta, GA.

Weiner, B. (1985). An attributional theory of achievement motivation and emotion. *Psychological Review, 92*, 548-573.

CHAPTER

9

Being and Feeling Interested: Transient State, Mood, and Disposition

MARY AINLEY
University of Melbourne

In the introduction to a special issue of *Educational Psychologist*, Pintrich (1991) asserted motivation could no longer be ignored in the educational research agenda. A decade later Schutz and Lanehart (2002) pointed to a similar shift in thinking turning attention to the role of emotions in education. It was pointed out that challenges for researchers included achieving greater understanding of the character of emotion and addressing the difficulties for traditional research methods posed by the fluid, changing nature of emotions (Schutz & DeCuir, 2002). It is our contention that some of the research problems arising from the fluid, changing nature of emotions and the limitations of traditional research methodologies can be addressed by focusing on the interface between student and academic tasks, and by inquiring into some of the positive emotions, especially interest, that are involved in students' achievement experiences. Consideration of the affective character of students' on-task experiences (Ainley, Hidi, & Berndorff, 2002), has prompted us to address issues of methodology with the aim of developing techniques that might give us better access to the content of students' affective states. This chapter will discuss some of the ideas, methods, and outcomes of our research program.

ON-TASK AFFECTIVE EXPERIENCE

Consider some observations that might be made from any secondary school classroom. It is a mathematics class. Some students are working with notebooks and a textbook; others are using computer notebooks. Two students are ignoring their books and computer screens and talking to each other. They exchange a few whispered words. One student bends down and searches through her bag. A few minutes later they are looking and pointing at a photograph and laughing. Nearby, another student has his eyes glued to the screen as he scans through a Web page. Alongside him, one of his classmates is slumped over his books and appears to be asleep. In another part of the room two students are conversing animatedly. One of the students points excitedly to the computer screen demanding the other look at the Web address where he has found the answers to all their homework mathematics problems. Yet another student appears to be engrossed in reading emails but quickly switches the screen to the mathematics problem and sighs when she notices the teacher watching her.

Each of these cameos represents a different interactive pattern with other class members, the teacher, and the task that has been set. Sometimes students are displaying emotion, and their facial expression leaves no doubt as to what they are feeling about the task, their classmate, or their teacher. On other occasions, feelings are less obvious from facial expression, but as with the student asleep at his desk, posture expresses feelings. Whether expressed overtly or covertly, the affective character of students' education experiences is of increasing interest to researchers.

As our title suggests, the perspective guiding our research program is consistent with theories and models that attribute behavior to both personal and situational factors, more accurately, to the interaction of personal and situational factors (e.g., Matthews & Zeidner, 2004). Disposition, mood, and transient state are perspectives representing different levels of generality—*disposition* because, whatever the educational context, students bring with them characteristic ways of processing experience. Whether they are described as social schemas and scripts, or cognitive organizations, whether the behavioral episode is a learning task or social interaction with classmates or teachers or both, individual students have typical ways of processing experience. *Mood* deserves our attention because it acknowledges that to understand students' experiences, we must acquire information concerning what went on before. *Transient state* represents the momentary, dynamic elements of students' experiences, and students' reports of their experiences are primary data for accessing these states.

In the literature, moods are distinguished from emotions by their extension over time; emotions are relatively transient states whereas moods are feelings that persist over time. Emotion and mood are also distinguished in terms of the specificity of their trigger; emotions have a specific target, a "highly accessible and salient cause" (Forgas, 2000, p.6), whereas moods with their extension over time may become relatively disconnected from the target (Erber, 1996; Frijda, 2000; Russell, 2003). Characteristically mood research distinguishes between positive and negative, or pleasant and unpleasant affect (Watson & Clark, 1994), and there is considerable debate over the relationship between these concepts (Green & Salovey, 1999; Tellegen, Watson, & Clarke, 1999; Watson, Wiese, Vaidya, & Tellegen, 1999). There is a significant focus in research on emotions in education on the way situations can trigger behavior consistent with disposition, or, consistent with current mood; the ways transient state may impact mood and how both may impact disposition. There are bidirectional influences operating that necessitate employing developmental perspectives to fully appreciate these patterns (von Eye & Schuster, 2000).

At the same time, we know that situations—whether in terms of task content, interaction with a peer, or interaction with the teacher—can trigger behavior that is out of character with known dispositions or predispositions. We know that situations can change mood by triggering inconsistent feeling states and that all of these possibilities might be happening within any one classroom, in any group of students. Understanding the affective character of disposition, mood, transient state, and the links between them will lead to greater understanding of the range of students' affective experiences. From this knowledge, ways to encourage and support student achievement are likely to emerge.

Monitoring and recording students' reports of transient states are central to the methodology we have developed to access the character of dynamic affective experiences. The software used is called *Between the Lines (BTL)* (Ainley & Hidi, 2002; Ainley, Hidi et al., 2002). Students report how they are feeling at a specific point during an achievement task. This is only part of what is encompassed by the term emotion. Contemporary theory and research includes biological and neurochemical perspectives; perceptual, cognitive, and higher-order processing; and phenomenological or subjective, conscious experiences (e.g., Cacioppo & Gardner, 1999; Frijda, 2000; Panksepp, 1998). However, we are focusing on the micro level of students' affective experiences, the phenomenology of achievement emotions, monitoring students' reports of their on-task feelings, and mapping them onto other task-related variables. This approach complements insights from research that focuses on larger behavioral episodes.

Achievement emotions can be addressed using different temporal perspectives: prospective, process, and reflective (Pekrun, Goetz, Titz, & Perry, 2002). When students are asked to predict how they are likely to feel in a future achievement task, the reports represent prospective states, anticipatory states, or expectancies. When students are asked how they are feeling while engaged on academic tasks, the perspective is concerned with ongoing process. Finally, when asked to reflect on their feelings after completing an academic task, students are either reporting how they are feeling about task completion (e.g., pride, relief) or giving a retrospective account of their on-task affective experiences (e.g., enjoyment, anxiety). Central to our approach is the focus on task processes. We are concerned with what students feel as they proceed through an academic task, from initial expectancies through to reflections on task completion. Mood and disposition can influence any point in this sequence.

AN INNOVATIVE METHODOLOGY

Schutz and DeCuir (2002) argued that capturing the dynamic and fluid character of students' emotions posed methodological difficulties. Typically questionnaires have been used to assess mood (e.g., positive and negative affect, PANAS, Watson & Clark, 1994) and disposition (e.g., trait anxiety, STAI, Spielberger & Krasner, 1988). However, transient state is more fleeting. The character and intensity of feelings may change as part of the developing relation between student and task. Laughter and enjoyment as the girls share the photograph are quickly replaced by embarrassment when the teacher notices and points out their inattention to the whole class.

Our methodology uses interactive computer software to monitor changes in specific states as students engage with academic tasks. Working online, students respond to probes, which are pop-up screens similar to those commonly seen in recreational technology that require a quick indication of feelings at a specific time. Research findings based on the range of these measures can be found in Ainley and Hidi (2002); Ainley, Hidi et al., (2002) Ainley, Hillman, & Hidi (2002); Ainley, Corrigan and Richardson (2005); Hidi, Ainley, Berndorff and Del Favero (in press), and, Pearce, Ainley and Howard (2005). Recording on-task affective experiences in this way preserves the temporal sequence of feelings. This is important for assessing the place of affective experience in self-regulation of learning where the sequence of planning, on-task responses, and reflections can be recorded. On-task probes provide a relatively quick form of measurement and can be completed in about half a minute with minimal intrusion into the ongoing task (see Ainley & Chan, 2006). This is critical when the focus of the research is a spontaneous reaction rather than a considered response. However, further investigation is needed concerning conditions where probes may deflect students from their task, or, give them a different perspective on the task, compromising the validity of these data. In addition, many of our measures of transient state are single item measures. As we have argued elsewhere (see Ainley & Patrick, 2006), reliability and validity of these measures has some support in the literature especially when the construct is very familiar to the respondent (Wanous, Reichers, & Hudy, 1997). Goetz, Frenzel, Pekrun and Hall (2006) have recently reported consistency between single-item and multi-item scales assessing enjoyment, anxiety, and boredom. Further work is needed to support reliability and validity of on-task process measures.

The same technology we use to monitor transient states includes on-line mood and disposition measures, allowing the on-task states to be anchored in a sequence that includes disposition, mood, or both. What have we found so far?

ON-TASK TRANSIENT AFFECTIVE STATES

A number of insights are emerging from our research documenting students' on-task affective experiences. First, a key component of engagement or involvement with academic tasks is experience of a state of interest. Second, at any one time there may be a number of additional emotions activated in students' on-task affective experience that reflect the character of students' responses to specific aspects of task content. We will discuss some of the findings that support these contentions.

The State of Interest

Earlier we described observations from a classroom where students were alert, concentrating, engrossed, or excited. Studies using our probes suggest that students who report feeling strong interest as they are progressing through a task are more likely to choose to continue when given the choice of continuing or quitting from the task (Ainley, Corrigan et al., 2005). An essential feature of the BTL methodology is that while students are making decisions about how they progress through the task, the software is recording the specific path they choose. In some of the tasks involving reading text, students were able to choose the order of texts, and, more importantly, at critical points were able to choose to stop reading a text and move to something else (Ainley, Hidi et al., 2002; Ainley, Hillman et al., 2002). In other studies, students did not have the option of quitting, and we tracked affective responses across the complete task and monitored changes in affective state associated with different text content. Records of time spent on each part of the task monitor such choices. Text topics and content in sections of text can generate different levels of interest, and the variability recorded has been consistent with the changing character of tasks. When the content in a text on ecotourism changed from a narrative about tourist victims of a crocodile attack to a section on crocodile habitat, interest decreased dramatically (Buckley, Hasen, & Ainley, 2004).

Our findings support a number of theoretical positions concerning the role of interest in learning and highlight the need to articulate more closely the nature of interest as a resource for learning. Researchers from both individual and situational perspectives stress that interest is not a characteristic of the person or of the object, but is a relation between person and object, whether the object is a task, a physical object, or an idea (Alexander, 2004; Hidi, 1990; Hidi, Renninger, & Krapp, 2004; Hoffmann, Krapp, Renninger, & Baumert, 1998). Further specification of the quality of the relation between person and object that justifies the term interest is required.

Hidi's writings concerning situational interest suggest that positive affect, negative affect, and even both positive and negative affect can define the relation between person and object that is active when situational interest has been triggered (see Hidi, 1990; Hidi & Harackiewicz, 2000). What we have been measuring as the state of interest can be generated from either individual or situational sources or both simultaneously. It essentially concerns a feeling of positive activation directed to a particular task. There is a target for the feeling, and the feeling is associated with focused attention and a willingness to engage with and investigate further. Hence from the perspective of the interested student, critical attributes of this relation between person and object include having a sense of connection with the object, having positive feelings of activation, wanting to explore, investigate, or engage with the object, and making decisions that promote and maintain the connection over time. The state of interest involves focused energy or activation. The student is alert and concentration is focused. This is a state of positive arousal, but it is not general arousal in the sense that it can be switched to any object. It is focused and relates to a specific object, whether a concrete task, an idea, or an anticipated activity. Hence, the state of interest is a key component in students' motivation to learn and achieve, and it underpins findings that have reported consistent associations between individual interest and learning (Alexander, 2004; Renninger, Hidi, & Krapp, 1992; Schiefele, 1996).

This perspective on interest as a positive affective state serving an important function in the motivation of learning and achievement is consistent with propositions advanced in prominent theories of emotion. Since Tomkins (1962) referred to the emotion of interest as "both what is necessary for life and what is possible" (p.345), there have been a number of emotion theorists who have included *interest* in their lexicon of emotions. At the neurophysiological level, Panksepp (1998) links states of interest and curiosity with a 'seeking' system. Izard and Ackerman (2000) define the function of interest as "motivation and energy mobilization for engagement and interaction" (p.257). Fredricksan expresses this in her *broaden-and-build* model of positive emotion (2000): "negative emotions (e.g., fear, anger, sadness) narrow an individual's momentary thought-action repertoire toward specific actions" serving a survival function. On the other hand, "positive emotions (e.g., joy, interest, and contentment) broaden an individual's momentary thought-action repertoire." Although not directly related to immediate survival purposes, there is an immediate expansion of knowledge and, over time, cognitive resources. In this way, positive emotions have an adaptive significance. More recently, Silvia (2005) has argued that interest meets the basic criteria for emotions; an organized and coherent set of phenomenological, expressive, physiological, motivational, and goal components.

Our approach does not deal directly with all of the levels that define emotions but concentrates on students' reports of their affective states. Thus we are able to explore some of the motivational, cognitive, and behavioral consequences that follow arousal of the state of interest. As we have argued earlier, feeling interested in an object supports specific decisions and action patterns: approaching, inspecting, exploring, and seeking information. In sum, the state of interest involves positive, focused, directed arousal, which prompts approach and engagement with the task. We watch the student engrossed in scanning his computer screen. Seeing his interest, we can confidently predict that he will continue examining the contents of the Web page and find out more about the mathematics topic.

COMBINATIONS OF AFFECTS

Interest is a positive activation dimension that is a central motivational factor in a wide range of achievement tasks. However, interest often occurs in combination with other feelings that pick up on specific aspects of the task, be it solving mathematics problems, reading a text, or preparing a position paper on a social issue. These specific feelings can be both positive and negative. Earlier we described how researchers focusing on individual interest referred to positive affect, and most frequently associate this with experiences of liking and enjoyment. On the other hand, Hidi (1990) suggests that situational interest can involve both positive and negative affect. Silvia (2005) has investigated the interest-enjoyment relationship further, showing that this varies as a function of different stimulus properties. For example, interest can be triggered in more complex situations than enjoyment or judgments of pleasantness.

In our early studies we used an emotion probe based on Izard's (1977) differential emotions theory, with one face icon representing each basic emotion (*sad, surprised, interested, scared, embarrassed, angry, happy, bored, disgusted*). We added *neutral* to allow students to report feeling nothing in particular. With academic tasks, *interested, neutral,* and *bored* were the most commonly reported states. Emotion probe screens were repeated, allowing students to choose more than one emotion.

In the first studies this was an unlimited choice. Few students went beyond three emotions, but many chose a second emotion. In response to this pattern and in light of well-documented empirical distinctions between activation and valence dimensions (e.g., Feldman, 1995), we designed a new form of the emotion probe that involved having students make separate reports of the interest (or activation) dimension and the valence of additional feelings at different points in the task (Buckley et al., 2004). One form of this probe is shown in Figure 1. The first screen is the interest or activation dimension (sometimes measured as a 5-point scale from bored [1] to interested [5], sometimes from not at all interested [1] to very interested [5]).

As shown in Figure 1, the second screen consists of a set of emotion icons and students report how they are feeling by clicking on an icon. An additional screen requires a rating of intensity of the chosen feeling. These screens can be repeated if the specific investigation allows students to choose more than one emotion, and the set of icons can also be varied in line with the specific research goals.

A wide range of positive and negative feelings has been recorded in combination with feeling interested. For example, when presented with a text concerning ecological issues on the survival of dolphins, students who reported that they expected to find the text very interesting were also likely to choose emotion icons representing feeling *sad* or *sorry* (Ainley, Corrigan et al., 2005). In a study using a text presenting case histories illustrating issues to do with euthanasia, a high state of interest was coupled with feelings of *sympathy* or *sadness* (Buckley et al., 2004). These feelings clearly represented students' responsiveness to the specific content of the texts. Hence while the state of interest represents the positive affect that is crucial for engagement, a diverse range of both positive and negative feelings can be aroused by different tasks and different achievement contexts.

Considered together, these findings suggest complex relationships between interest and other feelings activated in learning settings. They suggest that students' affective experiences are complex combinations in which interest plays a central role. For example, in our studies some emotion icons are consistently chosen alongside reports of high and increasing levels of interest. Others are reported along with low and decreasing levels of interest. In a recent study conducted with a sample of 8th grade girls, the problem task involved using sets of information to generate a piece of writing expressing an opinion on a topical issue (Ainley & Chan, 2006). Affective responses were

FIGURE 1
Example of emotion probe screens.

recorded three times across the task: pretask, midtask, and immediately following students' submission of their answers. Students' state of interest significantly increased across the task, from 2.94 to 3.42 on a 5-point scale. When presented with the first valence screen, more than half the students chose the *neutral* face icon, at the midtask this had decreased to just over one third of the students, and at final probe less than one-quarter of the students reported feeling *neutral*. The activation ratings from these students suggest the level of interest associated with *neutral* was on the low side of average (2.79, 2.80, and 3.05). *Anxious* was chosen less frequently as the task progressed and was associated with moderate interest (3.00, 2.80, and 3.00). On the other hand, more students chose *joyful* as the task progressed, and ratings of interest were higher (3.57, 3.90, and 4.15). Feeling *proud* was reported only at the end and was associated with high interest (4.00). That not all students were positive about the task is seen in the small numbers of students who reported feeling *hopeless* and who reported lower levels of interest as the task progressed (3.00, 2.30, and 1.25). Clearly there are important associations between levels of the state of interest and other on-task emotions that need to be investigated further. To do this we need to extend our methods for recording students' affective experiences to identify more precisely the character of these associations.

One interpretation of these associations referred to by Pekrun (2005) is that affects triggered by a specific feature of the task are transformed into interest. There are also bidirectional and reciprocal patterns to be considered. As triggering of interest opens the student to receiving more information, other affects may be aroused, which both persist across the task and maintain student interest in the task. The arousal of other affects may diminish interest. Despite these as yet unanswered questions, our findings do suggest that on-task experiences are likely to consist of complex organizations of feelings. However, identifying the role of such organizations of feelings in the development of attitudes and approaches to learning will require extending our measures in ways that can capture more of the character of these complex organizations of affect.

We have described our research approach and some recent findings monitoring students' transient affective states as they work through an academic task. But emotion or affect as transient state is only part of the story. Both mood and disposition contribute in significant ways to students' on-task affective experience.

MOOD AND TRANSIENT STATE

Students come to a classroom bringing with them specific mood states reflecting a wide range of personal experiences. Sometimes events that have been accumulating over an extended period of time mean that students are absorbed by these events and are unresponsive to classroom activities. Being asleep at the desk is more likely to be the result of an accumulation of events outside the classroom than just a response to the academic activity. Whether a classroom lesson or homework, mood states are part of the learning context and can influence the effectiveness of learning environments.

Mood has been shown to influence a range of cognitive processes, for example, memory (Eich & MacCaulay, 2000) and social judgments (Berkowitz, Jaffee, Jo, & Troccoli, 2000). Martin (2000) has proposed that positive and negative moods do not necessarily have "inherent relations" with memory and cognitive processing, but rather mood is one of a range of critical inputs influencing behavior in specific contexts. The individual continually draws inferences about their own feeling states as well as their situation. In the context of our research concerning transient, on-task affective states, this means that it is important to ask how students' moods relate to task responses, and whether working on academic tasks can change mood.

Efklides and Petkaki (2005) suggested that research into mood-cognition relationships is important for understanding self-regulation of learning. Using mood induction procedures they investigated relationships between mood and metacognitive experiences when students had to apply arithmetic operations quickly and accurately. Of particular importance for the content of this chapter is their finding that positive mood was significantly and positively related to self-reported interest and liking of the task. Whereas Efklides and Petaki (2005) investigated the effects of experimentally induced mood on students' metacognitive experiences, we assessed mood as students commenced a classroom task. To complement approaches that consider positive and negative affect as separate variables, we identified groups of students showing different combinations of positive and negative affect (Ainley, Flowers, & Patrick, 2005).

Mood items[1] adapted from a short form of the PANAS (Watson & Clark, 1994) were included at the beginning of the software to examine how mood might influence and might be altered by an academic task. Erber and Tesser (1992), using mood induction procedures, demonstrated that participants who were given a 10-minute task consisting of cognitively demanding algebra problems reported attenuation of their mood. A group who was given simple algebra problems and a group who simply waited out the time with no task, showed little change in self-reported mood on retest. Our research question was whether 7th and 8th grade students working on a task that required them to select and process information resources before writing on a topical local issue would bring about any change in mood (Ainley, Flowers, et al., 2005). Using the initial mood scale responses, separate dimensions of positive and negative affect were identified. These were then used as positive and negative affect scores in a cluster analysis that generated three groups of students who

[1] This consisted of 10 items: 5 negative affect items: scared, guilty, upset, nervous, and hostile; and 5 positive affect items: proud, enthusiastic, strong, determined, and alert.

came to the task with different mood profiles (labeled as *unhappy, anxious,* or *happy* groups). Forty percent of the 150 students were classified in the *happy* group. These students came to the task with relatively high levels of positive affect and low levels of negative affect. Most maintained their *happy* mood across the task. Following Martin's (2000) mood-as-input model, we would expect that students with a more positive mood would approach the task with more confidence and report a higher state of interest in response to the topic than the other groups. Both predictions were confirmed.

At the commencement of the task, 61 students were identified in the *unhappy* group (low negative affect and very low positive affect), but at the end when the same three mood profiles were identified, 20 of these students were identified by the clustering procedure as members of the *happy* profile group (32 remained in the unhappy group and 9 were part of the *anxious* profile group). They still reported low negative affect but their level of positive affect had increased and was now sufficiently above average for them to be grouped with the more positive mood students. The cumulative picture emerging from tracking responses of students who reported mood change was that initially they had more positive goal orientations than those who remained in their *unhappy* mood. When the task was presented, they were no more confident or no more interested in the task than those who maintained their *unhappy* mood. However, once into the task there were significant changes. The more positive edge to their general achievement goal orientation was picked up in higher salience of both on-task mastery and performance-approach goals. Interest tended to be higher than the no change group, and by the end of the task they expressed significantly higher interest in what they had done. Reflections on their experience indicated a stronger sense of achieving their goals and satisfaction with the quality of what they had written. Students who reported mood change differed in terms of goal orientation (disposition) as well as the way they responded to the task, suggesting a dynamic interdependency between disposition, mood, and transient state.

Of the 26 students who began the task with very high levels of negative affect (*anxious*), 17 remained that way, reporting similarly high levels of negative affect at the end of the task. These students did not respond positively to the cognitive challenge of the task. An important agenda for future investigations is to identify how moods characterized by lack of positive affect (the *unhappy* profile) and intense negative affect (the *anxious* profile) might be turned around to make learning environments more effective.

DISPOSITION AND TRANSIENT STATE

At all levels of education students bring with them characteristic ways of processing experience that extend beyond immediate mood. Some students enjoy learning and come to the classroom eager to learn. Others are disaffected and come reluctantly. The important questions concern how these relatively enduring personal characteristics, which are reflected in students approaches to learning, whether measured as achievement goal orientations, curiosity, or individual interests, influence on-task affective experiences.

Individual interest has been researched extensively as a relatively well-developed personal orientation towards specific domains or objects and individual interest has clear implications for the arousal of the state of interest. For example, Hidi and Renninger (2006) refer to individual interest as a "relatively enduring predisposition to reengage" (p.113). They argue that to be maintained, individual interest requires the support of ongoing engagements with the interest object. Having a well-developed individual interest in reading means that a student will be favorably disposed towards tackling many and varied texts. Texts are readily seen as something to choose to read. The student with a strong individual interest in science is unlikely to hesitate and will launch into the challenge of a new scientific puzzle or question. Students without the same well-developed interests are more likely to need contextual supports to make the same choices. They are more likely to be influenced by how the topic is presented, how the question is framed, and what special media and novelties are used to attract their attention (e.g., Cordova & Lepper, 1996). In a number of our studies, individual interest in relevant domains has been included in the design and measured prior to commencement of the academic task. Individual interest variables are predictive of how interesting students expect the task to be, and this effect flows through to the active on-task state (e.g., Ainley, Corrigan et al., 2005; Ainley, Hidi et al., 2002).

Another important disposition in relation to students' responsiveness to academic tasks is curiosity. Berlyne's (1960) work on perceptual and epistemic curiosity stimulated the development of trait measures of curiosity (e.g., Beswick, 1971), which focused on general levels of responsiveness to novel situations or problems. Like individual interest, trait curiosity develops in response to certain types of experience. In earlier work on curiosity (Ainley, 1987) two separate dimensions (breadth-of-interest curiosity and depth-of-interest curiosity) were identified and were defined by different types of novel activities and experiences that prompt approach and exploration. The depth-of-interest curiosity dimension, the tendency to approach novel and puzzling phenomena in order to understand the novel event, is particularly relevant to discussion of student's affective experience and approach to academic tasks. Our findings (Ainley, 1998) support a concept of curiosity as a complex affective-cognitive structure where the emotions associated with arousal of curiosity included interest, enjoyment, and surprise. This pattern was consistent in samples at 7th grade and at college level. In addition, scores on the depth-of-interest scale were found to be relatively stable in a group of female students who completed the scale three times, from 7th to 11th grade.

That dispositional variables affect the transient state is not surprising when the character of the disposition is considered. Individual interest and curiosity in common with other complex forms of human motivation have often been conceptualized as personal organizations of affect and cognition that energize and direct behavior. Dewey (1913) referred to interest as a "cognitive-affective synthesis." Izard, more than 50 years later in his work on infant emotions (Izard, 1977), described how basic emotions become elaborated into affective-cognitive structures. A similar position can be seen in recent writings exploring general relationships between affect, cognition, and motivation. For example, Matthews and Zeidner (2004) describe personality traits as self-regulatory mechanisms, organized units of emotion, cognition, and motivation.

The general point is that there are functional combinations of affective and cognitive processes, which persist across time as organized processing structures. In new situations, perceptual and appraisal processes draw on salient affective-cognitive organizations to generate specific states. With respect to interest, the affective state occurs in relation to an object. When the object is an achievement task, relevant prior knowledge structures are activated, and the task is appraised as offering some extension or expansion of current knowledge.

Questions about the development of relevant dispositions are key questions in understanding students' affective responses to academic tasks and need further investigation. Hidi and Renninger (2006) suggested that the predispositions represented in individual interests are developed through repeated experience of having situational interest triggered in the relevant domain. The student with a well-developed interest in mathematics or reading will have had repeated experiences where situations have triggered and maintained their level of interest. The same point is made in different ways by Alexander (2004) and Krapp (2005).

As with the dispositional characteristics of individual interest, there are important developmental issues to be addressed in relation to curiosity. Izard and Malatesta (1987) have linked infant facial expressions of the emotion interest with exploratory behavior. Significant relationships have been shown between exploratory behavior in early childhood and the development of later cognitive competencies (Raine, Reynolds, Venables, & Mednick, 2002; Trudewind, 2000). Other investigations have demonstrated the role of parents and teachers as supportive contexts for the development of curiosity (Endersley, Hutcherson, Garner, & Martin, 1979). More recently, attachment researchers (e.g., Grossmann, Grossmann, & Zimmermann, 1999) have pointed to the importance of the attachment relationship, with the hallmark of secure attachment being a "careful but curious orientation to reality."

These perspectives on individual interest and curiosity highlight that organizations of affect and cognition are activated by opportunity to engage with the interest "object," whether that be a broadly or more narrowly defined domain, or set of experiences. The compatibility between these conceptualizations of interest as disposition and contemporary theories of emotion suggest that investigation of the role of emotions in the dispositions and traits that contribute to the motivation of learning can draw further insights from closer scrutiny of contemporary emotion theories.

BEING AND FEELING INTERESTED

At the beginning of this chapter we advanced the proposition that interest has a central role in the positive affective experiences that support students' achievements. We have explored this proposition in relation to our own research, which has focused attention on developing and executing a methodology that examines the students' on-task reactivity, with specific reference to dynamic sequences of transient affective states, in particular the state of interest. We have reported findings supporting the contention that the positive directed attention characteristic of interest is related to students' decisions and choices about their engagement with academic tasks. It has also been shown that a range of additional positive and negative affects may occur in combination with the state of interest. Our focus on affective states that are transitory has acknowledged the important influences of mood, disposition, and situation on the character of students' affective reactions to academic tasks. We have also argued that these perspectives on the role of affect in education draw on and are consistent with a significant body of contemporary research and theory from education and from related areas of social and developmental psychology. The findings summarized here need to be expanded and supplemented with research into the developmental factors that contribute to positive dispositions and moods generating on-task states that facilitate achievement.

Glancing back into the mathematics classroom we notice that the two girls have put the photograph away and are now concentrating on what they are writing in their notebooks. The boy nearby still has his eyes glued to the screen. Alongside him a notebook is open showing lots of scribbled notes. The sleeping student is awake and is turning pages in his textbook as he looks quizzically around the room. The homework Web detectives have copied down all the answers and are grinning smugly. The student who was caught reading her emails is heatedly debating the correct answer to a mathematics problem with the student next to her. The dynamics of the class have changed and a new configuration of affective responses can be detected.

Acknowledgments

Acknowledgments go to the honors students who have worked on projects associated with the development of this research program. Where their work has been used in this chapter, reference has been made to conference papers, and published papers reporting their work.

References

Ainley, M. (1987). The factor structure of curiosity measures: Breadth and depth of interest curiosity styles. *Australian Journal of Psychology, 39*(1), 53-59.

Ainley, M. (1998). Interest in learning and the disposition of curiosity in secondary students: Investigating process and context. In L. Hoffmann, A. Krapp, A. Renninger & J. Baumert (Eds.), *Interest and learning: Proceedings of the Seeon Conference on gender and interest*. Kiel, Germany: IPN.

Ainley, M., & Chan, J. (2006). *Emotions and task engagement: Affect and efficacy and their contribution to information processing in during a writing task*. Paper presented at the Meetings of the American Educational Research Association, San Francisco, CA.

Ainley, M., Corrigan, M., & Richardson, N. (2005). Students, tasks and emotions: Identifying the contribution of emotions to students' reading of popular culture and popular science texts. *Learning and Instruction, 15*(5), 433-447.

Ainley, M., Flowers, D., & Patrick, L. (2005). *The impact of mood on learning—the impact of task on mood*. Paper presented at the Annual Conference of the European Association for Research in Learning and Instruction, Cyprus.

Ainley, M., & Hidi, S. (2002). Dynamic measures for studying interest and learning. In P. R. Pintrich & M. L. Maehr (Eds.), *Advances in motivation and achievement: New directions in measures and methods*. (Vol. 12, pp. 43-76). Amsterdam: JAI, Elsevier Science.

Ainley, M., Hidi, S., & Berndorff, D. (2002). Interest, learning and the psychological processes that mediate their relationship. *Journal of Educational Psychology, 94*(3), 545-561.

Ainley, M., Hillman, K., & Hidi, S. (2002). Gender and interest processes in response to literary texts: Situational and individual interest. *Learning and Instruction, 12*, 411-428.

Ainley, M., & Patrick, L. (2006). Measuring self-regulated learning processes through tracking patterns of student interaction with achievement activities. *Educational Psychology Review*.

Alexander, P. (2004). A model of domain learning: Reinterpreting expertise as a multidimensional, multistage process. In D. Y. Dai & R. J. Sternberg (Eds.), *Motivation, emotion and cognition: Integrative perspectives on intellectual functioning and development* (pp. 273-298). Mahwah, NJ: Lawrence Erlbaum Associates.

Berkowitz, L., Jaffee, S., Jo, E., & Troccoli, B. T. (2000). On the correction of feeling-induced judgmental biases. In J. P. Forgas (Ed.), *Feeling and thinking: The role of affect in social cognition* (pp. 131-152). Cambridge: Cambridge University Press.

Berlyne, D. E. (1960). *Conflict, arousal and curiosity*. New York: McGraw-Hill.

Beswick, D. G. (1971). Cognitive process theory of individual differences in curiosity. In H. I. Day, D. E. Berlyne & D. E. Hunt (Eds.), *Intrinsic motivation: A new direction in education* (pp. 156-179). Toronto: Holt, Rinehart & Winston of Canada.

Buckley, S., Hasen, G., & Ainley, M. (2004, November). *Affective engagement: A person-centered approach to understanding the structure of subjective learning experiences*. Paper presented at the Australian Association for Research in Education, Melbourne, Australia.

Cacioppo, J. T., & Gardner, W. L. (1999). Emotion. *Annual Review of Psychology, 50*, 191-214.

Cordova, D. I., & Lepper, M. R. (1996). Intrinsic motivation and the process of learning: Beneficial effects of contextualization, personalization, and choice. *Journal of Educational Psychology, 88*, 715-730.

Dewey, J. (1913). *Interest and effort in education*. Boston: Houghton Mifflin.

Efklides, A., & Petkaki, C. (2005). Effects of mood on students' metacognitive experiences. *Learning and Instruction, 15*, 415-431.

Eich, E., & MacCaulay, D. (2000). Fundamental factors in mood-dependent memory. In J. P. Forgas (Ed.), *Feeling and thinking: The role of affect in social cognition* (pp. 109-130). Cambridge: Cambridge University Press.

Endersley, R. C., Hutcherson, M. A., Garner, A. P., & Martin, M. J. (1979). Interrelationships among selected maternal behaviors, authoritarianism and preschool children's verbal and non-verbal curiosity. *Child Development, 50*, 331-339.

Erber, R. (1996). The self-regulation of moods. In L. L. Martin & A. Tesser (Eds.), *Striving and feeling: Interactions among goals, affect, and self-regulation* (pp. 251-275). Mahwah, NJ: Lawrence Erlbaum Associates.

Erber, R., & Tesser, A. (1992). Task effort and the regulation of mood: The absorption hypothesis. *Journal of Experimental Social Psychology, 28*, 339-359.

Feldman, L. (1995). Valence focus and arousal focus: Individual differences in the structure of affective experience. *Journal of Personality and Social Psychology, 69*, 53-66.

Forgas, J. P. (Ed.). (2000). *Feeling and thinking: The role of affect in social cognition*. Cambridge, UK: Cambridge University Press.

Fredrickson, B. L. (2000). Cultivating positive emotions to optimize health and well-being. *Prevention & Treatment, 3*, 1-26.

Frijda, N. H. (2000). The psychologist's point of view. In M. Lewis & J. M. Haviland-Jones (Eds.), *Handbook of emotions* (2nd ed.) (pp. 59-74). New York: Guilford Press.

Goetz, T., Frenzel, A., Pekrun, R., & Hall, N. C. (2006). The domain specificity of academic emotional experiences. *Journal of Experimental Education, 75(1)*, 5–29.

Green, D. P., & Salovey, P. (1999). In what sense are positive and negative affect independent? A reply to Tellegen, Watson and Clark. *Psychological Science, 10*(4), 304-306.

Grossmann, K. E., Grossmann, K., & Zimmermann, P. (1999). A wider view of attachment and exploration: Stability and change during the years of immaturity. In J. Cassidy & P. R. Shaver (Eds.), *Handbook of attachment: Theory, research and clinical applications*. (pp. 760-786). New York: Guilford Press.

Hidi, S. (1990). Interest and its contribution as a mental resource for learning. *Review of Educational Research, 60*, 549-571.

Hidi, S., Ainley, M., Berndorff, D., & Del Favero, L. (2006). The role of interest and self-efficacy in science-related expository writing. In S. Hidi & P. Boscolo (Eds.), *Motivation and interest in writing* (pp. 201–216). Amsterdam: Elsevier. Kluwer.

Hidi, S., & Harackiewicz, J. M. (2000). Motivating the academically unmotivated: A critical issue for the 21st century. *Review of Educational Research, 70*, 151-179.

Hidi, S., & Renninger, A. (2006). The four-phase model of interest development. *Educational Psychologist, 41*, 111-127.

Hidi, S., Renninger, A., & Krapp, A. (2004). Interest, a motivational variable that combines affective and cognitive functioning. In D. Y. Dai & R. J. Sternberg (Eds.), *Motivation, emotion and cognition: Integrative perspectives on intellectual functioning and development* (pp. 89-115). Mahwah, NJ: Lawrence Erlbaum Associates.

Hoffmann, L., Krapp, A., Renninger, A., & Baumert, J. (Eds.). (1998). *Interest and learning: Proceedings of the Seeon conference on interest and gender*. Kiel, Germany: IPN.

Izard, C. E. (1977). *Human emotions*. New York: Plenum Press.

Izard, C. E., & Ackerman, B. P. (2000). Motivational, organizational and regulatory functions of discrete emotions. In M. Lewis & J. M. Haviland-Jones (Eds.), *Handbook of emotions* (2nd ed.) (pp. 74-78). New York: Guilford Press.

Izard, C. E., & Malatesta, C. E. (1987). Perspectives on emotional development I: Differential emotions theory of early emotional development. In M. Lewis & J. M. Haviland-Jones (Eds.), *Handbook of infant development* (2nd ed.), (pp. 494-510). New York: John Wiley & Sons.

Krapp, A. (2005). Basic needs and the development of interest and intrinsic motivational orientations. *Learning and Instruction, 15*, 381-395.

Martin, L. L. (2000). Moods do not convey information: Moods in context do. In J. P. Forgas (Ed.), *Feeling and thinking: The role of affect in social cognition* (pp. 153-177). Paris: Cambridge University Press.

Matthews, G., & Zeidner, M. (2004). Traits, states, and the trilogy of mind: An adaptive perspective on intellectual functioning. In D. Y. Dai & R. J. Sternberg (Eds.), *Motivation, emotion, and cognition: Integrative perspectives on intellectual functioning and development* (pp. 143-174). Mahwah, NJ: Lawrence Erlbaum Associates.

Panksepp, J. (1998). *Affective neuroscience: The foundations of human and animal emotion*. New York: Oxford University Press.
Pearce, J., Ainley, M., & Howard, S. (2005). The ebb and flow of online learning. *Computers in Human Behavior, 21*, 745-771.
Pekrun, R. (2005). Progress and open problems in educational emotion research. *Learning and Instruction, 15*, 497-506.
Pekrun, R., Goetz, T., Titz, W., & Perry, R. P. (2002). Academic emotions in students' self-regulated learning and achievement: A program of qualitative and quantitative research. *Educational Psychologist, 37*, 91-105.
Pintrich, P. R. (1991). Editor's comments. *Educational Psychologist, 26*, 199-205.
Raine, A., Reynolds, C., Venables, P. H., & Mednick, S. A. (2002). Stimulation seeking and intelligence: A prospective longitudinal study. *Journal of Personality and Social Psychology, 82*, 663-674.
Renninger, A., Hidi, S., & Krapp, A. (Eds.). (1992). *The role of interest in learning and development*. Hillsdale, NJ: Lawrence Erlbaum Associates.
Russell, J. A. (2003). Core affect and the psychological construction of emotion. *Psychological Review, 110*, 145-172.
Schiefele, U. (1996). Topic interest, text representation, and quality of experience. *Contemporary Educational Psychology, 21*, 3-18.
Schutz, P., & DeCuir, J. T. (2002). Inquiry on emotions in education. *Educational Psychologist, 37*, 125-134.
Schutz, P., & Lanehart, S. L. (2002). Introduction: Emotions in education. *Educational Psychologist, 37*, 67-68.
Silvia, P. (2005). What is interesting? Exploring the appraisal structure of interest. *Emotion, 5*, 89-102.
Spielberger, C. D., & Krasner, S. S. (1988). The assessment of state and trait anxiety. In R. Noyes, M. Roth & G. D. Burrows (Eds.), *Handbook of anxiety: Classification, etiological factors, and associated disturbances* (Vol. 2, pp. 31-51). New York: Elsevier.
Tellegen, A., Watson, D., & Clarke, L. A. (1999). On the dimensional and hierarchical structure of affect. *Psychological Science, 10*, 297-303.
Tomkins, S. S. (1962). *Affect, imagery and consciousness*. (Vol. 1, The positive affects). New York: Springer.
Trudewind, C. (2000). Curiosity and anxiety as motivational determinants of cognitive development. In J. Heckhausen (Ed.), *Motivational psychology of human development* (pp. 15-38). Amsterdam, Netherlands: Elsevier Science.
von Eye, A., & Schuster, C. (2000). The road to freedom: Quantitative developmental methodology in the third millennium. *International Journal of Behavioral Development, 24*, 35-43.
Wanous, J. P., Reichers, A. E., & Hudy, M. J. (1997). Overall job satisfaction: How good are single item measures? *Journal of Applied Psychology, 82*, 247-252.
Watson, D., & Clark, L. A. (1994). Emotions, moods, traits and temperaments: Conceptual distinctions and empirical findings. In P. Ekman & R. J. Davidson (Eds.), *The nature of emotion: Fundamental questions* (pp. 89-93). New York: Oxford University Press.
Watson, D., Wiese, D., Vaidya, J., & Tellegen, A. (1999). The two general activation systems of affect: Structural findings, evolutionary considerations, and psychological evidence. *Journal of Personality and Social Psychology, 76*, 820-838.

Test Anxiety in Educational Contexts: Concepts, Findings, and Future Directions

MOSHE ZEIDNER
University of Haifa

Gregory Mendel, the noted pioneer and founder of classical genetics, was the son of peasant farmers, living in what is now Slovakia. Early on, his teachers recognized Mendel as an extremely talented and promising student. With his sterling academic record, he gained admission to the renowned University of Vienna to pursue his interests in the natural sciences. While he was there, he received a first-class education from some of the academic luminaries of his time. Unfortunately, however, Mendel evidenced a rather severe case of evaluation anxiety. Every time he had to face an important university exam, he became physically ill, taking months to fully recover and get back to his academic work. As a result of this serious and debilitating condition, he was unable to complete his academic work and was forced to leave the university without completing his degree. In order to subsist, he joined a monastery in the city of Brno, where he continued to pursue his interest in inheritance and conduct experiments on plants to help uncover the mechanisms in the inheritance of physical traits in plants. Although his theory and results were at first discredited by key members of the biological community, his work eventually gained worldwide recognition and acclaim. As attested by Mendel's experience, evaluative anxiety can have serious consequences for one's physical and mental health, as well as for one's educational achievements and occupational career. At the same time, not everyone with evaluative anxiety will also necessarily fail in life's tasks.

Tests and evaluative situations have emerged as a potent class of stressors in Western society, which bases many important decisions relating to an individual's status in school, college, and work on tests and other assessment devices. Test anxiety is frequently cited among the pivotal factors at play in determining a wide array of unfavorable outcomes for students, including: poor cognitive performance, scholastic underachievement, psychological distress, and ill health (Zeidner, 1998). In addition to taking its toll in human suffering and impaired test performance, test anxiety may also jeopardize assessment validity in the cognitive domain and constitute a major source of construct-irrelevant systematic variance in test scores (i.e., test bias). To the extent that anxiety influences performance in some substantial way, some examinees will perform worse than their ability or achievement would otherwise allow. Indeed, a student's performance on a classroom exam may be as much an indicator of the students' ability to cope with high levels of evaluative stress and anxiety in the classroom as a reflection of the ability or achievement the exam aims at measuring. Thus the measurement of any particular ability or proficiency will be confounded with anxiety.

This chapter examines current and recurrent issues in test anxiety theory, assessment, research, and intervention. This chapter begins with a brief description of the test anxiety construct, including basic issues and conceptualizations. I then move on to discuss key issues in the assessment of test anxiety, using both self-report measures as well as alternative assessment procedures. I then briefly examine the relationship between test anxiety and academic performance and discuss a number of key issues in test anxiety research, with a focus on personal and situational determinants of test anxiety. I then discuss clinical parameters, including coping strategies as well as interventions tailored to alleviate test anxiety. The chapter concludes by pointing out future trends and directions and the implications of current research for test anxiety theory, research, and practice in educational settings.

BASIC AND CONCEPTUAL ISSUES

The term *test anxiety* refers to the set of phenomenological, physiological, and behavioral responses that accompany concern about possible negative consequences or failure in an evaluative situation (Zeidner, 1998). Test anxiety is typically evoked in educational settings when a student believes that his or her intellectual, motivational, and social capabilities and capacities are taxed or exceeded by demands stemming from the test situation.

Much of the ambiguity and semantic confusion associated with the status of test anxiety as a psychological construct stems from the fact that different investigators have invested this term with quite divergent meanings. Thus test anxiety has been used to refer to several related yet logically very different constructs, including stressful evaluative stimuli and contexts, individual differences in anxiety proneness in evaluative situations (i.e., trait anxiety), and fluctuating anxiety states experienced in a test situation (i.e., state anxiety). Although the question still looms large whether test anxiety is best conceptualized as a relatively stable personality trait (individual difference variable) or an ephemeral emotional state, a widely accepted definition (see Spielberger & Vagg, 1995) construes test anxiety as a *situation-specific personality trait*.

A number of theoretical perspectives, surveyed by Zeidner (1998), have been suggested in the literature. Whereas cognitive-interference theories focus on the attentional demands of anxiety on the cognitive system and the debilitating effects of self-related cognitions on performance, deficit theories focus on the study and test-taking skill deficits of test-anxious students. Deficit theories of anxiety and competence are limited by their neglect of the interplay between the person's handling of environmental threats and their dispositional vulnerability.

Next, we discuss the dynamic interaction between person and situational demands, with reference to the *Self-Referent Executive Function* (S-REF) theory of emotional distress (Zeidner & Matthews, 2000; Zeidner & Matthews, 2005). The theory builds on earlier work on transactional stress processes (Lazarus, 1999) and cybernetic models of self-regulation (Carver & Scheier, 1989), to specify how executive processing of self-referent information generates anxiety and worry. This processing is shaped by declarative and procedural self-knowledge held in long-term memory. Dispositional or trait influences on anxiety are controlled by individual differences in the content of self-knowledge, consistent with evidence previously reviewed.

As shown in the S-REF model, as applied to test anxiety and graphically depicted in Figure 1, self-referent processing is generated initially by intrusions of threatening cognitions or images generated by external stimuli or internal cycles of processing. In the case of test anxiety, these would be thoughts of potential failure on the exam. The intrusions activate executive processing that seeks to initiate appropriate coping. Choice of a coping strategy is influenced by retrieval from long-term memory of self-referent knowledge and schematic plans for action. In the short-term, acute distress and worry are generated by accessing negative self-beliefs, that one lacks academic competence, for example, and by choosing counterproductive coping strategies, such as self-blame and avoidance, that focus attention on personal shortcomings in the academic domain. Of special importance are metacognitive beliefs that maintain negative self-referent thinking, for example, that it is important to monitor one's worries. In the longer term, distress may be maintained by dysfunctional styles of person-situation interaction. The well-adjusted person modifies self-knowledge to accommodate reality and learns of more effective coping strategies, such as resolving to study harder after a poor examination performance. However, preservative worry appears to strengthen and elaborate negative self-beliefs, such as being unable to cope

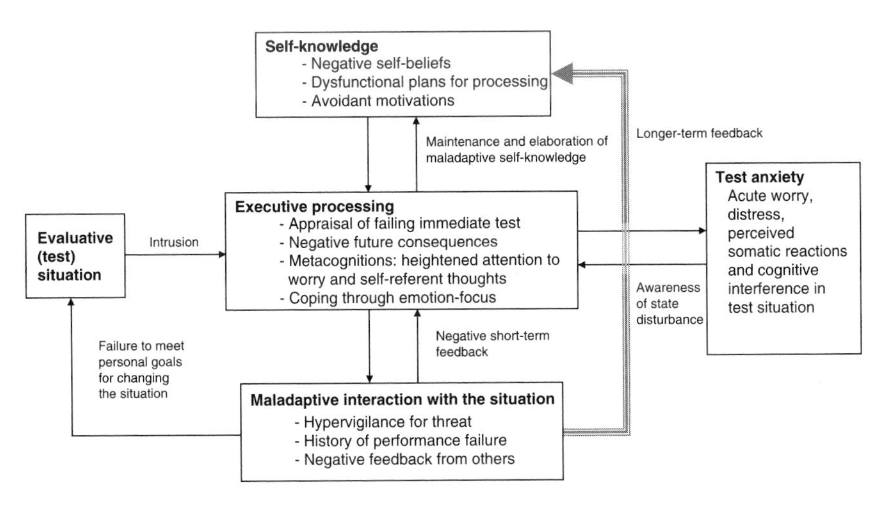

FIGURE 1

A prototypical self-regulative model of test anxiety (adapted from Zeidner & Matthews (2005).

with examinations. In addition, avoidant coping strategies lead to lack of exposure to situations that might enhance task-relevant skills. Thus the test-anxious student may be reluctant to study because the study situation focuses attention on the feared event.

MEASUREMENT AND ASSESSMENT OF TEST ANXIETY

Subjective reports include any direct report by the person regarding his or her own test anxiety responses, usually elicited via questionnaires, single-item rating scales, and think-aloud procedures or interviews before, during, or after an important exam. Self-report measures of *state* anxiety ask individuals to report which of the relevant symptoms of anxiety they are *currently* experiencing in a particular test situation, whereas trait measures ask subjects to report symptoms they *typically* or *generally* experience in test situations. Self-report inventories have been the most prevalent format for assessing test anxiety, largely because they are considered to provide the most direct access to a person's subjective experiential states in evaluative situations, possess good psychometric properties, are relatively inexpensive to produce, and are simple to administer, score, and interpret.

Fortunately, most of the more popular test anxiety inventories (e.g., Spielberger's *Test Anxiety Inventory [TAI]*, 1980; Sarason's *Reactions to Tests*, 1984; Suinn's *Test Anxiety Behavior Scale*, 1971; Benson & El-Zahhar's *Revised Test Anxiety Scale*, 1994; Wren & Benson's *Children's Test Anxiety Scale, 2004* have satisfactory reliability coefficients, typically from .85 to .95. During longer intervals between assessments, such personality traits as test anxiety may change, causing lower stability coefficients. Additional factors influencing reliability are test length, test-retest interval, variability of scores, and variation within test situation.

Most scales that have been constructed used exploratory factor analytic techniques (e.g., Spielberger's *TAI*). Confirmatory factor analysis was used early in the 1980s to test the adequacy of the indicator-factor relationship in the measurement model of test anxiety scales (e.g., Schwarzer, Jerusalem, & Lange, 1982), and has also recently been employed for purposes of item analysis and selection. Recent years have seen more sophisticated methods, such as confirmatory factor analysis and latent state-trait theory, in validating test anxiety scales and in decomposing the effects of person and occasion (e.g., Schermelleh-Engel, Keith, Moosbrugger, & Hodapp, 2004).

The fact that anxiety is such a complex construct, encompassing worry, self-preoccupation, physical upset, disruptive feelings, and maladaptive behaviors, makes it particularly difficult for researchers to sort out all these components. Researchers have found it particularly useful to differentiate between a *cognitive* facet (e.g., worry, irrelevant thinking) and an affective facet (e.g., tension, bodily reaction, perceived arousal). Thus test-anxious individuals may be characterized by their thoughts, somatic reactions, feelings, and frequently their observable behaviors in evaluative situations. In any test situation, test-anxious subjects may experience all, some, or none of these test anxiety reactions. The specific anxiety response manifested may vary, depending on the constitutional qualities and past experience of the individual, the nature of the problem to be solved, and various situational factors affecting the level of anxiety evoked. The Worry and Emotionality components of test anxiety are revealed to be empirically distinct, though correlated.

Zeidner and Nevo (1992) assessed the dimensionality of the Hebrew version of Spielberger's TAI via multidimensional scaling methods. Accordingly, the 16 scale items assessing the Worry (e.g., thoughts of doing poorly interfere with ability to concentrate on the exam) and Emotionality (e.g., feelings of confidence and relaxation during test) facets of the Test Anxiety Scale were administered to a student sample sitting for a college entrance exam. The intercorrelation matrix of the items was submitted to smallest space analysis. As shown in Figure 2, presenting the results of the analysis, the two-space is partitioned by the Worry (W) and Emotionality (E) facets of test anxiety, thus lending additional credibility to the reliability of the two-facet partition of the test anxiety space. It is noted, however, that the separation between the W and E items was a bit fuzzy, with some of the items (e.g., items 6 and 18) possibly presenting a mixture of W and E content.

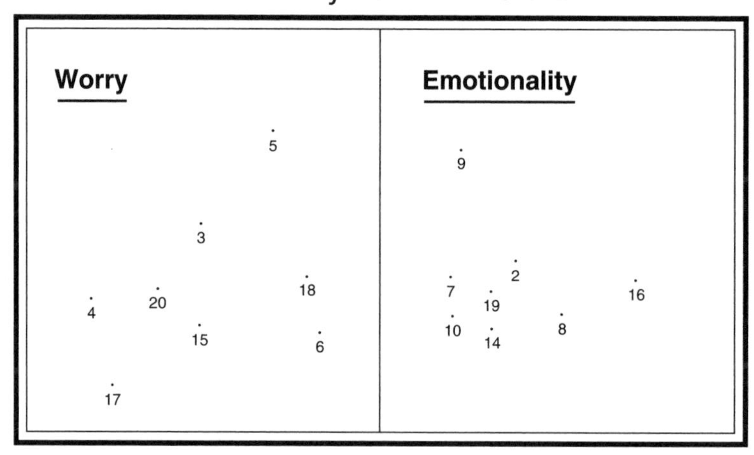

FIGURE 2
Multidimensional scaling of test anxiety inventory items.

Alternative Assessment Procedures

Although self-report inventories remain the most popular assessment tools, a variety of less frequently used assessments have been employed, including: think-aloud procedures (e.g., listing as many thoughts and feelings the student recalls having during this test), physiological measures designed to gauge changes in somatic activity believed to accompany the phenomenological and behavioral components of test anxiety (e.g., pulse, heart rate, respiration rate, skin resistance level), trace measures (e.g., accretion levels of corticosteroids, adrenaline products, free fatty acids), performance measures (e.g., examination scores, semester grade point averages, latency and errors in recall of stress-relevant stimulus materials), and unobtrusive observations of specific behaviors reflective of test anxiety in a test situation (perspiration, excessive body movement, chewing on nails or pencil, hand wringing, "fidgety" trunk movements, and inappropriate laughter when subjects were engaged in exam situations). Despite some important advantages, these alternative indices often suffer from a number of formidable methodological problems, including questionable construct validity, poor reliability, and low practicality in naturalistic field settings (Zeidner & Matthews, 2003). Overall, the assessment of test anxiety has not kept pace with the theoretical advances in conceptualizing the construct. Thus much of the construct domain (e.g., task irrelevant thinking, off-task thoughts, and poor academic self-concept) is underrepresented in current measures of test anxiety.

TEST ANXIETY AND COGNITIVE PERFORMANCE

Hundreds of studies have investigated the complex pattern of relations between anxiety and different kinds of performance. Test anxiety has been found to interfere with competence both in laboratory settings as well as in true-to-life test testing situations in school or collegiate settings (see Zeidner, 1998 for review). Processing deficits that relate to test anxiety include general impairments of attention and working memory, together with more subtle performance changes, such as failure to organize semantic information effectively.

Hembree's (1998) meta-analytic study, based on 562 North American studies, demonstrated that test anxiety correlated negatively, though modestly (about −.20) with a wide array of conventional measures of school achievement and ability at both high school and college level. Data collected on students from upper elementary school level through high school show that test anxiety scores were significantly related to grades in various subjects, although the correlation was typically about −.2. Cognitive measures (i.e., aptitude and achievement measures combined) correlated more strongly with the Worry than Emotionality component of test anxiety ($r = -.31$ vs. $-.15$). Similarly, Worry was slightly more strongly correlated with course grades than Emotionality (Worry: $r = -.19$; Emotionality: $r = -.19$). Effects sizes were higher for low-ability students than high-ability students. They were also higher for tasks perceived as difficult than tasks perceived as easy. Overall, evaluative anxiety appears to account for about 4% of the performance variance in a variety of evaluative settings, including math performance, sports, occupational settings, and social settings (Zeidner & Matthews, 2005). Thus the importance of test anxiety as a key construct in understanding sources of student distress, impaired test performance in classroom evaluative situations, and academic underachievement is now readily apparent. This situation demands that test anxiety be better understood through systematic assessment and research and appropriately dealt with (Sarason, 1980).

A number of studies have sought to identify moderator variables that accentuate or reduce deficits in performance. For example, evaluative settings, speeded timed conditions, and negative feedback appears to be especially detrimental to test-anxious subjects, whereas providing reassurance, a structured situation, and social support may eliminate the deficit (Zeidner, 1998).

In a true-to-life study among 378 Israeli college students sitting for their college entrance exams, I examined the moderating effects of phase of testing on the anxiety-performance relationship (Zeidner, 1991). Students were randomly assigned to one of two assessment conditions: (1) *pretest phase*, in which test anxiety was measured via the *TAI* prior to Scholastic Aptitude Test (SAT) administration, and (2) *posttest phase*, in which test anxiety was measured by the *TAI* following the SAT administration. Basically, time of assessment was shown to have a significant moderating effect on the anxiety-performance relationship, with correlations between test anxiety and SAT scores of −.11 and −.40, respectively, prior to testing and following testing. Thus it is critical to know when test anxiety was assessed to interpret the observed correlation between anxiety and performance. In fact, the inconsistencies reported in the literature in the anxiety-performance relationship may be due to differences among studies in the particular phase of testing at which test anxiety was measured.

These results substantiate previous theorizing in the literature (Folkman & Lazarus, 1985) that during the highly ambiguous anticipatory stage of testing (Time 1), the correlation between emotions associated with harm or threat appraisal (e.g., test anxiety) and test performance would be low, reflecting the high degree of uncertainty about both the emotions and the outcome. By contrast, the emotional and cognitive feedback provided to the examinee by the test experience at Time 2 is assumed to affect the accuracy and validity of the individual's performance expectancies and sense of competency, thus allowing the examinee to adjust his or her expectations and harm emotions accordingly. The negative test anxiety outcome emotions, which reflect appraisals about what has already transpired, tend to become increasingly negatively correlated with performance.

DETERMINANTS OF TEST ANXIETY

Interactional models of stress and anxiety (Endler & Parker, 1992; Lazarus, 1999) assume that situational anxiety in evaluative context is determined by the reciprocal interaction of personal traits (i.e., trait anxiety) and the characteristics of situations (i.e., social-evaluative). We next examine research on the role of personal and situational factors in test anxiety.

Personal Factors

The experience of evaluative anxiety is near universal across people differing in age, gender, and culture. A meta analysis of test anxiety data from 14 national sites (Seipp & Schwarzer, 1996) showed that, although mean test anxiety levels varied somewhat across cultures, test anxiety was a prevalent and relatively homogenous cross-cultural phenomenon.

The *differential hypothesis* of the interactional model (cf. Endler & Parker, 1992) claims that state anxiety will be experienced in an evaluation situation when there is a congruency or fit between the nature of a person's vulnerability (i.e., high evaluative trait anxiety) and the nature of the situation (evaluation/ego-threatening). Thus individuals high on evaluation anxiety are expected to show a higher increase in state anxiety than subjects low on evaluation anxiety primarily in a social evaluation situation (as opposed to, say, daily routine situation). The 'differential hypothesis' was tested by Zeidner (1998) in a study conducted among 198 Israeli college students (76% female) preparing for midterm exams. Specifically, it was predicted that significant differences in state anxiety would be found between high vs. low social evaluative trait-anxious students in evaluative conditions, and at the same time, nonsignificant differences in state anxiety would be observed between high vs. low social evaluative trait-anxious students in neutral conditions. Students were assessed for anxiety and coping during two phases: (1) a *neutral phase*, in which subjects were assessed during midsemester, and (2) an *evaluative phase*, in which subjects were assessed during an evaluative period, prior to midterm exams. State anxiety and situational coping served as criterion measures. Overall, the evidence supports the differential hypothesis of the interactional model of anxiety. Thus any account of determination of coping and anxiety in test situation needs to consider individual difference variables and situational variables.

Situational Parameters

In a series of studies we examined the effects of contextual and situational variables on test anxiety. Next, we present a number of exemplary studies to illustrate this programmatic research.

One line of research tested the effect of reference or comparison group, often called the *big-fish-little-pond effect* (Marsh, 1987; cf. Pekrun, Frenzel, Goetz, & Perry, 2006) with respect to test anxiety and academic self-concept (see Zeidner & Schleyer, 1999). Reference group theory posits that self-perceptions in educational settings, such as self-concept and evaluative self-cognitions, are shaped by the process of social comparison. Thus students compare their own attributes and attainments with their reference groups and use this relativistic impression as one basis for forming their self-perceptions and reaching conclusions about academic and social status. The central hypothesis, deduced from social comparison and reference group theory, was that gifted students enrolled in special gifted classes would perceive their academic ability and chances for success less favorably compared to students in regular, mixed-ability classes. Those negative self-perceptions, in turn, will serve to deflate students' self-concept and elevate their levels of evaluative anxiety and result in depressed school grades. The hypothesis was tested on a sample of 982 gifted students partaking in two types of classes: (1) special homogeneous gifted classes ($n = 321$), and (2) mixed-ability heterogeneous classes ($n = 661$), with a one-day pullout program. Students were administered an abridged version of the Test Anxiety Inventory ("Thoughts of doing poorly interfere with my concentration on the exam;" k = 12, alpha = .87). In addition, students were administered an academic self-concept scale (e.g., "I learn fairly easily;" k = 16, alpha = .85). Overall, our findings supported the big-fish-little-pond effect for test anxiety and academic self-concept. Thus both test anxiety and academic self-concept are shown to be of a dynamic character and shaped in part by social comparison processes. As shown in Figure 3, academic self-concept as well as both the Worry and Emotionality components of test anxiety were observed to be lower for gifted children in homogeneous gifted classes than in heterogeneous classes. The elevated test anxiety in gifted classes may be accounted for by a combination of factors, including: higher teacher and student performance expectations, fierce competition, and strong fear of failure. Thus the data are consistent with prior research showing that test anxiety varies with changes in students' social reference group.

To what extent does perceived control over the test situation impact anxiety and performance? On one hand, the literature focusing on decision-making under stress (Janis & Mann, 1977) would suggest that the constraints of having to choose among competing alternatives might plunge the individual into a conflict situation and increase subjective stress resulting in anxious behaviors and poorer performance. On the other hand, providing an individual with a choice among items may strengthen his or her sense of control over the situation and facilitate internal accommodation to outside events (Mills & Krantz, 1979), thus reducing stress and anxiety. The alternative hypotheses were tested by Keinan and Zeidner (1987) in a sample of 74 8th grade students, equally divided by gender. Students were informed they would be given a short math quiz and instructed to respond to three out of five items. Students were allocated to one of the two following testing conditions: (1) *decisional control*: students were given a short algebra quiz [e.g., 3 (X-2) + 3X = 60, X = ?] and instructed to respond to any three out of five items; or (2) *no decisional control*: same as above, except students were given the first three items on the exam and asked to respond to them. As shown in Figure 4, students tested under decisional control conditions were less anxious and also attained higher test scores. The data support the notion that provision of choice in an evaluative situation enhances examiner's perceived freedom of control over source of

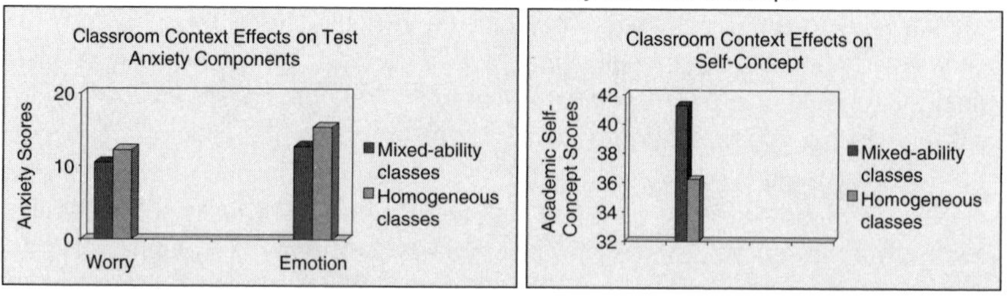

FIGURE 3
Test anxiety and academic self-concept, by educational program.

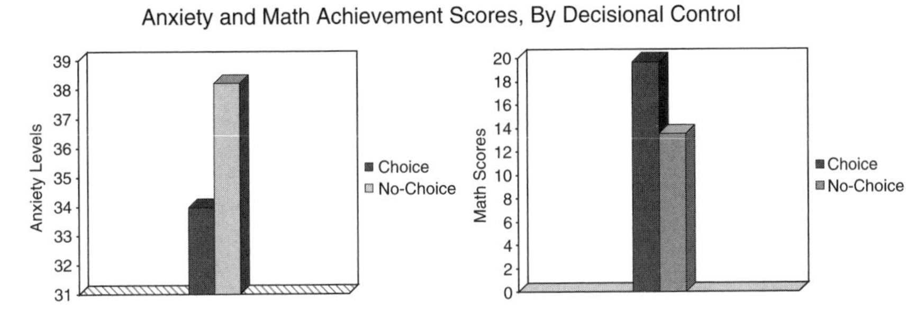

FIGURE 4

Anxiety and achievement, by decisional control.

threat. This in turn allows more favorable psychological adjustment of one's interior milieu to outside stimuli, lowering anxiety and elevating test attainment. Thus the results are more consistent with the hypothesis stating that provision of choice evokes less stress and anxiety relative to no choice.

CLINICAL PARAMETERS: COPING AND INTERVENTIONS

Coping With Test Anxiety

Test situations are currently viewed as a promising area of research for understanding how people cope with ego-threatening social encounters and how coping affects adaptational outcomes. An increasing number of studies over the past three decades have specifically focused on the ways students cope with stressful social evaluative encounters. Research by Folkman and Lazarus (1985) and Carver and Scheier (1994) provide support for the claim that problem-focused coping is adaptive in evaluative contexts, where such efforts will produce the desired outcomes. Problem-focused coping was shown to be especially adaptive in a student population during the anticipatory stage of an exam, when something can still be done to shape the outcome.

A review of the literature on coping with test anxiety (see Zeidner, 1998) concludes that it is meaningfully related to various forms of coping behaviors. Specifically, test anxiety relates positively to higher emotion-focus (e.g., trying to control anxiety symptoms) and greater avoidance (e.g., trying not to think of the test), but not to lower task-focus (e.g., focusing effort on task performance). Zeidner (1996) examined the coping strategies of 100 high school and 241 college students who were preparing for an important exam. Trait test anxiety and palliative coping strategies were both significantly predictive of state anxiety and situational coping in both groups. Furthermore, Zeidner (1994) reported that emotion-oriented coping responses were significant predictors of state anxiety among college students in close proximity to an important college exam.

Overall, research in evaluative situations concurs that some kinds of coping responses to some kinds of test situations and exigencies do make a difference, mainly with respect to affective outcomes. However, it is not entirely clear whether coping influences outcomes, coping merely covaries with adjustment to exam situations, or coping and distress are mutually intertwined reflections of something else.

Cognitive-Behavioral Interventions

A bewildering array of test anxiety treatment programs have been developed and evaluated over the past three decades (see Zeidner, 2004, for a review). Test anxiety intervention programs have flowered largely because of the salience of test anxiety in modern society and the general concern for the debilitating effects of test anxiety on the emotional well-being and cognitive performance of many. Treatment fashions and orientations have swayed sharply from the clinical to the behavioral, and more recently to the cognitive perspective—essentially mirroring the evolution of the behavior therapies.

Attempts to reduce debilitating levels of test anxiety and enhance test performance have typically focused either on treatments directed toward the emotional (affective) or cognitive (worry) facets of test anxiety (Spielberger & Vagg, 1995). Thus treatment programs include both *emotion-focused* treatments, designed largely to alleviate negative affect experienced by test-anxious persons, and *cognitive-focused* treatments, designed to help the test-anxious client cope with worry and task-irrelevant thinking and enhance their test performance.

Cognitive behavior modification (CBM), as applied to test anxiety intervention, is a multifaceted treatment designed to influence the various components of anxiety. The author and his coworkers (Zeidner, Klingman, & Papko, 1988) implemented an exemplary CBM primary prevention program among fifth and sixth grade elementary school students drawn from twelve classes in Israel. The five-phase treatment program was based primarily on Meichenbaum's (1977) cognitive modification model, implemented by those teachers whose homeroom classes participated in the study. The five phases of the program were: (1) *educational presentation*, providing students with a conceptual framework for understanding the nature of test anxiety by illuminating the nature, origins, and antecedents of test anxiety; (2) training in relaxation techniques and in the fundamentals of rational thinking; (3) *coping imagery and attentional focusing skills*, introducing students to coping imagery, which was then practiced with other previously taught techniques (positive self-statements, relaxation exercises, etc.); (4) *time management and work schemes*, focusing on the management of time both during and after the exam period and various test-taking strategies; and (5) *rehearsal and strengthening of coping skills*, aimed at rehearsing and fortifying the coping skills taught in previous sessions, primarily with the aid of guided coping imagery. Students were given instruction in using the coping techniques in future test situations. In conclusion, students summarized what they thought they had learned during the course of the training program. Evaluation of the effects of this proactive CBT program points to its effectiveness in meaningfully enhancing students' cognitive performance in test situations, with student performance meaningfully improving on three cognitive measures. As shown in Figure 5, the program was not successful in reducing test anxiety, although it may have taught some useful cognitive skills to students, thus accounting for their differential rise in test performance relative to the control. It is not unlikely that students in the experimental group may have become more aware of their test anxiety as a result of their experience, thus elevating their test anxiety scores. Although this particular program was not successful in significantly reducing students' test anxiety, reviews of the literature suggest that the combination of cognitive treatment and skills training targeted at reducing test anxiety is among the most successful types of interventions available to date (see Zeidner, 1998).

EDUCATIONAL IMPLICATIONS AND DIRECTIONS FOR FUTURE RESEARCH

In the next section I discuss the implications of our work in the area of test anxiety for psychoeducational practice. Furthermore, I note that although contemporary research has made important strides in mapping out the test anxiety terrain, there is still much uncharted territory that needs to be explored and more extensively mapped out by future research. I therefore highlight a number of these important areas, pointing out needed directions for future research.

Conceptualizations and Basic Issues

A variety of models and theoretical perspectives have been proposed over the past 50 years or so to account for various facets of test anxiety in educational

Experimental Effects for WISC[1] Performance and Test Anxiety Scores

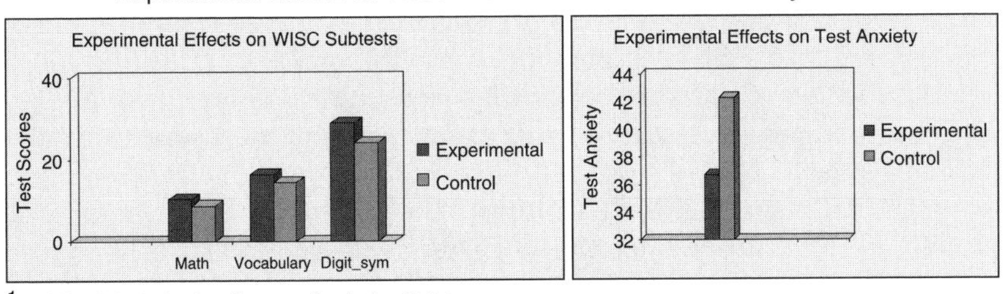

[1]WISC = Wechsler Intelligence Scale for Children

FIGURE 5

Effects of cognitive modification program on anxiety and cognitive performance.

settings, but no single unifying model is able to account for the multiple phenomena (antecedents, phenomenology, consequences) and the many complex empirical findings. Thus future test anxiety research would benefit from efforts directed at theory construction, and we have presented a provisional process model pointing in that direction. This may be achieved through broader integrative theoretical formulations, amalgamation of existing theoretical perspectives, identification of complementary approaches, common conceptual elements across theories, and so on.

Test anxiety is clearly not a unified phenomena, and I have identified a variety of different types of test anxious students (Zeidner, 1988). Development of a comprehensive taxonomy of test-anxious students would be useful for both theoretical, research, and intervention purposes. Furthermore, despite earnest efforts by practitioners to individualize treatments to the particular needs and problems of test-anxious students, we still do not have clear evidence to indicate which of the various intervention approaches is most effective for particular types of test-anxious students or for treating different manifestations of test anxiety. This stems, in part, from the absence of an established typology of test-anxious persons.

Methodology

There is a strong need for large-scale and systematic research relating to various facets of test anxiety, based on multiple observations of various target groups, at various time points, and in various contexts and cultural settings. Future research would benefit from application of sophisticated research designs—longitudinal and multivariate experimental designs, in particular. Data analysis would also benefit from application of state-of-the-art multivariate procedures, including hierarchical linear models, nonrecursive causal modeling, and multidimensional scaling techniques. Future conceptualizations and research should make more allowances for complex associations between variables, including reciprocal relationships and feedback loops as well as nonlinear relationships and interactions. More complex designs would certainly help in assessing the complex interactions between objective characteristics of the evaluative situation, personal variables, the expression of anxiety and related emotions, coping responses, and adaptive outcomes.

The key content facets represented in current test anxiety scales are rather limited and restricted in scope, with traditional scales ignoring the specificities of individual responses and situations. The response system, with focus mainly on cognitive and affective parameters, is often the only content facet represented in most current scale items. Seldom do test anxiety scales inform us about the various situational and personal factors eliciting test anxiety (e.g., anxiety proneness, inadequate preparation, over-stimulation), the full range of manifestations of test anxiety (e.g., cognitive, affective, behavioral), coping procedures and strategies, the consequences of test anxiety, or the dynamic fluctuations in test anxiety states across various phases of a stressful evaluative encounter. The restricted content scope can be improved by employing more systematic domain mapping procedures (e.g., through facet theory) and using better representation of additional facets in the test specification matrix, and subsequently on the test anxiety inventory.

When used for diagnosing and treating test anxious students, current instruments only allow measurement of the overall level of test anxiety or identification of a few of its key components. Prevalent measures are not very informative with respect to how anxiety is expressed in a student and in what situations. Future scales need to be more relevant for planning, execution, and evaluation of educational intervention through specification of the various antecedent conditions, manifestations, and consequences of test anxiety.

Finally, based on our work on the effects of decisional control, educational psychologists have often not looked favorably on free choice questions because of psychometric considerations (i.e., low reliability). However, as our research on decisional control suggests, considerations relating to examinee's emotional disposition during testing may be equally important and should therefore be given due weight and consideration by test specialists and teachers when deciding upon test administration policy.

Empirical Research in Educational Settings

Further research is needed on the specific school-related encounters that shape children's anxiety reactions and avoidance behaviors in evaluative situations. Research would benefit from more large-scale systematic and controlled studies that would pinpoint the effects of a wide array of classroom and school environmental variables (e.g., group climate and norms, evaluation and grading practices, tracking and streaming, transitional periods, teacher characteristics, teacher-student interactions, peer pressures, expectations) on the development of test anxiety in general, and different anxiety components (e.g., Worry vs. Emotionality), in particular. Additional research is also needed on the relationship between a child's failure-induced anxiety experiences in the preschool and elementary school years and their anxiety and cognitive performance later on in life (e.g., high school, college, and on-the-job performance). Also, the interaction between teacher test anxiety and student test anxiety is worthy of systematic investigation.

Further research is also needed to map out the specific effects of chronic evaluative stress on the physical and psychological health of school populations. Thus more research would help us better understand the effects of evaluative stress on maladaptive types of coping (e.g., alcoholic consumption, drug use), various forms of pathology (e.g., suicide, depression), and somatic illness in high-risk populations.

Although we have focused on highly test-anxious students in this paper, future research would benefit from examining the developmental and situational determinants of students who are on the low end of the test anxiety continuum. Thus more research is needed, focusing on resilient and low test-anxious students, who tend to view tests more as challenges than threats and who show adaptive coping responses to social evaluation situations. Furthermore, little research has been devoted to uncovering students' coping resources and factors that may serve to buffer negative emotions prior to and during stressful evaluative encounters in student populations. Future research would also benefit from examining the additive and interactive effects of test anxiety and other emotions (e.g., anger, sadness, guilt, pride, envy, joy) on a student's success and well being (cf. Pekrun, & Frese, 1992).

Our research on the impact of educational context on test anxiety in gifted students has a number of practical implications for placement of gifted students. Thus attending a selective educational framework may lead to higher school achievement, particularly for very gifted students. Yet, at the same time it may lead to reduced academic self-concept and higher test anxiety. Thus parents need to consider both the costs and benefits of sending a gifted child to a particular educational framework. On one hand, there may be little basis for assuring that students in selective classes will be advantaged on all fronts by attending selective classes. For some, the early formation of a poor self-image or the development of high test anxiety may be more detrimental than the possible benefits of attending a high ability school and concomitant higher school achievement. On the other hand, one may conceivably judge the positive academic outcomes of special classes in enhancing school achievement to be more important than evaluative anxiety or self-concept. Gifted students in a regular class whose academic self-concept was boosted relative to that for students in special classes may be in for a rude shock when they enter a profession in which there may be other equally gifted people. Also, there may be other gains that have long-term payoffs, such as the development of skills and lifelong friendships. Parents and counselors need to consider the drawbacks and advantages for a particular student before a placement decision is made.

Clinical Parameters: Coping and Interventions

About two decades ago most researchers in stress and coping would probably not have seriously questioned the assumption that coping is an important determinant of a person's emotional well-being during the various phases of a stressful transaction. Today, in contrast, researchers are asking whether coping helps; is it epiphenomenal or may it even interfere with outcomes such as emotional adjustment (Zeidner, 1988). Further research is needed to clarify how coping strategies resolve exam-related problems, relieve emotional distress, and prevent future difficulties in classroom evaluative situations.

As noted, a myriad of test anxiety intervention programs have been reported in the literature. Our research focused on assessing a school-based primary prevention program to help students cope with test situations.

The CBT program we implemented in the school system supports the provision of primary prevention programs to help students cope with test situations. Furthermore, the results appear to validate a number of assumptions derived from the tenets of psychological health education and primary prevention. First, psychological education and provision of test coping skills in the classroom context are believed to be as useful as the clinically oriented intervention by health professionals, implemented only after test anxiety has emerged as a full-blown classroom problem. We further believe that professional intervention, after repeated student failure or acute manifestations of test-anxiety reactions, can further heighten students' stress reactions. Therefore it would be more effective to provide students with relevant coping skills as part of a primary prevention program before acute test-anxiety levels are established.

A tacit assumption of many behavioral treatments is that the reduction of anxiety would release attentional and cognitive resources, thus enabling test-anxious examinees to devote a higher proportion of their capacity to learning and performing on evaluative tasks. However, as our experience dictates, procedures designed to reduce emotionality, while clearly useful in modifying subjectively experienced anxiety, by themselves appear to have little effect on cognitive performance. Overall, emotion-focused treatments appear to be relatively ineffective in reducing test anxiety, unless these treatments contain cognitive elements. It may therefore be necessary to combine such approaches with therapy modes focusing specifically on cognitive change to reliably elicit improvement in cognitive performance.

Most available studies of test anxiety intervention programs may be considered "outcome studies." Future research needs to assess differential types of treatment designed to assure maximum congruence between the test-anxious client and a particular form of intervention. Thus future research needs to provide a better answer to the question: What treatment works best for different individuals and under what conditions? Also, we currently need research to promote the development of interventions that would more reliably reduce test anxiety as well as improve academic performance. Current methods are more successful in modifying the former than the latter. Furthermore, current research suggests a number of ways in which teachers can help reduce test anxiety in the classroom (see Zeidner, 1998). Some of these procedures are summarized in Table 1. It is also noted, in passing, that a teacher's anxious behaviors may contribute to students' anxiety, and this is worthy of further research.

Finally, it is important to stress that test anxiety needs to be understood within the context of a person's life and social milieu, and requires appreciation of the possible multiple and interactional influences on anxiety scores. This includes the subject's past affective and academic history, and current social, emotional, vocational, and economic adjustments, as well as behavior during the exam. When a life history (no reported test anxiety in the past) is in disagreement with the test anxiety scale results, it is best to pause before

TABLE 1
Some Practical Suggestions for Optimizing Testing Conditions (based on Zeidner, 1998)

- Provide examinees with advance information about the test (e.g., content to be assessed, time limits, test format, and mode of administration).
- Strive to keep the average item difficulty level under control, incorporate a reasonable number of easy items, place them early in the exam, and avoid unnecessary use of extremely difficult or complex test material.
- Attempt to match the test format and mode of administration with students' preferences for specific test formats (e.g., multiple-choice or essay) and their prior experience (e.g., with computers and computerized testing).
- Assure greater examinee control of the test situation by allowing choice among items, use of open books, and adaptive testing.
- Provide examinees with the opportunity to blow off steam and comment on any facet of the test they so desire during testing.
- Create a non-threatening test atmosphere by providing examinees with task-oriented rather than ego-oriented instructions, avoiding emphasis on competition, eliminating threatening proctors, etc. Humor, soothing background music, and snacks may help to ease the tension for some examinees.
- Relax time pressures and limits whenever possible.
- Provide reassurance and emotional social support to test anxious examinees.
- Provide external memory aids and other supports.
- Provide appropriate facilities (e.g., recovery room) for anxious examinees who freeze up, to regain their composure and continue with the exam.

making a diagnosis or decision on the basis of the test anxiety scale alone, as the former is generally a more reliable criterion. Thus interpretation should only be made after examining the relevant information beyond test scores. A simple composite test anxiety score should never be used in describing, predicting, or explaining an examinee's behavior. Sound interpretation involves integrating various sources of data and assimilating them into an exposition that describes the examinee's functioning, details specific strengths and weaknesses, and predicts the specific behavioral manifestations one could be expected to see.

References

Benson, J., & El-Zahhar, N. (1994). Further refinement and validation of the revised test anxiety scale. *Structural Equation Modeling, 1*, 203-221.

Carver, C. S., & Scheier, M. F. (1989). Expectancies and coping: From test anxiety to pessimism. In R. Schwarzer, H. M. Van der Ploeg, & C. D. Spielberger (Eds.), Advances in test anxiety research (Vol. 6, pp. 3-11). Lisse, Netherlands: Swets & Zeitlinger.

Carver, C. S., & Scheier, M. F. (1994). Situational coping and coping dispositions in a stressful transaction. Journal of Personality and Social Psychology, 66, 184-195.

Endler, N.S., & Parker, J. (1992). Interactionism revisited: Reflections on the continuing crisis in the personality area. European Journal of Personality, 6, 177-198.

Folkman, S., & Lazarus, R. S. (1985). If it changes it must be a process: Study of emotion and coping during three stages of a college examination. Journal of Personality and Social Psychology, 48, 150-170.

Hembree, R. (1988). Correlates, causes, effects, and treatment of test anxiety. *Review of Educational Research, 58*, 7-77.

Janis, I. L., & Mann, L. (1977). Decision making: A psychological analysis of conflict, choice, and commitment. New York: Free Press.

Keinan, G. & Zeidner, M. (1987). Effects of decisional control on test anxiety and achievement. AQ25 Personality and Individual Differences, 8, 973-975.

Lazarus, R.S. (1999). Stress and emotion: A new synthesis. New York: Springer. Marsh, H. W. (1987). The big-fish-little-pond effect on academic self concept. Journal of Educational Psychology, 79, 280-295.

Meichenbaum, D. (1977). Cognitive-behavior modification: An integrative approach. New-York: Plenum Press.

Mills, R. T., & Krantz, D. S. (1979). Information, choice and reactions to stress: A field experiment in a blood bank with laboratory analogue. Journal of Personality and Social Psychology, 7, 608-620.

Pekrun, R., & Frese, M. (1992). Emotions in work and achievement. International Review of Industrial and Organizational Psychology, 7, 153-200.

Pekrun, R, Frenzel, A., Goetz, T., & Perry, R. P. (April, 2006). Control-value theory of academic emotions: How classroom and individual factors shape students affect. Paper presented at the annual meeting of the American Education Research Association, San Francisco, CA, April, 2006.

Sarason, I. G. (Ed.). (1980). Test anxiety: Theory, research and applications. Hillsdale, NJ: Erlbaum.

Schermelleh-Engel, K., Keith, N., Moosbrugger, H., & Hodapp, V. (2004). Decomposing person and occasion-specific effects: An extension of latent state-trait (LST) theory to hierarchical structures. Psychological Methods, 9, 198-219.

Sarason, I. G. (1984). Stress, anxiety, and cognitive interference: Reactions to tests. *Journal of Personality and Social Psychology, 46*, 929-938.

Schwarzer, R., Jerusalem, M., & Lange, B. (1982). A longitudinal study of worry and emotionality in German secondary school children. In R. Schwarzer, H. M. Van der Ploeg, & C. D. Spielberger (Eds.), Advances in test anxiety research (Vol. 1, pp. 67-81). Lisse, Netherlands: Swets & Zeitlinger.

Seipp, B., & Schwarzer, C. (1996). Cross-cultural anxiety research: A review. In C. Schwarzer & M. Zeidner (Eds.), Stress, anxiety, and coping in academic settings (pp. 13-68). Tubingen, Germany: Francke-Verlag. Spielberger, C. D., & Vagg, P. R. (1995). Test anxiety: A transactional process. In C. D. Spielberger & P. R. Vagg (Eds.), Test anxiety: Theory, assessment, and treatment (pp. 3-14). Washington, D.C.: Taylor & Francis.

Spielberger, C. D. (1980). *Test Anxiety Inventory: Preliminary Professional Manual.* Palo Alto, CA: Consulting Psychologists Press.

Suinn, R. M. (1971). *Suinn Test Anxiety Behavior Scale.* CO: RMBSI.

Wren, D. G., & Benson, J. (2004). Measuring test anxiety in children: Scale development and internal construct validation. *Anxiety, Stress, and Coping, 17,* pp. 227-240.

Zeidner, M. (1991). Test anxiety and aptitude test performance in an actual college admission testing situation: Temporal considerations. Personality and Individual Differences. 12, 101-109.

Zeidner, M. (1994). Personal and contextual determinants of coping and anxiety in an evaluative situation: A prospective study. Personality and Individual Differences, 16, 899-918.

Zeidner, M. (1996). How do high school and college students cope with test situations? British Journal of Educational Psychology, 66, 115-128.

Zeidner, M. (1998). *Test anxiety: The state of the art.* New York: Plenum.

Zeidner, M. (2004). Test anxiety. In C. Spielberger (Ed.), Encyclopedia of applied psychology (Vol. 3, pp. 545-556). San Diego: Academic Press.

Zeidner, M, Klingman, A., & Papko, O. (1988). Enhancing students test coping skills: Report of a psychological health education program. Journal of Educational Psychology, 80, 95-101.
Zeidner, M., & Matthews, G. (2000). Intelligence and personality. In R. J. Sternberg (Ed.). Handbook of intelligence (2nd ed.) (pp. 581-610). New York: Cambridge University Press.
Zeidner, M. & Matthews, G. (2003). Test anxiety. In R. Fernandez-Ballesteros (Ed.), Encyclopedia of psychological assessment (Vol 2, pp. 964-969). Beverly Hills, CA: Sage.
Zeidner, M., & Matthews, G. (2005). Evaluative anxiety. In A. Elliot & C. Dweck (Eds.), Handbook of competence and motivation (pp. 141-166). New York: Guilford Press
Zeidner, M. & Nevo, B. (1992). Test anxiety in examinees in a college admission testing situation: Incidence, dimensionality, and cognitive correlates. In K. Hagtvet (Ed.), Advances in test anxiety research (Vol. 7, pp. 288-303). Lisse, Netherlands: Swets & Zeitlinger.
Zeidner, M. & Schleyer, E. (1999). The big-fish-little-pond effect for academic self-concept, test anxiety, and school grades in gifted children. Contemporary Educational Psychology, 24, 305-329.

Students' Emotions: A Key Component of Self-Regulated Learning?

PETER OP 'T EYNDE, ERIK DE CORTE, & LIEVEN VERSCHAFFEL

University of Leuven

> "Emotions are not just the fuel that powers the psychological mechanism of a reasoning creature, they are parts, highly complex and messy parts, of this creature's reasoning itself." (Nussbaum, 2001, p. 3)

Since the 1980s, self-regulation has taken a prominent place in our thinking about learning and instruction. In line with constructivist views on learning, it is pointed out that learning is not something that happens *to* students but happens *by* students. More specifically, it is seen as

> ...the self-directive process through which learners transform their mental abilities into task-related academic skills. This approach views learning as an activity that students do for themselves in a proactive way, rather than as a covert event that happens to them reactively as a result of teaching experiences. (Zimmerman, 2001, p. 1)

Originally, self-regulation was almost exclusively perceived as the regulation of cognitive processes resulting in an emphasis on 'higher order' information processing and metacognition. Motivational and affective factors were considered minor components in explaining students' learning behavior and results (see Pintrich, Marx, & Boyle, 1993; Schutz & Davis, 2000).

Although the conception of (self-regulated) learning and competence has broadened over the years to include conative (i.e., motivational and volitional) and affective components next to cognitive ones, the research field is still struggling to come to a balanced understanding of the nature and role of these different components (e.g., Boekaerts, Pintrich, & Zeidner, 2000; Zimmerman & Schunk, 2001). In this respect, the editors of the *Handbook of Self-Regulation* (Boekaerts et al., 2000) argue the following with regard to emotional processes:

> An intriguing question for future research on the structure and process of self-regulation is "How should we deal with emotions or affect?" Some experts... view emotions as part and parcel of the self-regulatory process... By contrast, other models... do not assign a functional role to affect. (p. 754)

Traditionally, the role of specific affective variables (e.g., emotions, feelings, moods) in school learning has been hardly studied, with the exception of anxiety (see Pekrun, 2000). In the last 15 years, however, several scholars from all over the world have reported research that analyzes the role of emotions and feelings in school and academic contexts (e.g., Boekaerts, 1997; Efklides & Volet, 2005; Meyer & Turner, 2002; Pekrun, Goetz, Titz, & Perry, 2002; Schutz & Lanehart, 2002).

Inspired by the work of these scholars, we initiated a research program to investigate the role of emotions in students' (self-regulated) learning in the mathematics classroom (see Op 't Eynde, De Corte, & Verschaffel, 2001). In this chapter we outline the central characteristics of this research and discuss the main empirical findings. First, we briefly discuss how the integration of a socio-constructivist perspective on (mathematics) learning and a component systems approach of emotions provides a comprehensive theoretical framework for the study of the role of emotions in the mathematics classroom. Next, we describe how this theoretical perspective has informed our specific research approach and methodology. Then we give an overview of the different studies and their main results. Finally, we discuss the implications of these findings for educational practice and future research.

THEORETICAL FRAMEWORK

A Socio-Constructivist Perspective on Learning

Reacting against the strictly individual nature of learning depicted in traditional constructivist approaches, new perspectives on cognition and learning were introduced during the 1990s that stressed the importance of features of the socio-historical context, as notified in the work of Vygotsky and others (e.g., Wenger, 1997). In line with Vygotskian theory, a socio-constructivist account of learning stresses that learning is characterized by a reflexive relation between *the context and the individual* (e.g., Op 't Eynde et al., 2001).

Consequently, we take a perspective on students' learning that acknowledges the (socio-historically) situated *as well as* the constructivist nature of cognition and learning (e.g., Cobb & Bowers, 1999).

But such a socio-constructivist perspective is not only characterized by its focus on the interplay between the social context and the individual, but also points to the close interaction between cognitive, conative (i.e., motivational and volitional), and affective factors as constitutive for students' learning (see Op 't Eynde, De Corte, & Verschaffel, 2006). Grounded in and bounded by the specific as well as the broader socio-historical context, affective processes, like emotions and moods, in close interaction with cognitive, motivational, and volitional processes, constitute students' learning behavior.

A Dynamical Component Systems Approach Toward Emotions

Taking a socio-constructivist view as the theoretical basis for our research, we have found a dynamical component systems approach of emotions to represent a promising conceptualization of emotions that further clarifies the situated nature of emotional processes, as well as their intricate relations with cognitive and conative processes (see Op 't Eynde, De Carte & Verschaffel, 2006). A component systems approach defines an emotion to consist of multiple interacting systems or processes. Aligned with Scherer (2000), we identified the following component systems of emotions:

- the cognitive system (appraisal)
- the Autonomic Nervous System, or ANS (affect)
- the monitor system (affect)
- the motor system (action)
- the motivational system (action)

Such an approach stresses the process-character of emotions. They are perceived as emerging on-line (during a particular event) in a specific context through the interactions between cognitive (appraisal), affective (ANS and monitor system), and conative (motor and motivational) systems. The appraisal-driven modifications and the ensuing feedback interactions between the different subsystems constitute the typical pattern characterizing the emotion.

Within a component systems approach, students' appraisal processes play a crucial role. Students' appraisals of their ongoing goal-directed interactions with the world initiate and direct the emotional process (see also Lazarus, 1991). Schutz and Davis (2000, p. 246) point out that these appraisals from which emotions emerge are a function of "the underlying organization of our knowledge, goals, beliefs and experiences in a particular domain."

Indeed, there is a broad consensus today that at the level of the individual, next to physiological processes, students' domain-specific and metacognitive knowledge as well as self-efficacy, control, and value beliefs are significantly related to their academic emotions (e.g., Boekaerts & Niemivirta, 2000; Pekrun, 2000). Also, in mathematics education, students' emotional reactions toward mathematics are thought to be the outcome of consciously or subconsciously activated personal *evaluative* cognitions or beliefs about mathematics, the self, and mathematics learning situations. Consequently, students' mathematics-related belief systems as well as their mathematical knowledge and skills are identified as central mental structures underlying students' experienced emotions and functioning in the mathematics classroom (see Malmivuori, 2001; Op 't Eynde, De Corte, & Verschaffel, 2002).

However, emotions are a function of *student-environment* interactions. This implies that, in interaction with individual characteristics, features of the specific task and classroom context shape the emotions experienced by students in the classroom. In this respect, students' emotions are fundamentally *situated* in a twofold way. First, students' appraisals of a learning event are situated within a time-space frame characterized by a specific classroom, task, teacher, the instructions and comments given, the phase in the problem-solving process, and so on. Second, students' emotions are also situated as the appraisal processes are colored by students' knowledge and belief systems that are themselves social in nature and embedded in the specific and broader sociohistorical context. Indeed, the immediate classroom context, the classroom culture, the former mathematics classrooms and activities in which students participated, the general ideas about mathematics and mathematics learning held in the broader society, all determine in fundamental ways students' specific knowledge and belief systems (see Op 't Eynde, De Corte, & Verschaffel, in press Turner, Meyer, Cox, Logan, DiCintio, & Thomas, 1998).

RESEARCH APPROACH AND METHODOLOGY

The aforementioned socio-constructivist perspective on learning and emotions functions as the theoretical framework for our research. It sets the stage for the general research approach taken, the theoretical models developed, and the studies carried out.

If learning is perceived as essentially an interaction between cognitive, conative, and emotional processes, a *multidimensional approach* that addresses these respective processes and the interactions among them is required. Although different in nature, there are complex and close interactions among these processes. On one hand, the emotional experience itself consists of multiple interactions among affective, cognitive (appraisal), and motivational processes. On the other hand, within learning activities, students' emotional experiences are intricately linked to their learning goals and their action control behavior (i.e., volition). But they are also related to students' cognitive and metacognitive knowledge and strategies.

When focusing on the emotions themselves, it is clear that one should study as much as possible the *different component systems* in relation to each other (appraisal-affect-action system). Therefore the use of facial coding systems and registration systems of physiological parameters that grasp the evolutions in the motor system and the ANS should complement information about students' appraisal processes. However, since appraisal processes are the driving force behind the interactions between the different systems and are at the core of an emotion, it is not possible to study an emotion independent of a specific individual and context. Therefore researchers should always take an *actor's perspective* rather than an observer's perspective to really grasp the meaning structure behind students' appraisal processes and emotions (see Cobb & Bowers, 1999; Op 't Eynde, De Corte, & Verschaffel, 2006).

Since emotions emerge on-line, or in other words during transactions within a specific context, it is essential to study them in the *classroom context*. Next to analyzing the dynamic on-line interactions between the respective processes characterizing learning, it is, however, also necessary to address the more stable factors that are involved. It is important to distinguish between the more state-like perceptions and processes (e.g., task-specific perceptions) and the more trait-like student characteristics (e.g., beliefs, knowledge) that determine emotional processes and learning processes in general (see also infra, Fig. 1). Although survey studies and questionnaires might be suitable methods to measure trait-like variables, on-line questionnaires, experience sampling methods, video-based stimulated recall interviews, and instruments to measure physiological parameters are examples of more appropriate methods and instruments that may be used to assess the continuous flow of motivational and emotional processes in specific situations (e.g., Boekaerts, 2002; Op 't Eynde & Hannula, 2006; Prawatt & Anderson, 1994).

Based on the theoretical framework described above and the general methodological principles that follow from it, we developed a hypothetical model of mathematical problem solving (see Figure 1) to guide our research on the role of emotions in mathematical problem solving.

Starting from a general model of mathematical problem solving that differentiates between orientation, organization/implementation, and evaluation phases in the problem-solving process, Figure 1 represents the concrete student and context characteristics involved (e.g., Boekaerts, 1992; Kuhl, 1994; Schoenfeld, 1985). Also, it clarifies the respective trait- and state-like student characteristics and processes that constitute the problem-solving process. The studies discussed below all address some of the components and relations represented in the framework, depending on the specific research questions of the respective studies. At a more general level, however, the three studies aim to clarify one central research question: *What is the role of emotional processes in mathematical problem solving and which student characteristics and classroom context factors determine these processes?*

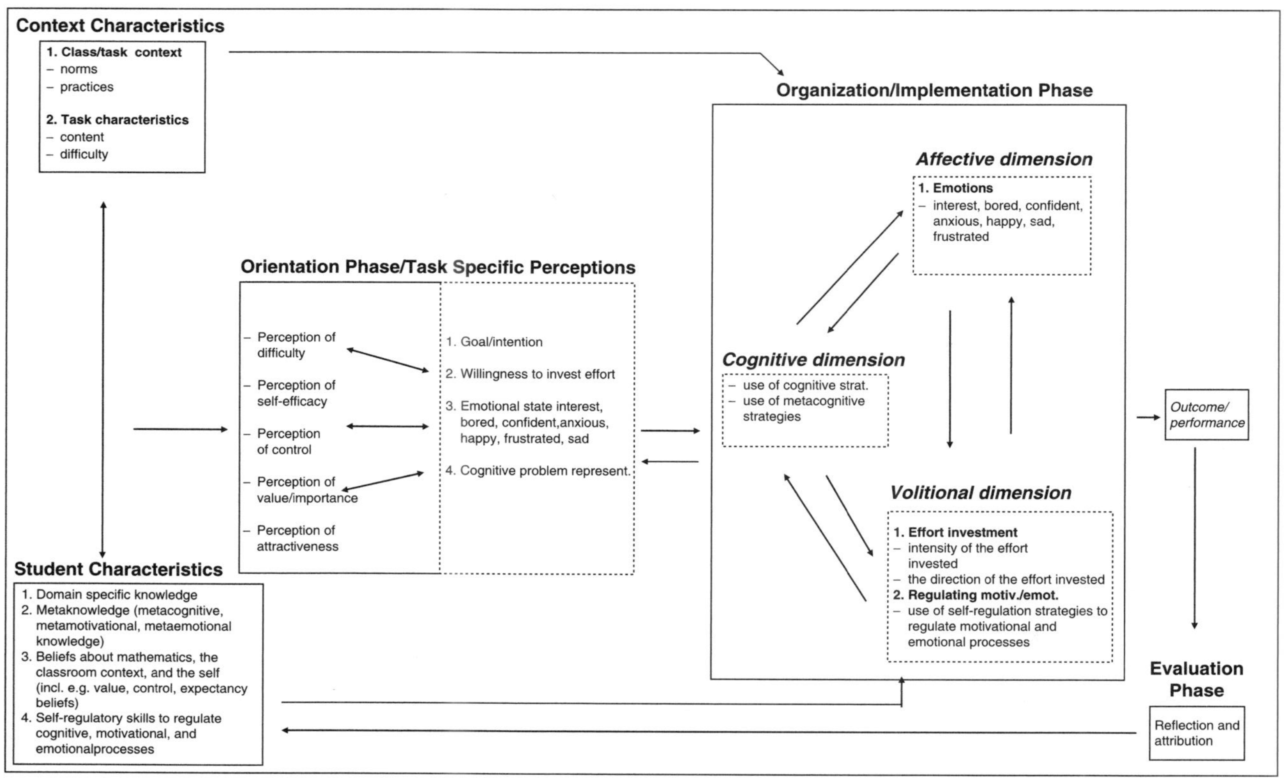

FIGURE 1

A multidimensional model of mathematical problem solving

STUDENTS' EMOTIONS IN THE MATHEMATICS CLASSROOM: EMPIRICAL FINDINGS

Study 1: Students' Mathematics-Related Belief Systems

Before analyzing the interactions among the emotional, motivational, and cognitive on-line processes that characterize mathematical problem solving, we needed to get a better understanding of some of the more trait-like student characteristics that in interaction with the classroom context are thought to influence these processes. Taking into account those students' beliefs and especially their control and value beliefs, which have been found to have an important impact on the appraisal processes, we first analyzed the nature and structure of students' mathematics-related belief systems.

Research Design and Methodology

To measure mathematics-related belief systems, we conducted a survey study in which students were presented a self-developed Mathematics-Related Beliefs Questionnaire (MRBQ) designed to measure in an integrated way their beliefs about mathematics education, about the self in relation to mathematics, and about the social context in their specific mathematics classroom (see Op 't Eynde et al., 2002; in press). The MRBQ was administered to a sample of 365 Flemish junior high students (age 14 years). This sample involved 21 classrooms spanning the different tracks in the second year of Flemish secondary education (i.e. the classical track, the humanities track, and the vocational track).

Results

A principal component analysis indicated that a four-factor model best represents the structure of students' mathematics-related belief systems. It consists of

- Beliefs about the class context, specifically the role and the functioning of the teacher (F1);
- Beliefs about the (intrinsic) value of and perceived competence in mathematics (F2);
- A view of mathematics as a social activity that refers to the usefulness of mathematics in real life and, more generally, to the fact that mathematics is grounded in human practices and is perceived as a dynamic discipline (F3); and
- A view of mathematics as a domain of excellence allowing students, through its exact and absolutist nature, to show one's competence compared to others (F4).

We found that, in general, Flemish junior-high students have a rather dynamic view of mathematics. They do not perceive mathematics primarily as a subject in which they want to prove themselves as better than others. Most students consider mathematics a rather valuable, interesting subject, and feel pretty confident about their capabilities. However, the variance analyses point to important differences among groups of students, depending on the track level they are in, their gender, and their achievement level. In particular, students' track level appears to be significantly related to their mathematics-related belief system.

On the one hand, this result certainly substantiates the influence of the social context on students' belief systems. Track level can be interpreted as a composite variable that includes a variety of classroom and other characteristics such as the actual content of the mathematics classes (the level of difficulty aimed at), the teaching style, class grouping, disciplinary problems, parental expectations, and so on. On the other hand, since students choose tracks or are directed towards specific tracks based on their former school results, their choice is in many cases related to their intellectual level (see Van Damme, Van Landeghem, De Fraine, Opdenakker, & Onghena, 2004). Thus the relevance of track level for students' belief systems may refer as much to a relation between students' intellectual level and their beliefs as to a close connection between beliefs and the classroom and school context.

A K-means cluster analysis showed that it is possible to distinguish between four significant types of students with regard to their mathematics-related beliefs profiles (see Tables 1).

In general, the differences between the respective profiles are characterized by an increase in the scores on the first three factors (i.e., F1, F2, and F3). Students with a negative belief profile do not value mathematics and do not feel very competent in it (F2). They also have a negative view of the teacher and his functioning (F1). Further, they are reasonably convinced of the social and dynamic nature of mathematics (F3) and hardly perceive mathematics as a domain of excellence (F4). Students with a mildly positive belief profile are more convinced of the value of mathematics and feel more competent in it than students with a negative belief profile, but their score on F2 is still rather low. Typically these students, similar to those with a negative belief profile, have a negative view of their teacher. Students with

TABLE 1
Factor Means of Cluster Profiles (Max Score on Each Factor = 5)

	N	F1 Classroom	F2 Competence	F3 Social	F4 Excellence
Negative belief profile	34	1,50	1,45	2,63	1,77
Mildly positive belief profile	67	1,55	2,92	3,58	2,27
Positive belief profile	168	3,16	3,24	3,74	1,56
Highly positive belief profile	110	3,43	3,53	3,94	2,88

a positive and highly positive belief profile score respectively moderately high and high on the first three factors, illustrating a very positive and dynamic view of mathematics in the classroom. Notwithstanding the clear influence of track level on the different beliefs, we found students from the four belief profiles in all tracks.

Although identifying those clusters has some relevance in itself, this was mainly done to have an indicator of the differences between students regarding their mathematics-related beliefs systems. This is important since we wanted to include in our multiple case studies a variety of students and classrooms that represented to the largest extent possible relevant differences in students' beliefs that are supposed to be important determinants of the emotions they experience in the mathematics classroom.

Study 2: The Role of Emotions in Mathematical Problem Solving

As explained before, an analysis of the role of emotions in mathematical problem solving from a socio-constructivist perspective forces us to study the on-line interactions between emotional, conative, and cognitive processes, addressing trait as well as state aspects of learning in a specific classroom context. The framework of mathematical problem solving represented in Figure 1 offers a representation of the main variables involved. In the next section, we specify which variables are systematically addressed in this study and describe the methodology used.

Research Design and Methodology

Since emotions clearly are a function of person-environment interactions, we opted for a design that, on the one hand, addresses differences in students' mathematics-related belief profiles, and on the other hand, takes into account differences in classroom contexts. Therefore we carried out a multiple case study in the second year of junior high school with students (age 14 years) of four classes in four different schools. The classes had basically the same curriculum for math but differed in the track of secondary education they followed. One was a classical Latin/Greek class (high level), one a humanities class (moderate level), one a humanities class (following a specific student-centered Freinet pedagogy), and the last one a vocational class (low level). We selected the four classes that, based on their scores on the MRBQ, could be considered representative examples of classes belonging to these respective tracks. Within each of these classes, we then selected four students whose mathematics beliefs systems best matched the four belief profiles found in the cluster analysis. This resulted in a group of 4×4 students.

In four successive lessons, all students of the four classes were asked to solve four (one for each lesson) rather complex, realistic mathematical problems, each consisting of several subtasks. Although the problems were complex, they only included mathematical operations that were representative of the primary school level. The complexity derived from their realistic nature and—related to it—the unusual way of presenting the problems and subtasks. For example, one problem consisted of a one-page story about the Kosovan war:

> A group of Kosovan refugees tried to go to Albania through the mountains. In the mountains, a woman gives birth to a baby that appears to be ill and urgently needs specialized medical care. There are two possibilities, one on foot and by car, and another with a delta plane of the Red Cross. The students had to calculate the fastest way, given the different speeds of the respective means of transport and the distances they have to travel. Basically the problem consisted of four subtasks that had to be solved successfully to find the correct solution, i.e., the fastest way. The four subtasks involved:
>
> 1. Calculating the travel time on foot from the mountains to the first village;
> 2. Calculating the travel time by car from the village to the city where the hospital is situated. Adding 1 and 2;
> 3. Calculating the weight to be carried by the delta plane and checking if there is no overweight; and
> 4. Calculating the travel time with the delta plane from the mountains to the hospital, comparing this result with the sum of 1 and 2.

Only the sixteen target students were asked to fill in the first part of the "Online Motivation Questionnaire (OMQ)" after they had skimmed the problem and before they actually started to work. The OMQ has been found to be a reliable instrument to measure some of the students' task-specific perceptions mentioned in Figure 1: task attractiveness, self-efficacy, usefulness, learning intention, and anxiety (see Boekaerts, 2002). One item was added to the OMQ in which we asked students to indicate the action goal they wanted to reach by solving the problem (work avoidance goal, task accomplishment goal, instructional goal, knowledge building goal). Knowledge building goals typically are representative of a mastery orientation, whereas task accomplishment goals more closely relate to a performance orientation. The exact orientation of instructional goals depends on the classroom context (i.e., the nature of the expectations of the teacher regarding problem solving as embedded in classroom norms and practices). Thus, instructional goals can be either mastery or performance oriented.

Every target student was also asked to think aloud during the whole problem-solving process, which was videotaped. Immediately after finishing, the student accompanied the researcher to a room adjoining the classroom where a Video-Based Stimulated Recall Interview took place (Prawatt & Anderson, 1994). This enabled students to describe in their own words how they tackled the task and experienced problem solving (see taking an actor's perspective in research).

Data Analysis

The analysis of the data gathered involved a cyclic procedure. In a first phase, we studied the thinking aloud and problem-solving protocol for all students, their answers to the On-Line Motivation Questionnaire already defined, the videotape of the problem-solving process, and the interview transcript. We used these data to describe the different (mental) experiences and activities that characterized the problem-solving process. This resulted in sixteen rich narratives of the way students' handled and experienced the problem. Although attention was paid to (meta)cognitive and motivational as well as emotional processes, the focus of the subsequent analysis was on the last category.

In a second phase, emotions were labeled and identified through a triangulation of the following data sources: (1) facial actions observable on the videotape tape; (2) vocalizations on the tape; (3) bodily actions on the tape; and (4) interpretations and appraisals explicated during the interview. The data were coded making use of some aspects of existing coding systems (e.g., the maximally discriminative facial movement coding system; Izard, 1983). The core of the labeling process, however, consisted of an analysis of students' interview data to interpret their appraisals of certain events. Starting from these appraisals and taking into account the information regarding the facial, bodily, and vocal expressions, the emotion was then defined.

In the third phase of the analysis, the focus moved from describing and interpreting phenomena to explaining them. The data, specifically students' task-specific perceptions in the orientation phase, were reanalyzed to unravel and explicate relations between these perceptions and what actually happened during problem solving. The interview transcripts were also further investigated to look for relations between students' mathematics-related beliefs and their problem-solving behavior. Finally, after this vertical analysis of each student's problem-solving process, a horizontal approach was taken by looking for recurrent patterns and/or fundamental differences between students that might deepen our understanding of what happens during problem solving and more specifically of the role of emotions in this process.

We have discussed the research design and data analysis procedures characterizing this study at some length here, because the approach taken most directly addresses our central research question and incorporates in the best way possible the methodology implied in our theoretical perspective (see for more details Op 't Eynde et al., 2006; Op 't Eynde & Hannula, 2006). Although it is our view that this approach and the different instruments used constitute a valid and reliable way to analyze emotions in classroom learning in its full complexity, there clearly are some limitations. At the moment, we still are confronted with a lack of solid research instruments that allow us to capture physiological and neurological aspects of emotional processes *in the classroom*. Consequently, a full analysis of all the component systems of emotions in the classroom was not yet feasible, leaving room for development of more appropriate and reliable measures of emotional processes. Clearly, the approach taken is also very time-intensive, implying that only a small number of students could be studied in this way. Other studies in similar and different contexts are needed to confirm our findings and to further advance our understanding of the role of emotions in mathematical problem solving and self-regulated learning.

Results

Analyzing closely the different processes characterizing the orientation phase of problem solving (see Figure 1), and comparing the task-specific perceptions (i.e., task attractiveness, self-efficacy, usefulness, learning intention, anxiety, and the nature of the action goal) of students with different belief profiles (see Table 1), indicates that the mathematics-related beliefs students hold are closely related to their perceptions and their emotions when confronted with a mathematical problem in the classroom.

Given the limited number of students in the study, investigating the relations between the respective variables using correlation analysis would not enable us to fully grasp the nature of the relations. Therefore we will use the results of a nonparametric correlation analysis as a first indicator of the interrelations and complement it with a more detailed inspection of the patterns represented in Table 2.

The correlations indicate that students in higher tracks considered the tasks less attractive than students in lower tracks (−.53). Moreover, a significant correlation was found between the perceived attractiveness of the task and the action goal (.51). A qualitative inspection of the patterns suggests those students' action goals were also closely related to their mathematics-related beliefs profile. Three out of the eight students with a lower beliefs profile (i.e., a negative profile or mildly positive profile) indicated that they wanted to avoid the problem rather than solve it. This is entirely different from students with a positive or highly positive beliefs profile, where no one had an avoidance goal. On the contrary, half of these students had a knowledge-building goal when starting with the problem. With one exception, students with knowledge-building goals were not found among the lower beliefs profiles.

The students with a negative belief profile also scored lowest on self-efficacy (mean of 2.19). This is true especially in the humanities reform and the classical class. Apparently, the classroom context in these two classrooms was not very supportive for the self-efficacy of the student with a negative beliefs profile. This was not unexpected. Based on observation data and the teacher's answers to a short questionnaire, teaching in this class was characterized by a strong performance-oriented classroom culture.

TABLE 2
Task-Specific Perceptions by Mathematics-Related Belief Profile (Max Score = 6)

	Task attractiveness	Self-efficacy	Usefulness	Learning intention	Anxiety	Action Goal
Negative Profile						
Imaram[v]	2.50	2.75	3.00	1.00	0.67	Task accomplish.
Adam[ht]	2.00	2.50	4.00	4.50	3.33	Knowledge building
Edwin[hr]	1.50	1.50	2.00	4.50	4.00	Avoidance
Margot[cl]	1.50	2.00	4.00	4.00	3.00	Avoidance
Mean	1.88	2.19	3.25	3.5	2.75	
Mildly Positive Profile						
Joanna[v]	4.50	3.75	4.00	5.00	3.33	Instructional
Belinda[ht]	2.00	2.75	3.00	3.50	3.33	Task accomplish.
Frederick[hr]	1.50	3.00	2.00	4.50	0.67	Instructional
Naomi[cl]	1.50	1.25	2.00	2.50	1.00	Avoidance
Mean	2.38	2.69	2.75	3.88	2.08	
Positive Profile						
Kurt[v]	2.00	2.75	2.00	4.00	5.00	Task accomplish.
Geoffrey[ht]	3.00	2.50	5.00	5.00	3.67	Knowledge building
Frank[hr]	3.00	4.25	4.00	4.00	2.67	Knowledge building
Otto[cl]	1.50	3.00	3.00	5.00	1.67	Task accomplish.
Mean	2.38	3.13	3.50	4.50	3.25	
Highly Positive Profile						
Laurence[v]	2.00	2.00	3.00	4.50	5.00	Knowledge building
Daniel[ht]	2.00	2.25	2.00	3.50	4.00	Task accomplish.
Hector[hr]	3.00	3.25	4.00	3.00	2.00	Knowledge building
Rebecca[cl]	1.50	2.25	3.00	5.50	2.33	Instructional
Mean	2.13	2.44	3.00	4.13	3.33	

Track indicators: [v]vocational track; [ht]humanities traditional; [hr]humanities reform; [cl]classical.

As far as the humanities reform class is concerned, these results seem to indicate that the more student-oriented way of working compared to the traditional humanities class does not really stimulate the self-efficacy of the student with a negative beliefs profile. The opposite is the case for the students with a positive or highly positive beliefs profile. They were a lot more self-efficacious in the humanities reform class than in the traditional humanities class.

With regard to anxiety, there seems to be a relation with the classroom context independently of students' beliefs profile (−.42), with an exception for the students with a negative belief profile. The students in the vocational and humanities traditional class systematically were more anxious when confronted with the problem than students in the humanities reform and classical class. It is not clear how this can be explained. The difference in intellectual aspirations in mathematics between vocational education and more general education at the individual as well as the classroom level might explain why the vocational students experience more anxiety than others when confronted with the same problem. However, this is hardly the case in the two humanities classes that are of the same intellectual level. Apparently, other differences at the individual and/or classroom level come into play here. Although we did a limited analysis of classroom practices, it clearly was insufficient to reveal all relevant aspects. Similar studies with more subjects are needed to further unravel these issues.

We next looked at the interactions between emotional, cognitive, and conative processes. When analyzing students' problem-solving behavior from a multidimensional perspective, addressing emotional, cognitive, and conative processes, we found that there is an individually changing flow of emotional experiences that derives from students' interpretations and appraisals of the events that occur during mathematical problem solving in class (see also Op 't Eynde et al., 2006). Emotions were very much part of problem solving in the four mathematics classrooms. Students felt at times annoyed, frustrated, angry, worried, anxious, relieved, happy, and nervous. Mostly, students experienced unpleasant emotions during problem solving, with frustration and nervousness as the most frequently observed emotions (for a more detailed discussion, see Op 't Eynde, De Corte, & Mercken, 2004).

This is not surprising. If a problem really is challenging, then students by definition do not know the answer immediately (see Mayer & Wittrock, 1996). So they get stuck and frustrated at certain points in the process. Indeed, unpleasant emotions usually were experienced at moments when students were not able to solve the problem as fluently as they had anticipated. Experiencing the inadequacy of the cognitive strategies used is apparently as much an emotional as a (meta)cognitive process (see also Mandler, 1989). Emotions then trigger students to redirect their behavior and to look for alternative (cognitive) strategies (i.e., problem solving as a coping strategy).

Experiencing emotions during problem solving, however, not only determines if students stick to the use of a certain cognitive strategy or look for alternative approaches, but also influences their conative processes. Unpleasant emotional experiences can lead students to reconsider their motivational task specific perceptions, including the action goal they intended to realize when solving the problem. Belinda, who was not that confident about her mathematical competence in general (mildly positive beliefs profile, see Table 2), but was convinced that she would be able to solve the problem, started doubting her capacities at the first cognitive blockage. Although she originally had a task-accomplishment goal, experiencing feelings of frustration and anger induced a reevaluation of her goals, resulting in very low motivation and unsuccessful behavior.

In a similar way, however, emotions can confirm the original task-specific perception and further weaken or strengthen students' motivation for the task. Edwin had a negative mathematics-related beliefs profile and initially an avoidance goal. When confronted with a block during problem solving he became quite apathetic, stating: "You see, I already knew that it would be too difficult. It does not matter what I do." His motivation went down even more, resulting in hardly any concentrated or task-focused behavior. Geoffrey, on the other hand, having a positive beliefs profile and being very motivated to engage in the problem with the ambition to really learn something from it (a knowledge-building goal), became even more involved in solving the problem when he happily managed to find the correct answer to the first subtask.

In summary, our data do not only point to students' beliefs and task-specific perceptions as important antecedents of emotions, they also suggest how emotions interact with cognitive, motivational, and volitional processes to determine the course of the problem-solving process. We found many students who experienced unpleasant emotions during mathematical problem solving. Some were able to effectively deal with those unpleasant emotions in a task-focused way, others got lost in ruminating thoughts and avoidance behavior. Students' competence to self-regulate these unpleasant emotions in effective ways might be an important determinant of successful mathematical problem solving. Therefore in the next study we started to explore students' knowledge and use of strategies to self-regulate their emotions in the mathematics classroom.

Study 3: Students' Use of Emotional Regulation Strategies

Research shows that, in general, students use different coping strategies to regulate their emotions in stressful learning situations (e.g., Schutz & Davis, 2000). Typically, a distinction is made between problem- (or task-) and emotions-focused coping (e.g., Carver, Weintraub, & Scheier, 1989). In the multiple case study discussed previously, we saw students using problem- or task-focused coping. But escape-avoidance strategies, seeking distraction, or seeking social support were also used to deal with such emotions as frustration (see Op 't Eynde et al., 2004). In Study 3, we investigated in a more systematic way the kind of emotional (self-) regulation strategies students use in learning mathematics in school contexts (see also, Op 't Eynde, De Corte, & Mercken, 2005).

Research Design and Methodology

We conducted a survey study focused on the coping behavior of 393 Flemish second-year (age 14 years) and fourth-year (age 16 years) secondary school students in different stressful, mathematical school situations. In addition, we analyzed the relations between their coping behavior, on the one hand, and their gender, age, educational track (general, technical, or vocational education), motivation for mathematics, and achievement in mathematics, on the other hand. Also, we investigated how students' coping behavior is related to the specific kind of stressful school situation they are confronted with and their familiarity with it.

To assess the kind of regulation strategies students employ when managing their emotions, we developed a Flemish version of Carver, Weintraub, and Scheier's (1989) COPE-questionnaire. This multidimensional coping inventory incorporates 13 conceptually distinct scales, and thus represents a wide variety of coping strategies. Students were asked to indicate on a 4-point Likert scale the regulation strategies they draw on to manage their emotions in three stressful situations related to mathematics learning: (1) confronted with a difficult mathematics test, (2) dealing with difficult mathematics homework, and (3) facing a difficult mathematics lesson. Students' answers on the items were scored as followed:

- Never use this strategy 0
- Rarely use this strategy 1
- Sometimes use this strategy 2
- Often use this strategy 3

An exploratory factor analysis was performed to identify the different categories of emotional regulation strategies used by students. Next, variance analyses clarified the relations between the strategies used and the three different mathematical school situations, students' familiarity with the stressful nature of these situations, their track, achievement level, age, and gender.

Results

Our data indicate that, in general, students do use coping strategies from time to time. However, the means mentioned in Table 3 are never above 2 (out of a maximum of 3), which illustrates that they do not systematically use coping strategies in stressful mathematical school situations.

Based on the frequency with which students make use of the six different categories of coping strategies, a cluster analysis identifies three main clusters. The first cluster includes the factors *active coping* and *humor and acceptance,* which are most frequently used by adolescents, with mean scores ranging from 1.39 to 1.77 on a scale of 0 (never used) to 3 (often used). The factors *social-emotional coping* and *abandoning and negation* encompass the second cluster

TABLE 3
Overall Factor Means and by Situation (Max. Score 3)

Factor	Test situation	Home work situation	Lesson situation	Overall
1. Active coping	1.74	1.77	1.76	1.76
2. Humor and acceptance	1.45	1.41	1.39	1.42
3. Social-emotional coping	1.15	1.12	1.06	1.11
4. Abandoning and negation	0.98	1.04	1.00	1.09
5. Religion	0.70	0.65	0.61	0.65
6. Alcohol and drug use	0.16	0.15	0.16	0.16
Overall	1.03	1.03	1.00	1.02

(range from 0.98 to 1.15). Secondary school students rarely reported using such coping strategies as *religion* and *alcohol and drug use*. The mean scores of coping strategies belonging to this cluster range from 0.15 to 0.70.

Although students mostly use active coping strategies that allow them to remain focused on the task and solve the problem, the variance analyses indicate that students of the lower tracks, students with little motivation, and students who are frequently confronted with stressful situations typically use fewer adequate coping strategies. We found, for example, that students of the vocational track use significantly fewer active coping strategies and social-emotional coping strategies in the different stressful situations. The finding that the more students are familiar with a stressful situation, the more they use less-adequate coping strategies, such as abandoning and denial, indicates that they do not spontaneously learn to tackle stressful situations in an effective way. More likely, they tend to end up in a negative spiral in which low motivation and the use of inadequate coping strategies result in experiencing more stressful situations with only one way out: abandoning and denial. Since students in vocational tracks in general have been found to be less motivated than students in other tracks, they seem to be extremely vulnerable to stressful situations.

CONCLUSION

In line with related research, the results of our studies indicate that students clearly experience emotions during mathematical problem solving that interact in different ways with cognitive, conative processes, and as such determine the course of the problem-solving process. At the individual level, students' mathematics-related belief systems, including their value and competence beliefs, directly and indirectly (through the task-specific perceptions) determine the appraisal processes that are at the core of the emotions they experience during problem solving in the mathematics classroom. But our data show clearly that aside from students' belief systems, characteristics of the classroom context, often related to track level, also have an impact on the development of students' beliefs and on the way these beliefs influence learning.

Whereas experiencing negative or stressful emotions does not necessarily have to be detrimental to learning and problem solving, for many students it appears to be the case. Consequently, it is important to find ways of organizing the classroom context in general and instruction in particular in a way that students either experience fewer unpleasant emotions or know how to deal with them in an effective task-focused way. Two lines of educational interventions seem appropriate here. One is directed towards the promotion of more positive beliefs, another is focused on equipping students with the necessary skills and strategies to (self-) regulate their emotions in effective ways.

Typical instructional practices that have been found to allow for the development of more positive beliefs and appraisals are, among others, autonomy support rather than a very strict control of student learning, co-operative work rather than competitive work, giving constructive feedback and allowing students to learn from errors rather than evaluation per se, teachers who explicitly show their motivation and interest in the subject field, and so on (e.g., Perry, Schonwetter, Magnusson, & Struthers, 1994). Classroom contexts characterized by these practices enable students to develop more positive and availing beliefs, resulting in more positive feelings and a greater involvement in learning.

Next to stimulating the development of more availing beliefs through implementing adequate classroom practices, education should stimulate students to acquire the necessary strategies to self-regulate their emotions. Gottman, Katz, and Hooven (1997) refer in this respect to classroom environments that enable students to master the necessary metaemotional knowledge and skills. A teacher who is functioning as an emotional coach and organizes classroom interactions accordingly characterizes such a classroom context. Gottman et al. (1997) identify three key dimensions of (emotionally) coaching educational environments: (1) a lack of derogation; (2) presence of warm interpersonal relations; and (3) a focus on cognitive as well as emotion scaffolding. This view is in line with Meyer and Turner's (2002) findings on emotions in the classroom. They point out that teachers' affective responses are important both at the academic or cognitive level and at the interpersonal level.

More research is needed that addresses the affective dimensions of learning from a learner's as well as a teacher's perspective. Only in that way can a research-based body of knowledge become available that allows instructional designers and teachers to develop powerful learning environments that adequately address emotions as a key component of self-regulated learning.

References

Boekaerts, M. (1992). The adaptable learning process: Initiating and maintaining behavioral change. *Applied Psychology: An International Review, 41*, 377-397.

Boekaerts, M. (1997). Capacity, inclination, and sensitivity for mathematics. *Anxiety, Stress, and Coping, 10*, 5-33.

Boekaerts, M. (2002). The on-line motivation questionnaire: A self-report instrument to assess students' context sensitivity. In P. R. Pintrich & M. L. Maehr (Eds.), *Advances in motivation and achievement, Vol. 12, New directions in measures and methods* (pp. 77-120). Oxford, UK: Elsevier Science Ltd.

Boekaerts, M., & Niemivirta, M. (2000). Self-regulated learning: Finding a balance between learning goals and ego-protective goals. In M. Boekaerts, P. R. Pintrich, & M. Zeidner (Eds.), *Handbook of self-regulation* (pp. 417-450). San Diego: Academic Press.

Boekaerts, M., Pintrich, P. R., & Zeidner, M. (Eds.). (2000). *Handbook of self-regulation*. San Diego: Academic Press.

Carver, C., Weintraub, J, & Scheier, M. (1989). Assessing coping strategies: A theoretically based approach. *Journal of Personality and Social Psychology, 56*, 267-283.

Cobb, P., & Bowers, J. (1999). Cognitive and situated learning: Perspectives in theory and practice. *Educational Researcher, 28*, 4-15.

Efklides, A., & Volet, S. (Eds.). (2005). Feelings and emotions in the learning process. [Special issue]. *Learning and Instruction, 15*, 377-515.

Gottman, J. M., Katz, L. F., & Hooven, C. (1997). Introduction to the concept of meta-emotion. In J. M. Gottman, L. F. Katz, & C. Hooven (Eds.), *Meta-emotion. New families communicate emotionally* (pp. 3-8). Mahwah, NJ: Lawrence Erlbaum Associates.

Izard, C. E. (1983, revised). *The maximally discriminative facial movement coding system*. Newark: University of Delaware, Instructional Resources Center.

Kuhl, J. (1994). A theory of action and state orientations. In J. Kuhl, & J. Beckmann (Eds.), *Volition and personality: Action versus state orientation* (pp. 9-46). Seattle: Hogrefe & Huber Publishers.

Lazarus, R. S. (1991). *Emotion and adaptation*. New York: Oxford University Press.

Malmivuori, M.-L. (2001). *The dynamics of affect, cognition, and social environment in the regulation of personal learning processes: The case of mathematics (Research report 172)*. Helsinki: University of Helsinki.

Mandler, G. (1989). Affect and learning: Causes and consequences of emotional interactions. In D. B. McLeod, & V. M. Adams (Eds.), *Affect and mathematical problem solving: A new perspective* (pp. 3-19). New York: Springer-Verlag.

Mayer, R. E., & Wittrock, M. C. (1996). Problem-solving transfer. In D.C. Berliner & R.C. Calfee (Eds.), *Handbook of educational psychology* (pp. 47-62). New York: Simon & Schuster Macmillan.

Meyer, D. K., & Turner, J. C. (2002). Discovering emotion in classroom motivation research. *Educational Psychologist, 37*, 107-114.

Nussbaum, M. C. (2001). *Upheavals of thought: The intelligence of emotions*. Cambridge, UK: Cambridge University Press.

Op 't Eynde, P., De Corte, E., & Mercken, I. (2004). *Pupils' (meta)emotional knowledge and skills in the mathematics classroom*. Paper presented at the Annual Meeting of the American Educational Research Association, San Diego, California, April 12-16.

Op 't Eynde, P., De Corte, E., & Mercken, I (2005). *Students' self-regulation of emotions in mathematics learning: How do they cope?* Paper presented at the 11th Biennal Conference of the European Association for Research on Learning and Instruction, Nicosia, Cyprus, August 23-27.

Op 't Eynde, P., De Corte, E., & Verschaffel, L. (2001). "What to learn from what we feel?": The role of students' emotions in the mathematics classroom. In S. Volet & S. Järvelä (Eds.), *Motivation in learning contexts: Theoretical and methodological implications* (pp. 149-167). Oxford, UK: Elsevier Science Ltd

Op 't Eynde, P., De Corte, E., & Verschaffel, L. (2002). Framing students' mathematics-related beliefs: A quest for conceptual clarity and a comprehensive categorization. In G.C. Leder, E. Pehkonen, & G. Törner (Eds.), *Beliefs: A hidden variable in mathematics education?* (pp. 13-37). Dordrecht, The Netherlands: Kluwer Academic Publishers.

Op 't Eynde, P., De Corte, E., & Verschaffel, L. (2006). "Accepting .emotional complexity:" A socio-constructivist perspective on the role of emotions in the mathematics classroom. *Educational Studies in Mathematics*.

Op 't Eynde, P., De Corte, E., & Verschaffel, L. (in press). Beliefs and metacognition: An analysis of junior high students' mathematics-related beliefs. In A. Desoete, & M. V. J., Veenman (Eds), *Metacognition in mathematics*. Hauppauge: Nova Science Publishers.

Op 't Eynde, P., & Hannula, M. (2006). The case of Frank. *Educational Studies in Mathematics*.

Pekrun, R. (2000). A social-cognitive, control-value theory of achievement emotions. In J. Heckhausen (Ed.), *Motivational psychology of human development*. (pp. 143-163). Oxford, UK: Elsevier.

Pekrun, R., Goetz, T., Titz, W., & Perry, R. P. (2002). Academic emotions in students' self-regulated learning and achievement: A program of qualitative and quantitative research. *Educational Psychologist, 37*, 91-105.

Perry, R. P., Schonwetter, D., Magnusson, J. L., & Struthers, W. (1994). Use of explanatory schemas and the quality of college instruction: Some evidence for buffer and compensation effects. *Research in Higher Education, 35*, 349-371.

Pintrich, P. R., Marx, R.W., & Boyle, A. (1993). Beyond cold conceptual change: Motivational beliefs and classroom contextual factors in the process of conceptual change. *Review of Educational Research, 63*, 167-199.

Prawat, R. S., & Anderson, A. L. H. (1994). The affective experiences of children during mathematics. *Journal of Mathematical Behavior, 13*, 201-222.

Scherer, K. R. (2000). Emotions as episodes of subsystem synchronization driven by nonlinear appraisal processes. In M. D. Lewis & I. Granic (Eds.), *Emotion, development, and self-organization: Dynamic systems approaches to emotional development* (pp. 70-99). Cambridge, UK: Cambridge University Press

Schoenfeld, A. H. (1985). *Mathematical problem solving*. Orlando, FL: Academic Press.

Schutz, P. A., & Davis, H. A. (2000). Emotions and self-regulation during test taking. *Educational Psychologist, 35*, 243-256.

Schutz, P. A., & Lanehart, S.L. (Eds). (2002). Emotions in education [special issue]. *Educational Psychologist, 37*, 67-134.

Turner, J. C., Meyer, D. B., Cox, K. E., Logan, C., DiCintio, M., & Thomas, C. T. (1998). Creating contexts for involvement in mathematics. *Journal of Educational Psychology, 90*, 730-745.

Van Damme, J., Van Landeghem, G., De Fraine, B., Opdenakker, M.-Ch., & Onghena, P. (2004). *Maakt de school het verschil?: Effectiviteit van scholen, leraren, en klassen in de eerste graad van het middelbaar onderwijs* [Does the school make a difference?: Effectivity of schools, teachers, and classgroup in the first two years of secondary education]. Leuven, Belgium: Acco.

Wenger, E. (1997). *Communities of practice: Learning, meaning, and identity*. Cambridge, UK: Cambridge University Press

Zimmerman, B. J. (2001). Theories of self-regulated learning and academic achievement: An overview and analysis. In B. J. Zimmerman & D. H. Schunk (Eds.), *Self-regulated learning and academic achievement: Theory, research, and practice* (pp. 1-37). New York: Springer

Zimmerman, B. J., & Schunk, D. H. (Eds.). (2001). *Self-regulated learning and academic achievement: Theoretical perspectives* (2nd ed.). Mahwah, NJ: Lawrence Erlbaum Associates.

CHAPTER

12

The Impact of Race and Racism on Students' Emotions: A Critical Race Analysis

JESSICA T. DECUIR-GUNBY
North Carolina State University

MECA R. WILLIAMS
Georgia Southern University

In recounting the events at a school assembly at Wells Academy's, Barbara, an African-American student, described her response to a civil rights leader's speech as the following:

> It was really an *emotional* day. [The Civil Rights Leader] was only supposed to speak for forty-five minutes. He ended up staying for over two hours. Classes were cancelled, students stayed in the auditorium to argue with him, [and] people started to forget how they were supposed to respect their elders.

Barbara's description of the African-American history assembly as *emotional* is important since the discussion of race and racism in the school context can often stir up a barrage of emotions, particularly unpleasant emotions. The school context is a ripe arena for issues of race and racism because of the history of racial prejudice and discrimination in the United States (U.S.), particularly in the educational system (Ladson-Billings & Tate, 1995). During segregation and Jim Crow, education became the default laboratory for experimentation on controlling racial attitudes and the emotions associated with them. In addition to the historical implications of race, emotion, and schools, the emotional nature of the developing adolescent, and the school context is an environment of overflowing emotions (Sadowski, 2003). This includes the emotions that result from racial encounters. Despite the relationship between racism and emotions, little research examines how issues of race and racism impact emotions within the school context. This chapter attempts to fill some of this research void.

As part of a larger body of research on racial identity development, racism, and the educational environment, our work evolved into the discourse of emotions and the significant role emotions play in the educational experiences of African-American adolescents.[1] Although this chapter addresses racism within the United States, we acknowledge that racism is a global problem and that people around the world manage similar emotions involving racial issues. Thus our discussion describes the emotions associated with racism in students in the United States.

The purpose of this chapter is to address the connection between emotions and issues of race and racism. Little research has described students' emotions in relation to the persistence of racism and their experiences with racial episodes. This chapter advances the discussion of race and emotions by communicating how racial messages habitually manifest unpleasant emotions in the lives of African-American students. In doing so, we use a Critical Race Theory (CRT) lens to comprehensively communicate how racial messages manifest their presence in our daily lives. We begin the chapter with discussions of emotions, racism, and the relationship between emotions and racism within the school context. Next, we describe the CRT framework and use it to explore how a race-based event elicited emotions in students at a predominantly White, elite, private school. To do so, we provide an example of how CRT and emotions research disentangle issues related to racism and intense emotions associated with racist acts. Last, we provide suggestions for future research as well as offer suggestions for change within the school context.

EMOTIONS

Emotions are identified as the language of a person's internal state of being and are typically based on or tied to physical and sensory feelings (Lazarus, 1999). Often described in both psychological and physiological terms, emotions impact our internal state of being and our physical responses. Emotion is the realm where thought and physiology are entangled, and where the *self'* is incapable of severing from perceptions of value and judgments (Zembylas, 2003). Consequently, emotions do not happen in isolation. They are responses to meaningful situations in the life of an individual, they motivate us to actions, and they impact our daily lives (Massey, 2002). Emotion "remains a strong and independent force in human affairs, influencing perceptions, coloring memories, binding people together through attraction, keeping them apart

[1] The original study concerned African American students' negotiation of racial identity in the private school context.

through hatred, and regulating their behavior through guilt, shame, and pride" (Massey, 2002, p. 20). In other words, emotions have the power to set the pace and the path we decide to take throughout life.

Emotions are profoundly commonplace within the social sector; however, emotions remain illusive in research literature. Currently, as this volume suggests, more researcher are making an effort to investigate emotions. Emotions became a substantive topic in sociology in the mid-1980s (Thoits, 1989). Only recently, emotions research has branched to education, discussing teachers' emotional involvement in their work (Carlye & Woods, 2002; Day & Leitch, 2001; Golby, 1996; Hargreaves, 1998; 2000), counselors and anger management (Besley, 1999), and students' discussions on shame and guilt (Ferguson, 1991; Ferguson, Stegge, Miller, & Olsen, 1999).

One reason there exists a paucity of research concerning emotions in schools is that emotions are constructed as antithetical to reasoning. Many professionals stress that public displays of emotions in schools are recognized as a disruption rather than a functioning part of the social atmosphere (Boler, 1999). The stigma attached to an emotional person delivers the perception that the individual in question either suffers from a serious mental illness or has emotional maladjustments. Discourse of emotions and emotional events are usually directed to school counselors who advise students on managing problems. Students who silently restrain their anger and hostility to racism and/or racist policies that demand they comply with the economic, political, and national interests of the majority often develop an increased anxiety about their social setting and the threat of a social conflict (Boler, 1999; Massey, 2002). Their emotions can become prime targets of social control because students sometimes feel ill at ease when publicly discussing their emotions. Because of this lack of discussion on race and racism, most students are unaware that many people experience a similar sting of uneasiness in opening a discussion about race and their associated emotions about racism.

RACISM

Racism is a powerful hegemonic tool that allows for the categorization and prejudgment of humans based on phenotypic characteristics (Shujaa, 1994). Categorizations help to form a system of advantages and disadvantages (Omi & Winant, 2005). Prejudgment occurs when the relationship between racial groups becomes one of superordinates and subordinates. The superordinate group, those exerting more power and control, set the parameters within which the subordinates are to operate (Tatum, 1997). In the case of the United States, the superordinate group is Anglo-Americans and the subordinate groups are people of color. The thoughts and behaviors of superiority by Anglo-Americans are immortalized in systems of standards that disadvantage people of color and subsequently maintain the hierarchical system of racism.

Racism can be conceptualized in various ways. Racism is often defined as overt actions conducted by fanatical groups; however, it is also defined as subtle daily occurrences (i.e., covert racism). There are three levels of racism that describe racist episodes and behaviors that are exhibited throughout society. These levels include institutionalized racism, personally-mediated racism, and internalized racism (Jones, 2000). First, institutionalized racism occurs when there is limited access to goods, services, and opportunities for people in a minority group. For example, in schools, this level of racism is manifested through racial tracking, which at times inhibits the placement of students of color in advanced placement courses (Oakes, 1985; Wise, 2005). Second, personally-mediated racism concerns assumptions made about the qualifications and intentions of others, according to their race. An example is when individuals believe African-American students receive awards due to quotas rather than competence. Last, internalized racism occurs when people of color accept the negative messages seen regarding their looks, abilities, and beliefs. This is manifested when African-American students embrace the *superiority of Whiteness* while engaging in self-deprecating internal and external speech.

These various levels of racism are important because they illustrate that racism is not always a hostile occurrence. On the contrary, racism often occurs quietly, yet consistently. However, in order to best address racism, it is important to expose how the levels of racism are manifested. It is imperative that we recognize the permanence of racism that is woven within the social fabric of U.S. society (Bell, 1992), which includes the need for students to counter racism within the school context.

EMOTIONS AND RACISM

As previously mentioned, racism is pervasive. Because of this, many African-American students wrestle with an array of issues involving racism within the school context. Some understand the limited possibilities and opportunities available for them as African Americans. While in this peculiar negotiation of self-conceptualization, some students may fall victim to internalized racism or attach themselves to negative stereotypical images (McKown & Weinstein, 2003; Perry, Steele & Hilliard, 2003). The acceptance of these images can affect self-perception. One way in which these negative images are presented within the school context is in the form of speech.

Racist beliefs often appear in our daily conversations and speeches. Matsuda, Lawrence, Delgado, & Crenshaw (1993) state that prejudiced words and stereotypes are often used under the guise of free speech and are harmful to the victim's emotional well-being. More specifically, there is an instinctive, defensive psychological and emotional reaction on the part of the victim of racist speech. Past experiences with racism, including those associated with racist speech, can impact how African-American students choose to respond to persistent everyday stressors, how they approach their goals, how they interact with family members, and how they harbor unpleasant emotions (Walsh, 2004). Emotions are intricately embedded in how individuals perceive speeches laced with racist beliefs because such beliefs are related to their self-perceptions. These emotions range from shock, fear, anger, and anxiety, which may all happen before rational thinking takes place.

CRITICAL RACE THEORY FRAMEWORK

We use CRT to examine the transactions among emotions and racism. The CRT movement emerged in the 1970s as a response to the U.S. legal system's inadequate approach to addressing the effects of race and racism in U.S. society (see Delgado & Stefancic, 2001). CRT combines the historical context of race and its relationship to racism, power, legal reasoning, and principles of constitutional law. The purpose of CRT is to respond to the stalled progress of traditional civil rights litigation to produce meaningful racial reform (Tate, 1997; Taylor, 1998). An additional goal of CRT is to challenge as well as find solutions for the systemic manifestations of *White privilege* that subordinate people of color. To do so, as described by DeCuir and Dixson (2004), CRT involves five key tenets: (1) *permanence of racism* (Bell, 1992, 1995); (2) *whiteness as property* (Harris, 1993); (3) *interest convergence* (Bell, 1980); (4) *critique of liberalism* (Gotanda, 1991); and (5) *counterstorytelling* (Delgado, 1989).

Specifically, the five tenets work to disentangle how race continues to be a significant factor in maintaining inequity in the United States as well as around the world. Each tenet adds a unique component to understanding how racism impacts the experiences of people of color. *Counterstorytelling* works to give voice to people of color, enabling them to describe their lived experiences of oppression and feelings of domination. *Permanence of racism* offers an evaluation of how racist views are created and maintained within our everyday lives. *Critiques of liberalism* consider the theories of race and race consciousness, and identify the slow progress and innumerable delays in resolving racial inequalities in the quest for racial justice. *Whiteness as property* displays the historical evidence of the advantages to White identity that are flagrantly seen in laws and policies that benefit Whites in political and social environments such as voting, citizenship, and education. *Interest convergence* details how the interest of Blacks in achieving racial equality has been accommodated only when this goal has converged with the interest of powerful Whites. (For a more in-depth discussion of the tenets of CRT, see DeCuir & Dixson, 2004).

Although the tenets of CRT emerged from the U.S. legal system, they have been embraced in other disciplines in the United States, including educational research (e.g., DeCuir & Dixson, 2004; Duncan, 2002; Ladson-Billings & Tate 1995; Solorzano & Yosso, 2002; Villalpando, 2003). As such, this chapter attempts to contribute to the body of CRT work in educational research by conveying the lived experiences of African-American students. We will address the tenets of the *permanence of racism* and the *critique of liberalism,* centering on color blindness, through the narrative component of counter-storytelling.

Permanence of Racism

CRT begins with the assertion that racism is a "normal, not aberrant" social construction; it appears customary and routine (Delgado, 1989). Because of hegemonic practices and the slow pace of litigation demanding racial justice, CRT presupposes that there will be an ever-present reality of racism and a burden of racial subordination on African Americans. Further, CRT presupposes that racism is an "integral, permanent and indestructible component of this society" (Bell, 1992, p. ix). Racism is so closely woven into our political, economic, and cultural institutions, it exists as "American as apple pie" (deMarrais, Wilson, & Williams, 2003; Powell, Jacob, & Kimberly? 2003). Acknowledging the idea of the permanence of racism involves embracing the ideas of *racial realism* (Bell 1992, 1995). The *racial realism* perspective involves realizing the systematic structure of racism. Specifically, it focuses on the institutionalization of racism throughout all economic, social, and political systems of the United States (Bell, 1992, 1995).

Racism is influenced by shared cultural experiences, which in turn influence ideas, attitudes, and beliefs. However, addressing such individual notions is an arduous task because of our inability to interrupt the racist thought processes that lead to racist actions. It can be said that the process of "acknowledging and understanding the malignancy [of racism] are prerequisites to the discovery of an appropriate cure" (Lawrence, 1987, p. 321). This suggests that addressing the permanence of racism involves attending to the difficulty of understanding how racism is individually perpetuated.

Critique of Liberalism

Similar, to the Permanence of Racism, the Critique of Liberalism enables a challenge to White privilege. Specifically, it addresses the notion of classical liberalism (Gotanda, 1991). The U.S. legal system is built upon the notion of classical liberalism, which involves the ideals of equal opportunity, individualism, and meritocracy. These ideals are guided by the notion of race neutrality or color blindness. One aspect of the critique of liberalism is the challenge of color blindness that contributes to White privilege and inequity.

Counterstorytelling

The neutrality of the law or colorblindness assumes equality and equal opportunity through race-neutral means. It involves the nonrecognition of race or the "recognition of racial affiliation followed by the deliberate suppression of racial considerations" (Gotanda, 1991, p. 6). In other words, color blindness involves "noticing but not considering race" (Gotanda, 1991, p. 16). Williams (1997) suggests that the notion of color blindness is an illusion. In not considering race, one has to consider its presence. The promotion of color blindness can have detrimental effects for people of color. It contributes to the devaluing of cultural heritage, the ignoring of racial group distinctiveness, and the promoting of White hegemonic culture.

In order to address the notions of the permanence of racism and the critique of liberalism (color blindness), symbolic representations such as stories are useful tools. Stories have an apocalyptic effect that unveils the injustices many African Americans face because of the racist beliefs of others. As such, narrative stories are often used as a vehicle to articulate the everyday incidences of racism and their impact on the African-American community (e.g., Bell, 1992; Crenshaw, Gotanda, Peller, & Thomas, 1995; Ladson-Billings, 1998). Counterstorytelling offers an alternate reality that "can open new windows into reality, showing us that there are possibilities for life other than the ones we live" (Delgado, 1989, p. 2414). The use of counterstorytelling allows for the challenging of dominant discourse and enabling otherwise subjugated voices to be heard. Counterstorytelling serves a dual role in that it is both a means of telling untold stories as well as a method of analysis (Delgado, 1989; Solórzano & Yosso, 2002). Counterstorytelling not only serves as a tool to describe the oppression of minorities, but also serves to provide "deconstructors and a reconstruction of human agency" (Ladson-Billings & Tate, 1995, p. 55). In addition, counterstorystelling is a useful way to increase feelings of pride and belongingness among African-American students. As a model for human agency, counterstorytelling gives African-American students the opportunity to speak out against the inequalities in schools and provides them a chance to critically analyze the issue of race and its prominence in America (Taylor, 1998).

COUNTERSTORIES AT WELLS ACADEMY

Using the CRT framework, we focus on the permanence of racism and the critique of liberalism to examine the counterstories of six African-American students from Wells Academy, regarding a school event featuring a prominent civil rights leader. Our analysis concentrates on the emotions associated with the participants' perceptions of the civil rights leader's talk, White classmates' reactions, and African-American students' reactions.

The majority of students in this K-12 school are White, with students of color making up approximately 10% of the total school population, and African-American students making up 8% of the high-school population. The school has a negative history of race relations and a troubled racial climate (see DeCuir & Dixson, 2004). We concentrate our discussion on six African-American students' perceptions of their White classmates' reactions. These particular students range in age from 14 to 17 years, sophomores to a recent graduate.

Perceptions of White Students' Reactions

During African American History Month at Wells, there is a tradition to invite an African-American speaker to discuss his or her sentiments regarding issues in the African-American community. After addressing the entire high school, the speaker is invited to participate in a question-answer session with a smaller number of students. Because Wells considers itself to be "liberal," it is not uncommon to invite speakers who would discuss issues of race and racism. During the 2001–2002 school year, Wells' Diversity Club recommended a local, prominent civil rights leader. The goal of his visit was to expose issues of race and racism within society as well as within the Wells community. Although the question-answer sessions are not always well attended, the session involving the civil rights leader appeared to be of interest to many students. It featured more than 100 students, nearly a fifth of the student body. This question-answer session was as spirited as, if not more than, the original assembly. Students were able to voice their opinions—mainly their differences—with the perspective of the civil rights leader.

According to the participants, many of the White students vehemently disagreed with the civil rights leader. This contributed to unpleasant interactions among the students, particularly between the White and African-American students. Some students even directly challenged the civil rights leader's views. The group that was perhaps the most upset was that of self-identified as U.S. Southerners. These students felt the civil rights leader was insulting their Southern heritage. Jasmine gave a description of these students:

> You had this section of White boys, who were all about Southern *pride*. They've got the confederate flags on their belts and they've got *Dixie* on their cell phones. They're these football boys. They didn't like that you were saying that their grandfather was a bad person. They didn't like that you were attacking the confederate flag and that you could have Southern *pride*.

According to the African-American students, these students felt attacked and their Southern roots were dishonored. Their identity and beliefs were shaken to a volatile and weak state. They could not defend their heritage, beliefs, or who they were. From the perspective of the African-American students, this caused them to question their perspectives of race and racism and, in doing so, caused an emotional uproar. These students attempted to defend their position of Southern pride and engaged in several interchanges with the civil rights leader. Barbara gave a description of such an attempt:

> This young white man was talking about the Confederate flag and how it meant [Southern] *pride* to him. The civil rights leader [asked him]: 'What do you have to be *proud* of?' This student couldn't answer; his friends couldn't answer it. The boy would say, 'But it is Southern heritage.' And the civil rights leader [said], 'Well, explain your heritage . . . You don't know how to answer that question. The only way students at this school have verified the Confederate flag was by saying it's for Southern heritage; it's for *pride*. And [you] don't have anything to back that up with.'

These self-indentified Southern students openly discussed their concerns in an effort to defend a historical perspective that gives them a sense of pride and revelry. However, the words of the assembly speaker made some of these students respond to the flinching sting of negative remarks about America's bitter history. The civil rights leader noted that the Confederates were a defeated coalition that represented rebellion and the desire for continuing to enslave people of African descent. Images of the confederate flag are part of a Southern tradition that identifies a dream that never was truly realized. Some Whites cling to that dream and make it a part of their identity. To discuss aspects of Southern identity as negative or confederate icons as racist hangs a heinous stigma on their forefathers and themselves.

As described by the participants, many of the students had such negative reactions because they misconstrued the civil rights leader's message. He was attempting to describe the permanence of racism and how Whites perpetuate it, as well as benefit from it. However, many of the students personalized his message because they viewed confederate icons as a part of their identity. They believed that the civil rights leader was attacking them personally, rather than attacking the system of racism. Jasmine provided an example: "A White girl was very *upset* and was *crying*. She was *upset* that he was saying that she was a bad person simply because she was White. I think that was what everyone was thinking he was saying." Michael believed that such personalization was the result of having a closed mind. Michael added: "There were some things that were misunderstood. It is really easy to misunderstand something that you are [being] accus[ed] of. When you go home defensive, you stop listening, start selective hearing and you hear what you want to hear." Malcolm, however, felt that many of the White students were uncomfortable with the civil rights leader's speech because they lacked exposure to alternative views on race and racism, and it led some students to feelings of White guilt. He explained:

> It was really kind of *sad* because no matter what, they [the White students] just walked away. They will remember [the civil rights leader] for a good little bit because this was the first time that they had to sit down and listen to these,

> what seemed to them, astronomical point[s] of views. They were acting like toddlers, two-year-olds. They were sitting in their seats, waving around. They couldn't stay still. They were raising their hands, walking out, *huffing* and *puffing, turning red*. You could tell that this was the first time that they've ever had to meet an opposing view.

As beneficiaries of White privilege, according to the participants, many of the White students could not understand the civil rights leader's message. They had difficulty realizing the function of race (Gotanda, 1991). Because these students embraced the beliefs of race-neutrality or color blindness, they could not understand the influence of race and the practice of racism within U.S. structures. More specifically, they could not comprehend how the system of racism is both beneficial to and perpetuated by White people. The White students particularly could not comprehend how they, as members of the White race as a whole, were benefiting from racism. It seemed perplexing to them that their ancestors could have contributed to the system of racism. Instead, many of them interpreted the message to be a personal insult. As a result, they refused to listen to alternative views. However, there were some White students who were receptive. They understood the concept of racism, as well as how they have benefited from it.

DISCUSSION

A mix of emotions is often elicited when discussing race and racism in racially mixed audiences. The sensitivity of the topic serves to encourage negative reactions. This sensitivity is particularly heightened in Southern U.S. contexts, in that racial discrimination is deeply rooted in the history of the U.S. South. As such, the difficulty of discussing issues of race and racism at Wells, a Southern, upper class school was not surprising. In fact, it was somewhat expected.

At the African-American history assembly, the civil rights leader attempted to evoke unpleasant emotions in the students. He wanted the students to realize the permanence of racism and question the manner in which people of color have been treated in the United States. In short, he wanted students to become *angry*, *upset*, and *mad* at the implementation and effects of racism. However, instead of being angry at the meaning of the message, many students internalized the message and became *angry*, *upset*, and *mad* at the messenger. It is common for students to engage in "ignoring, rejecting, excluding, reinterpreting" and "holding anomalous information in abeyance" when their beliefs are challenged (Ashton & Gregoire, 2003, p. 103). This reaction was not surprising since deeply entrenched beliefs are difficult to change (Ashton & Gregoire, 2003), particularly racial beliefs. Such reactions helped to create a schism among the students along color lines while simultaneously promoting dialogue concerning race and racism. The conversations on this hotly debated topic opened lines of communication that had not been explored before. Students opened emotionally-sensitive aspects of their lives and tried to translate how their identities and relationships were affected by the images and perceptions they constantly encountered in their school and community. The civil rights leader accomplished his mission in advocating for discussions on racism and its persistence in our society.

Since the participants were aware of issues relating to racism and how it is manifested within society, including school, they were aware of their White classmates' privilege and lack of racial awareness. As such, the participants were observant of their White classmates' reactions to the civil rights speaker. In particular, the participants noticed the unpleasant emotions that the White students exhibited (e.g., *anger, upset*). In the participants' views, many of the White students elicited reactions that stemmed from their mutilated egos, bruised feelings, and prejudiced attitudes. Their White classmates were confronted with statements that questioned their Southern heritage and its fundamental flaws. These students could not articulate why they were so proud of their heritage even when bluntly asked by the civil rights leader. As a result of this battle on pride and prejudice, the emotional reactions from the White students included *crying, huffing* and *puffing*, and *turning red*.

It also seemed that for the African-American students, this experience of sharing in their history that promotes survival, despite a negative social system, helped them identify with African Americans through nostalgic memories and feelings of pride (McCarthy, Pretty, & Catalano, 1990). The students' resiliency to persist despite the emotional problems associated with racism fostered a sense of pride for each individual and a greater sense of belongingness through a shared emotional connection or identity among the African-American students in their school environment. This emotional connection was created through the African-American students' opportunity to openly describe the events and associated feelings through counterstorytelling. Their narratives presented a shared history that has also become a triumphant story of accomplishments. Most of the African-American students identified with the pain and promise in their school environment that was understood but never clearly stated. Having an opportunity for counterstorytelling legitimatized and validated their feelings concerning the racist attitudes of their peers.

CONCLUSION AND IMPLICATIONS

The legacy of unresolved racial dilemmas continues to infest the United States' cultural and social existence. Race-related differences are highly documented in educational research. Most of these studies examine how race and racism play a critical role in educational practices. However, few studies examine the impact of race-related issues on students' emotions. The dialogue on race and racism is often one of difficulty. It often elicits many emotions, especially unpleasant emotions. In public settings, people of color who attend gatherings where negative stereotypes regarding people of color are discussed may feel annoyed, angry, powerless, and pained. In addition, it has been found that raising questions about racism are uncomfortable to Whites and considered divisive (Tatum, 1997). This discomfort helps to create unpleasant emotions for both people of color and Whites. To understand the impact of race and racism, it is imperative that we understand the emotions that are frequently associated with such topics. Our research suggests that researchers, counselors, and educators must do significant work to reach a better understanding of the impact of race and racism on emotions.

First, researchers need to address issues of race and racism in the school context. Although issues of race and racism permeate schools, this issue is not often explored in research. Next, researchers should examine emotions in the school context. As previously stated, adolescents often exhibit a barrage of emotions. As such, the school context is particularly conducive to the examination of emotions. Last, it would be particularly helpful if more research addressed the relationship between issues of race and emotions. In doing so, using a CRT framework (see DeCuir & Dixson, 2004) would provide a deeper discussion of the issues.

Next, it seems clear that African-American students need opportunities to discuss their experiences (this need may be even greater in predominantly White schools). It is imperative that all students feel that they belong in the school setting (Osterman, 2000). Thus a multicultural environment needs to be created. In creating a multicultural environment, African-American students should be provided with opportunities to talk about their experiences and the emotions associated with them. This may include the incorporation of multicultural counseling, mentoring (peer), focus groups, or cultural workshops and activities (see Reynolds, 1999).

Last, schools should begin to encourage constructive dialogues among racial groups regarding race and racism, instead of shrinking away from such difficult conversations. It is imperative that students have the opportunity to express the emotions that are associated with such issues, even if those emotions are unpleasant. The presence of unpleasant emotions will decrease as constructive dialogues occur, while simultaneously increasing the presence of pleasant emotions. However, to begin the dialogue, schools need to recognize the issues of race and racism that exist in their respective school environments and society. Schools have to be willing to create school-based interventions that address racism, discrimination, and prejudice (see McKown, 2005).

FUTURE RESEARCH

Although the initial purpose of this research study was to examine the racial identity of African-American adolescents, we soon discovered that emotions were essential to their experiences, particularly in their discussion of race and racism. As such, our future research plans include further examining the impact of race and racism on emotions experienced in the school context. In particular, we would like to investigate five key issues involving students and the school context. First, we would like to understand how different racial groups within the United States experience emotions when discussing race and racism. We hypothesize that students in general experience unpleasant emotions when discussing race and racism. However, we strongly feel that the source of unpleasant emotions will differ among Anglo-Americans and students of color; Anglo-American students have difficulty in discussing the role of Anglo-Americans in the origins and maintenance of racism.

Next, we would like to discover what students learn when they discuss race and racism in their classrooms and how they respond to these open class discussions. More specifically, we would like to see if students are more or less responsive to their teachers who are open to discussing issues of racism. This would lead to examining how such discussions impact students' social worlds and schools. Similarly, we will also explore the relationship between race and racism on the emotions of pride and belongingness in students. Such an investigation would require gaining an understanding of how discussions of race and racism create a sense of racial connectedness among racial groups. We would also like to explore the long-term effects of emotions concerning race and racism that are experienced in the school context. More specifically, we want to explicate how emotions relate to beliefs and belief changes related to preserving or reducing racism within the school context.

Last, we would like to conduct cross-cultural studies concerning students of color in various other countries (e.g., Great Britain, France, China). With growing racial diversity throughout the world, it is important to better understand how issues of race and racism are discussed along with the emotions associated with such conversations. Examining the relationship between race and emotions from various perspectives and in different cultural contexts will help serve as an additional means to improving race relations and understanding emotions within our social world.

References

Ashton, P. & Gregoire, M. (2003). At the heart of teaching: The role of emotion in changing teachers' beliefs. In J. Raths & A. McAninch (Eds.), *Advances in teacher education* (pp. 99-121). Greenwich, CT: Information Age Publishing

Bell, D. A. (1980). *Brown v. Board of Education* and the interest convergence dilemma. *Harvard Law Review, 93*, 518-533.

Bell, D. A. (1992). *Faces at the bottom of the well: The permanence of racism*. New York: Basic Books.

Bell, D. A. (1995). Racial realism. In K. Crenshaw, N. Gotanda, G. Peller, & K. Thomas (Eds.), *Critical race theory: The key writings that formed the movement* (pp. 302-312). New York: The New Press.

Besley, K. (1999). Anger management: Immediate intervention by a counselor coach. *Professional School Counseling, 3*(2) 81-90.

Boler, M. (1999). *Feeling power: Emotions and education*. New York: Routledge

Carlye, D. & Woods, P. (2002). *Emotions of teacher stress*. Stoke on Trent, UK: Trentham Books.

Crenshaw, K., Gotanda, N., Peller, G., & Thomas, K. (Eds.). (1995). *Critical race theory*. New York: The New Press.

Day, C., & Leitch, R. (2001). Teachers' and teacher educators' lives: The role of emotion. *Teaching and Teacher Education, 17*, 403-415.

DeCuir, J.T., & Dixson, A. (2004). "So when it comes out, they aren't that surprised that it is there": Using critical race theory as a tool of analysis of race and racism in education. *Educational Researcher, 33*(5), 26-31.

Delgado, R. (1989). Storytelling for oppositionists and others: A plea for narrative. *Michigan Law Review, 87*(8), 2411-2441.

Delgado, R., & Stefancic, J. (2001). *Critical race theory: An introduction*. New York: New York University Press.

deMarais, K., Wilson, J., & Williams, M. (2003, April). *The language of race: Power and the professional in the anger narratives of women teachers*. Paper presented at the annual meeting of the American Education Research Association. Chicago, Illinois.

Duncan, G. (2002). Beyond love: A critical race ethnography of the schooling of adolescent black males. *Equity & Excellence in Education, 35*(2), 131-143.

Ferguson, T. (1991). Children's understanding of guilt and shame. *Child Development, 62*(4) 827-839.

Ferguson, T., Stegge, H, Miller, E., Olson, M. (1999) Guilt, shame and symptoms in children. *Development Psychology*, 35(2), 347-357.

Golby, M. (1996). Teachers' emotions: An illustrated discussion. *Cambridge Journal of Education, 26*(3), 423-436.

Gotanda, N. (1991). A critique of "Our constitution is color-blind." *Stanford Law Review, 44*, 1-68.

Hargreaves, A. (1998). The emotional practice of teaching. *Teaching and Teacher Education, 14*(8), 835-854.

Hargreaves, A. (2000). Mixed emotions: Teachers' perceptions of their interactions with students. *Teaching and Teacher Education, 16*, 811-826.

Harris, C. (1993). Whiteness as property. *Harvard Law Review, 106*(8), 1707-1791.

Jones, C. (2000). Levels of racism: A theoretical framework and a gardener's tale. *American Journal of Public Health, 90*(8), 1212-1216.

Ladson-Billings, G. (1998). Just what is critical race theory and what's it doing in a nice field like education? *International Journal of Qualitative Studies in Education, 11*(1), 7-24.

Ladson-Billings, G. & Tate, W. (1995). Toward a critical race theory of education. *Teachers College Record, 97*(1), 47-68.

Lawrence, C. R. (1987). The id, the ego, and equal protection: Reckoning with unconscious racism. *Stanford Law Review, 39*(2), 317-388.

Lazarus, R. (1999). *Stress and emotions: A new synthesis*. New York: Free Association Books.

Massey, D. (2002) A brief history of human society: The origin and role of emotions in social life. *American Sociological Review, 67*, 1-29.

Matsuda, M., Lawrence, C., Delgado, R., & Crenshaw, K. (1993). *Words that wound: Critical race theory, assaultive speech, and the first amendment*. Boulder, CO: Westview Press.

McCarthy, M., Pretty, G., & Catalano, V. (1990). Psychological sense of community and burnout. *Journal of Community Psychology, 18*, 211-216.

McKown, C. (2005). Applying ecological theory to advance the science and practice of school-based prejudice reduction interventions. *Educational Psychologist, 40*(3), 177-189.

McKown, C. & Weinstein, R. (2003). The development and consequences of stereotype consciousness in middle childhood. *Child Development 74*(2), 498-515.

Oakes, J.(1985). *Keeping track: How schools structure inequality*. London: Yale University Press.

Omi, M. & Winant, H. (2005). The theoretical status of the concept of race. In C. McCarthy, W. Crichlow, G. Dimitriadis & N. Dolby (Eds.), *Race, identity, and repression in education* (pp. 3-12). New York: Routledge.

Osterman, K. F. (2000). Students' need for belonging in the school community. *Review of Educational Research, 70*(3), 323-367.

Perry, T., Steele, C., Hilliard, A (2003). *Young, gifted and black: Promoting high achievement among African American students.* Boston: Beacon Press.

Powell, C., Jacob, A., Kimberly, R. (2003) Relationship between psychosocial factors and academic achievement among African American students. *Journal of Educational Research, 96*(3), 7-15.

Reynolds, A. (1999). Working with children and adolescents in the schools: Multicultural counseling implications. In R. H. Sheets & E.R. Hollins (Eds.), *Racial and ethnic identity in school practices: Aspects of human development* (pp. 213-229). Mahwah, NJ: Lawrence Erlbaum Associates.

Sadowski, M. (2003). *Adolescents at school: Perspectives on youth, identity, and education.* Cambridge, MA: Harvard Education Press.

Shujaa, M. (1994). Afrocentric transformation and parental choice African American independent schools. In M. Shujaa (Ed.), *Too much schooling too little education: A paradox of black life in white societies* (pp. 362-376). New Jersey: African World Press.

Solórzano, D. G., & Yosso, T. J. (2002). Critical race methodology: Counter-storytelling as an analytical framework for education research. *Qualitative Inquiry, 8*(1), 23-44.

Tate, W. F. I. (1997). Critical race theory and education: History, theory and implications. In M. Apple (Ed.), *Review of research in education* (Vol. 22). Washington, D.C.: American Educational Research Association.

Tatum, B. (1997). *Why are all the black kids sitting together in the cafeteria? and other conversations about race.* New York: Basic Books.

Taylor, E. (1998). A primer on critical race theory: Who are the critical race theorists and what are they saying? *Journal of Blacks in Higher Education, 19,* 122-124.

Thoits, P. (1989). The sociology of emotions. *Annual Review of Sociology, 15,* 317-342.

Villalpando, O. (2003). Self-segregation or self-preservation? A critical race theory and Latina/o critical theory analysis of a study of Chicana/o college students. *Qualitative Studies in Education, 16*(5), 619-646.

Walsh, R. (2004). *Through white eyes: Color and racism in Vermont.* New York: Book Surge UC.

Wise, T. (2005). *White like me: Reflections on race from a privileged son.* Brooklyn, NY: Publishers Group West.

Williams, P. J. (1997). *Seeing a color-blind future: The paradox of race.* New York: Farrar, Straus, and Giroux.

Zembylas, M. (2003). Interrogating "teacher identity": Emotion, resistance and self-formation. *Educational Theory, 53*(1), 107-127.

PART

IV

Teachers' Emotions in Educational Contexts

CHAPTER

13

Teacher Identities, Beliefs, and Goals Related to Emotions in the Classroom

PAUL A. SCHUTZ

University of Texas at San Antonio (UTSA)

DIONNE I. CROSS, JI Y. HONG, JENNIFER N. OSBON

University of Georgia

> I always feel overwhelmed. I really do. I always feel like I have a lot to do . . . a lot of important things to do . . . I have the extra work of getting to know the students and understand what they need and think about how I can give them what they need academically and emotionally. I think about who doesn't have friends and how I need to maybe connect them with someone who's a little on the popular side - popular for good reasons. So I just kind of feel overwhelmed. (Ms. Walker quoted in Williams, Cross, Hong, Aultman, Osbon, & Schutz, 2006)

The above quote is but one example of how teachers see their emotional life in the classroom. For teachers, the classroom can be an intense environment where emotions vary from the extreme joy of an exciting lesson to the heart-wrenching sorrow of knowing a student is being abused or bullied by his or her classmates. For anyone who has spent time in a classroom, it is clear: the classroom is an emotional place!

The potential influences of emotions on the schooling of our youth, as well as the diversity and intensity of emotions in the classroom, are some of the reasons that we, as well as other scholars, have been investigating the nature of emotions in the classroom. To discuss our approach to inquiry on emotions in education, we will begin by discussing our theoretical perspective related to teacher identities, beliefs, and goals about emotions the classroom. Throughout this discussion, we infuse our own research findings associated with emotions in education. Next we discuss our methods and methodological approaches to the investigation of teachers and emotions and how we develop rigor in our research. Finally, we will discuss some implications related to how we see our research informing educational practice.

In this chapter we will be foregrounding teachers' experiences related to emotional transactions during classroom events. However, it is clear that in any discussion of classroom transactions it is difficult to extract one individual from the rest of the participants in the classroom. Classrooms involve both teachers and students who come to that activity setting with their own personal histories that have emerged from transactions within their own social-historical contexts. Therefore, in essence, our attempt will be to acknowledge this transactional interdependence while discussing classroom activities from teachers' perspectives.

THEORETICAL PERSPECTIVE

We will use Figure 1 as a model of our current thinking about the transactions among teachers' identities, beliefs, and goals about emotions in the classroom. At this point, we do not present it as a causal model. We simply use the model as a way of conceptually diagramming the transactions we see as the core constructs for our discussion of emotions in the classroom.[1] We will first discuss social-historical contextual influences and related constructs (outer box in Figure 1), then the activity setting and related issues (inner box in Figure 1), and finally some of our preliminary ideas about how these transactions relate to confirming or challenging old beliefs and building new beliefs.

Social Historical Contextual Influences

In Figure 1, the outer box labeled Social-Historical Contextual Influences represents various nested systems that influence the classroom activity setting. The contextual nature of teaching and the goals, beliefs, and emotions associated with teaching suggest that when considering a particular classroom activity, it must be kept in mind that classroom events occur within a web of social-historical contextual influences. To organize our thinking on this, we use Bronfenbrenner's ecological model (1986). This model represents developmental transactions as functions of a series of nested systems where, in this case, teachers and the various social contexts associated with classroom education transact. In Bronfenbrenner's model, and from the teachers' perspective, the *microsystem* refers to their closest personal relationships and activities. For example, it would include teachers' families, their friends,

[1] Although each construct in the model is presented separately to explain their transactional relationships in the classroom, we see them as being overlapping and closely intertwined.

students, and the transactional activities they are involved in at school and home. For our purposes, we focus on the *microsystem* that represents the face-to-face transactional activities among teachers and students.

In Bronfenbrenner's (1986) model, the various *microsystems* teachers are involved in are nested within a *mesosystem* that represents the transactional activities among teachers' various microsystems (e.g., a teacher's family or self-influencing classroom transactions). This *mesosystem* is nested within an *exosystem* that includes the social-historical influences that affect teachers, even though teachers may not be directly involved in the process (e.g., a community where the school is located, a school system, or the local newspaper). The *exosystem* in turn is embedded in a *macrosystem*, which represents the larger society and the potential social-historical influences on the other systems (e.g., cultural values, the federal government's proposed educational mandates, or the historical treatment of various groups within the society). Finally, the *chronosystem* adds the dimension of time as it relates to teachers' transactions (e.g., a teacher at age 21 may react differently to a particular classroom event than at age 50). We use this model to remind us that a particular classroom and the activities that occur in that classroom do not occur in a vacuum. They are part of a complex social-historical contextual web.

Our focus here will be on the transactions that occur at the *microsystem* level. We use actual events or activities that occur in the classroom as the

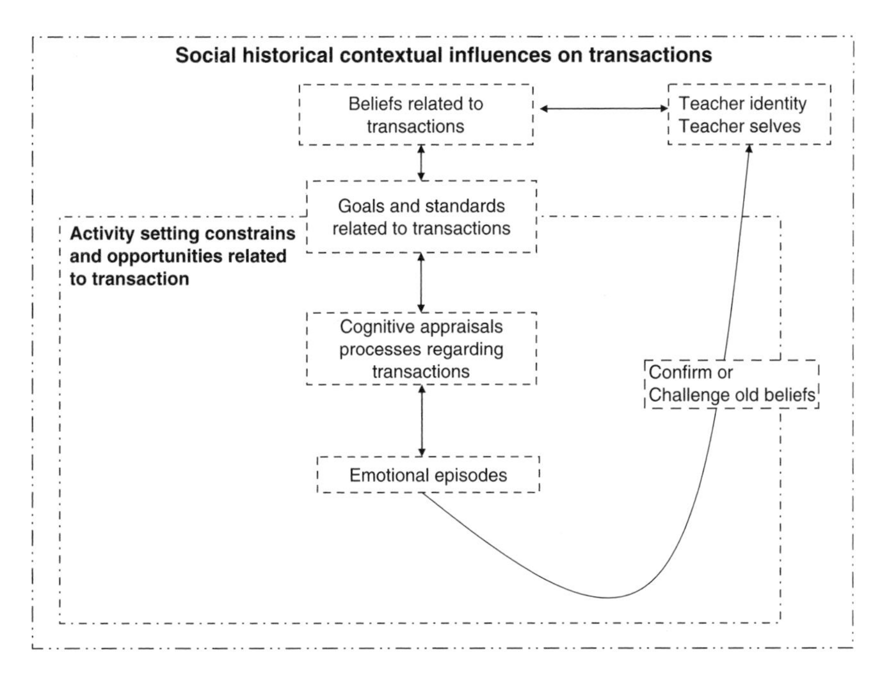

FIGURE 1

Transactions among teachers' indentities, beliefs, and goals.

focal point for our discussion. For teachers in an activity setting, we see the nature of their identities, beliefs, and goals as emerging from transactions among biological, environmental, and social-historical influences (Schutz, Crowder, & White, 2001). For example, Pajares (1992) refers to Van Fleet's (1979) work and his classification of this cultural transmission into three components: enculturation, education, and schooling. They encompass the learning that individuals undergo both inside and outside of the home throughout their lives. They describe both the formal and informal processes teachers assimilate into the cultural elements of their personal world, which guide their behavior. These three components are also used to explain the development of teachers' current knowledge about the external world and exposure to the ideas and theories present within it.

This socialization process influences the way teachers perceive and evaluate events during person-environment transactions. It serves as a filter through which they assess situations and determine the most appropriate emotional response relevant to the social-historical context. For teachers, their identities, beliefs, and goals tend to emerge from various combinations of these components of cultural transmission, including but not limited to their early student experiences (Pajares, 1992; Virta, 2002), parent-child relationships (Massey & Chamberlin, 1990), formal teacher training (Zeichner & Gore, 1990), and their initial teaching experiences (Knowles, 1992).

For example, the goal to become a teacher and the role of being a teacher is played out within particular social-historical contexts. As such, being a teacher has different meanings depending on the time and place that goal has emerged and is enacted. One illustration of these social-historical influences was demonstrated by Schutz, Crowder, and White (2001), who interviewed college students during their training to become a teacher and explained how these college students, with the goal to become a teacher, were either encouraged or discouraged in that pursuit because of social-historical influences, such as gender or their parents' beliefs about a teacher's role in society at that point in time. As such, being a teacher in the United States has a somewhat different meaning than it did 50 years ago, or than what it currently means to be a teacher in South Korea or Jamaica. The process of becoming a teacher, and the meaning of that goal to the individual, is represented in Figure 1 by the construct "teacher identity."

Teacher Identities, Selves, and Emotion in the Classroom

We see teacher identity as an overarching construct including beliefs, goals, and standards. In addition, we think of identity as the ways teachers perceive themselves as teachers and the way they portray themselves to their students. For instance, Ms. Jones, a third-year elementary school teacher highlighted by Aultman, Williams, Garcia, and Schutz (2006), perceives teaching as a "very selfless job." She thinks she has to invest all of her time and energy in her students—at least while she is at school. Her personal life and issues should be totally pushed aside until the bell rings.

In considering teacher identities, it is important to keep in mind that, although there is a certain amount of stability in how teachers see themselves as teachers, there is also continuous change as teachers transact among particular social-historical contexts (Zembylas, 2003; Danielewicz, 2001). This construction and reconstruction of teacher identity is based not only on the continually changing self-knowing of teachers, but also on teachers' perceptions of the profession itself. Thus what teachers know of themselves, their perception of the characteristics and nature of the teaching profession, and their beliefs about their roles are all interrelated in forming teacher identity.

Teacher identity establishes a framework from which teachers transact with others (Aultman et al., 2006), and that framework guides teachers' behavior and the ways they deal with emotions during classroom transactions. In other words, how teachers view themselves in their professional role influences their beliefs about pedagogy, how they portray themselves, and it guides their actions and emotional experiences during transactions. Ms. Walker, who teaches third and fourth grade reading, wants her students to see her as "a teacher who cares about their progress and cares about them as an individual" (Williams, et al., 2006). Such a perception guides her to take the "coach" role to motivate and encourage students.

As teacher identity and emotions are inevitably related to each other, teacher identity is often conveyed and expressed through emotions, whether it is unconscious or conscious. In other words, the emotions teachers experience reflect their sense of identity. Ms. Walker, who wants to portray a teacher self of respect and control, makes an effort not to express certain emotions in the classroom (Williams et al., 2006). She consciously excludes certain emotions that can make her look silly or out of control because those emotions may hinder her from carrying a desired teacher self.

As Nias (1996) noted, teaching is not just a technical job; rather, teachers invest their selves into their work. This means that teachers are continuously developing their identity through the interpretation of their experiences within the context they adopt. Thus teachers' identity not only influences their actions and emotions, but also their professional identity formation. Simply put, teacher identity and emotion are not linear or unidirectional; rather, they are inextricably related to each other through an ongoing, multidirectional, transactional process. Accordingly, when teachers experience unpleasant emotions, those emotions may threaten their identity by challenging their existing beliefs.

Teacher Beliefs and Emotion in the Classroom

In Figure 1, teachers' beliefs related to classroom transactions are shown as emerging out of transactions within social-historical contexts and playing a key role in what occurs in that activity setting. As indicated by Pajares (1992), beliefs tend to be a difficult construct to define. Part of the definitional problems involves the attempt to distinguish beliefs from knowledge. In this regard, Calderhead (1996) clarifies the distinction somewhat by stating that beliefs refer more to suppositions, commitment, and ideology, while knowledge encompasses factual propositions and understandings that inform skillful action. The "truth" value and objectivity are ascribed to knowledge, whereas beliefs are more considered individual subjectivity (Clandinin & Connelly, 1987; Frijda, Manstead, & Bem, 2000; Green, 1998). Rokeach (1968) incorporates these views and states that whereas beliefs can be viewed holistically, they comprise three components: cognitive (knowledge), affective (emotional), and behavioral (action). These components are often not clearly distinctive (Woolfolk-Hoy, Davis, & Pape, 2006). Because of its complex nature and the composition of belief systems, they tend not to be observable, leaving researchers to make inferences from individuals' speech, intentions, or actions.

The distinction between knowledge and beliefs becomes useful when we attempt to discuss the relationship between beliefs and emotions. As discussed by Shavelson and Stern (1981), individuals often resort to beliefs when there is no relevant information or knowledge available. This is important because the teaching profession is rich with interpersonal transactions that may respond to impulse and intuition rather than reflection (Pajares, 1992). Situations arise for which there are no easy solutions; hence, the teacher has no schema or cognitive resources to activate and so they are unsure of the necessary information or appropriate behavior to apply. In these situations, where they lack useful knowledge or strategies to use, they may resort to beliefs that can be riddled with problems and inconsistencies.

Williams et al. (2006) describe the relationship between teachers' beliefs about their classroom roles and its influence on how they negotiate classroom contexts and respond to classroom emotional events. In that study, there seemed to be connections between what the teachers believed their role should be in the classroom, how they structured their classroom activities, and the relationships they sought to develop. One example is a teacher who believed she needed to create an environment where all her students would achieve. This required her to regularly wear the "cheerleader" hat, motivating her students to learn. She recalls one incident where the students responded very negatively to a lesson she had prepared. Attributing their response to tiredness (unstable but controllable attribution, Weiner, 1994), she was able to respond positively, avoiding feelings of frustration that would have been a warranted response. Instead she put on her cheerleader hat and refocused the students. In this case, her judgments about the situation lead to feelings of hope, which encouraged her to try another approach that worked.

Goals, Standards, and Emotion in the Classroom

In addition to teacher identity and beliefs, teacher goals and standards are key dimensions of classroom transactions. That is, we see goals and standards as key organizing constructs for understanding the nature of emotions in the classroom (Boekaerts, 2007; Carver & Scheier, 2000; Powers, 1971; Schutz & Davis, 2000; Schutz & DeCuir, 2002; Turner & Waugh, 2007). Goals represent ways individuals, as members of social groups, think the world should and could be, as well as the way they would not like it to be (Ford, 1992; Markus & Nurius, 1986; Schutz, Crowder, & White, 2001). As such, we see goals and standards as reference points used, in this case, by teachers and the social-historical contexts they are embedded in to guide their thoughts and actions. In that role, goals and standards tend to provide directionality to teachers' thoughts and activities (Frijda & Mesquita, 1994; Lazarus, 1991, 1999; Solomon, 1976).

The nature and type of teachers' goals tend to emerge from and are influenced by their particular social-historical contexts (Markus & Kitayama, 1994; Ratner, 2000; Schutz, et al. 2001; Schutz, Hong, Cross, & Osbon, in press). Thus there are reciprocal transactional relationships among goals, standards, beliefs, and one's identity as a teacher. In other words, the goal to be a caring teacher emerges out of transactions within particular social-historical contexts. This goal, along with sufficient personal confidence and commitment, facilitative beliefs and social-historical contexts that encourage the goal, may develop into an identity or defining way of being. For teachers, the classroom activity setting represents the primary place where personal goals, beliefs, and identities related to teaching are transacted with their students.

Classroom Activity Setting

In Figure 1, the inside box represents the classroom activity setting. These activity settings tend to be patterned and repetitive places where individuals transact (Tharp & Gallimore, 1988). The classroom activity setting is defined by who is present and available, what they are doing and why, and where and when they are doing it (Tharp & Gallimore, 1988). From our perspective, we see the goal, beliefs, and professional identities that teachers embody as the lens through which teachers transact as participants in those classroom activities. In addition, we see judgments or appraisals teachers make regarding how things are going during a particular activity as key to the nature of the transactions within an activity setting.

Cognitive Appraisals Processes

We think of cognitive appraisals as judgments about the personal significance of a person-environment transaction (Lazarus, 1991). As such, they are related to individuals' understanding of how a particular event is unfolding and how it may affect their current and future goals and standards. They are not generated by factors in the environment or isolated mental processes, but by individuals' assessment of the person-environment transactions. These assessments may vary with respect to time and context (Lazarus, 1991) and produce a rich variety of emotions. This variation occurs because situations can be interpreted in different ways and they often arise from a combination of appraisals (Lazarus, 1991), socialization history (Kemper, 1987), and cultural expectations (Ratner, 2000; 2007). We see emotions and the potential for emotional regulation (Gross, 1998) as beginning with these appraisals or judgments related to teachers' identities, beliefs, goals, standards, and individuals' perceptions of how a goal pursuit is going in a particular activity setting. These judgments can occur rapidly and without awareness, yet we see them as essential for emotions to emerge (Frijda, 1993; Lazarus 1991; Pekrun, Frenzel, Goetz & Perry, 2007; Schutz & Davis, 2000; Smith, 1991; Sutton, 2007). For example, for anger to emerge during a teaching activity, a teacher must judge the activity as being important to their teaching goals. If the teacher does not see the activity as relevant, the potential for emotion diminishes. In addition, the teacher must appraise the activity as not going well or as goal incongruent. Finally, if the teacher judges the problem with the activity to be the fault of the student's attitudes or some mandated curriculum change, anger could emerge (Smith, 1991; Smith & Ellsworth, 1987). So, even though the appraisals may not be accurate or reflective, a teacher may choose to view the situation in a manner that is conducive to experiencing anger.

Emotional Episodes and Emotional Labor in Classroom Activity Setting

We see emotional episodes or experiences as emerging during judgments regarding the perceived success related to goal attempts (Schutz, et al. in press). Stein and her associates (e.g., Stein, Liwag, & Wade, 1996; Stein, Trabasso, & Liwag, 1993) suggest a dynamic theory of emotion episodes in which changes in the status of valued goals lead to emotional experiences that then evoke goal-directed behavior aimed at maintaining or reaching desired outcomes, or avoiding undesired outcomes. Depending on the nature of the transaction and the type of emotion experienced, a goal may be generated to maintain or change the current state. The process is dynamic in that new precipitating events can initiate another cycle of emotional experiences, which then lead to planning processes in a continuous goal-action-outcome sequence (Ainley, 2007; Op 't Eynde, De Corte, & Verschaffel, 2007; Turner & Waugh, 2007).

For example, Williams and her colleagues (2006) found that many teachers have deemed it necessary to suppress or avoid the display of emotions in the classroom. In an illustration of Stein's theory, these teachers view personal expressions from themselves or the students as a modification in the goals of the classroom. If acknowledged, this change in focus (for instance, two students exhibiting openly hostile behavior due to a disagreement outside the classroom) will set off an emotional "chain reaction" among all students that will eventually switch the priority of that classroom setting to resolving the issue rather than learning. Rather than addressing the occurrence, some teachers maintain their valued goal of teaching and suggest that the students either suppress their emotions until a more appropriate time or refer them to resources outside the classroom (e.g., principal or guidance counselor).

These results are similar to other research suggesting that within the context of classroom activity settings, teachers are expected to display emotions in particular ways depending on the nature of the events (Ekman, 1973; Morris & Feldman, 1996; Zembylas, 2003, 2005). We see these display rules as standards for appropriate emotional expression during classroom transactions. For example, in most transactions with students, teachers are expected to show pleasant emotions and suppress their unpleasant emotions (Schaubroek & Jones, 2000).

The expectation that teachers are obliged to follow particular display rules has been associated with the idea of emotional labor (Grandey, 2000; Hochschild, 1990; Morris & Feldman, 1996; Zembylas, 2005). For example, Morris and Feldman (1996) define emotional labor "as the effort, planning, and control needed to express organizationally desired emotion during interpersonal transactions" (p. 987). In the area of occupational psychology, researchers have suggested that emotional labor under some circumstances is related to emotional exhaustion (a key component of burnout), job satisfaction, and health symptoms (Hochschild, 1979, 1990; Morris & Feldman, 1996; Schaubroek & Jones, 2000). These findings suggest that emotional labor is an important issue in understanding emotions in the classroom.

In 1996, Morris and Feldman explicated four dimensions of emotional labor: frequency of emotional displays, attentiveness of required display rules, variety of emotions required, and emotional dissonance. For teachers, variations on these dimensions are thought to be associated with the amount of emotional labor. For example, teachers who work in prekindergarten to high school classrooms spend a number of hours per week in contact with students, staff, administrators, and parents. This provides the opportunity for not only a large number of emotional displays but also the need to display a variety of different emotions. Researchers have suggested that as the number and variety of displays increases, more emotional labor is needed to present and regulate appropriate emotion displays (Grandey, 2000; Morris & Feldman, 1996).

Morris and Feldman (1996) suggest that, in addition to frequency and variety, the required attentiveness to display rules is also important in regard to emotional labor. This dimension involves both the duration and the intensity of emotional displays. So, typically someone working in the lunchroom transacts with each student for a short period of time each day and with relatively low intensity. However, teachers in that same school spend considerably longer and more intense time with the students in their classes. As indicated, the frequency, variety, and attentiveness to display rules have been shown to be associated with emotional exhaustion (Jackson, Schwab, & Schuler, 1986; Maslach, 1982; Morris & Feldman, 1996).

Emotional dissonance, the fourth dimension of emotional labor discussed by Morris and Feldman (1996), refers to the potential conflict between what teachers actually feel during a classroom event and the perceived display rules they are expected to follow. As indicated, teachers come to the classroom with a variety of beliefs about what should be happening in the classroom. These beliefs may be in conflict with the expected display rules of the school (e.g., during school reform). This may result in the disequilibrium that has been associated with increased emotional exhaustion and a decrease in job satisfaction, potentially leading to quitting the teaching profession (Morris & Feldman, 1996).

Confirming or Challenging Old Beliefs and Building New Beliefs

As suggested, teaching tends to be an emotional profession. Teachers' beliefs about themselves and their teaching influence their goals and the appraisals they make during classroom transactions. As such, we recognize the influence of beliefs on emotions. However, theorists also have begun to examine the reverse relationship and have observed a multidirectional relationship between emotions and beliefs (Frijda et al., 2000). Thus emotions are also seen as influencing the content and strength of an individual's beliefs, often making them more resistant to change (Frijda et al., 2000; Lazarus & Smith, 1988).

Emotions are thought to influence beliefs in two ways: (1) beliefs that were previously nonexistent may be generated, or (2) existing beliefs may be altered or changed. These modifications may involve a reduction or increase in the strength of the belief (Frijda & Mesquita, 2000). Early philosophers (e.g., Hume, 1739/1969 as cited in Frijda et al., 2000) went further to state that emotions not only influenced beliefs but they also were integral to the manifestation of these thoughts into action. In this regard, beliefs were considered as "that upon which one is prepared to act," but without emotion or passion, this action would never be actualized. Therefore, while beliefs are precursors and sometimes guides for our actions, they are never sufficient to provoke action (Frijda et al., 2000). Although this reverse relationship has received increased interest, there is still little formal evidence to support this phenomenon (Frijda & Mesquita, 2000).

Ashton and Gregoire (2003) also talk about the relationship between affective and cognitive constructs. They refer to both Piaget's (1974) and Vygotsky's (1986) belief in the inseparable nature of cognition and emotion, where affect is thought to be the energizer or instigator of intellectual activity or thought. Referring to Piaget's theory of equilibration or cognitive conflict, they posit that emotions are integral to belief change, in that some emotions tend to emerge during mismatches between one's goals and/or beliefs and teachers perceptions of what is occurring. This mismatch is what is referred to as disequilibrium in Piaget's theory.

This reverse relationship, whereby emotions influence beliefs, was also observed in research on teaching (Williams et al., 2006). In that study, one teacher revealed that in her early years of teaching, she believed that teachers should be authoritative. As such, she was quite rough on the students because she felt that it was in their best interest. She felt satisfied that her detached demeanor was effective until she received a letter from a student. The letter described the impact she had on the student's life and evoked tears of joy. It was through this experience that the teacher recognized how her students saw her and the important role she played in their lives. The emotions elicited by the letter caused some internal conflict, which needed to be resolved. This led her to reassess and change her beliefs about her teacher role to one that was more compassionate, allowing her to express her own emotions more freely in the classroom and to more effectively address the emotional situations that arose.

In this case, emotions have the potential to motivate belief change, because dissatisfaction with current beliefs ultimately influences change in beliefs, which may in turn affect teachers' identities. The role of emotion in changing and maintaining beliefs has been recently emphasized, while uniting the issue of teacher identity and emotions (Carlyle & Woods, 2002; Van den berg, 2002). The relationship between unpleasant emotions and the loss of teacher identity are frequently reported in those studies. For instance, Morag, a 25-year-old arts teacher, said, "The stress of teaching splits you up so you don't know who you are any more" (Carlyle & Wood, 2002, p. 77). Thus teachers' identities can be reconstructed through the change in beliefs, which is caused by unpleasant emotions, such as dissatisfaction or burnout.

Closely tied to the nature of the emotional experiences and the influence of emotions on belief change or confirmation are the attributions teachers make regarding their explanation of the causes of the emotional experience (Clore & Gasper, 2000; Weiner, 1994, 2007). From the perspective of Weiner (1994), attribution theory would suggest that the difference between teachers' attributing an unsuccessful academic activity to an unsuccessful strategy rather than unmotivated students could have different consequences for confirming or changing their beliefs. In the one case, attributing the cause to the need for different strategies may result in reflection about their belief related to the content, their teaching philosophy, or the students they are teaching. On the other hand, attributing the unsuccessful academic activity to the students may simply reaffirm teachers' beliefs about a particular group of students.

Theoretical Perspective Summary

As indicated, we used Figure 1 as a model of our current thinking about the transactions among the constructs we see as important to our discussion of teachers' identities, beliefs, and goals about emotions in the classroom. The model suggests that teachers enter classrooms enmeshed within particular social-historical contexts. As such, teachers "bring with them" shared personal (e.g., identities, beliefs, and goals) and cultural (i.e., beliefs and values about schooling and learning) backgrounds that have emerged through transactions within historical contexts. In addition, as Bronfenbrenner's (1986) model suggests with his concept of the chronosystem, these contexts are continually changing.

We used the model in Figure 1 to conceptually diagram transactions among what we see as some of the key aspects related to teachers and emotions in the classroom. Our current program of research is focused on developing an understanding of how teachers think and talk about emotions in the classroom. How we go about investigating those issues is discussed next.

METHOD AND METHODOLOGICAL APPROACHES AND ISSUES OF RIGOR

Methodologically, our approach to inquiry tends to be pragmatic in nature. As such, we approach research as a problem-solving process. We develop our research questions based on our current understanding of problems, and then our methods of inquiry are developed to answer those questions (Schutz, Chambless, & DeCuir, 2004). As a result, we tend to use multiple research methods. This multimethods approach occurs not only within a particular study, but also, as indicated by Schutz, Chambless, and DeCuir (2004), across our program of research.

The focus of this chapter has been our research on teachers' identities, beliefs, and goals about emotions in the classroom. Our original problem was a concern with the emotional nature of the classroom and how teachers thought about and attempted to deal with emotions in the classroom. As a result, some of our earlier research questions focused on how teachers talked about their own and their students' emotions and how they talked about regulating student emotions in the classroom. Due to the nature of our questions, we began with what Schutz and DeCuir (2002) refer to as investigating process and meaning.

Investigating process and meaning in this case involved concentrating on the processes related to teachers' experiences with emotions in the classroom and the meaning they attached to those experiences (Schutz & DeCuir, 2002). As a result, we began our inquiry using a phenomenological approach. We were interested in what the experience of emotions in the classroom meant to the teachers themselves (Patton, 2002). During this phase, we chose interviews as our primary method of data collection. We thought interviews would provide us the best opportunity to get at the essence of this phenomenon from the teachers' own perspectives.

To ensure the accuracy of the thoughts and beliefs of teachers and to make sure that our analyses were credible, we used a variety of strategies to ensure rigor. For example, we conducted the interviews in the teachers' classrooms to obtain a more accurate picture of the context of the school and to obtain ideas of what the teachers envisioned for their classrooms. In addition, interviews were audiotaped, transcribed, and sent to the participants for member checks. Again, the goal here was to make sure that we had an accurate representation of what the teachers thought and wanted to share with us.

During the analysis phase, we were concerned with the credibility of our findings. Therefore, in the studies discussed in this chapter, we coded simultaneously, with each researcher coding the same transcript. The transcripts were then compared and the codes discussed among the research group. Hence, the development of categories was an ongoing, iterative process based on common codes across transcripts. As each transcript was coded, both the codes and the categories were reevaluated and refined to mirror participants' descriptions of the essence of the phenomenon. We developed the themes inductively from the data through examination of the categories generated. Multiple readings of the transcripts were completed to ensure the codes, categories, and themes reflected the overall context as well as the meaning-making of each participant (LeCompte & Preissle, 1993; Thompson, Locander, & Pollio, 1989). Through our interactions with the data, we created categories and themes. Once the codes were developed, we tested the codes' appropriateness deductively by examining deviant narratives that seemed not to fit the categories. In this confirmatory stage, we also developed hypothetical models by interpreting the relationships among the categories. It was through this rigorous iterative process that we attempted to first obtain accurate transcripts of the thoughts and beliefs of the teachers we interviewed, and then worked with the goal of making our analyses credible representations of the meanings of our participants regarding emotions in the classroom.

As indicated, we see inquiry as a multimethod, iterative process (see Pekrun & Schutz, 2007). As such, within our program of work we go though cycles of theory development and theory testing. It is clear that, at this point in our research on teachers' emotion, we iterate towards theory testing as we continue the process of developing our ideas about teachers' emotions in the classroom. Our goal is to work to the development of interventions for preservice and in-service teachers that will provide them with ways of being successful in their chosen line of work.

IMPLICATIONS FOR EDUCATIONAL PRACTICE

Emotions permeate educational contexts; therefore the study of emotions within that context may help to provide a more sophisticated understanding of the problems and potential solutions. As we discussed earlier, emotional experiences and actions often emerge from appraisals about the event. These judgments are influenced by the individual's personal goals, as well as their beliefs about themselves and their position within the particular activity settings. In the educational context, this has implications for the way we structure the learning environment, the training and education we provide for our preservice teachers, and the support and professional development for in-service teachers.

One area where emotion research can be useful is related to the concept of "emotional labor" (Hochschild, 1990; Morris & Feldman, 1996; Zembylas, 2003). As indicated, emotional labor refers to the teachers' or students' effort to express, repress, or manufacture emotions based on a perceived need during particular transactions (Williams, et al., 2006). For instance, if a teacher's felt emotions and pedagogically desired emotions are not congruent, then the teacher needs to put forth great effort to display the desired emotions (Morris & Feldman, 1996). Without looking into the nature and process of emotional labor, we cannot really understand how teachers construct emotions and why they express emotions the way they do.

Kemper (1993) and Hochschild (1990) noted that emotional labor and emotional regulation occur to maintain conformity with normative standards. In other words, what teachers perceive as "pedagogically desired emotions" and the teachers' decisions to regulate emotions emerge from certain ideologies to maintain the social norm. So, by understanding how the social consensus has been constructed and given meaning, and how the emotional dimension is related to the social norm, teachers may be able to problematize or embrace certain ideologies in school, and so learn how to regulate emotions in empowering ways (see Zemblyas, 2007).

In many countries, the high attrition rate of teachers is an increasing concern; hence, research exploring emotions and emotional labor are paramount. A number of studies identified unpleasant emotions, such as anger, stress, and anxiety, as core factors influencing teachers' decisions to dropout (Gaziel, 1995; Wilhelm, Dewhurst-Savellis, & Parker, 2000; Wisniewski & Gargiulo, 1997) or experience burnout. According to Hughes (2001), teacher burnout is often associated with the coping strategy of escape or the desire to escape.

As Folkman and Lazarus (1988a, 1988b) claimed, emotions and coping have multidirectional relationships. For example, when teachers assess a person-environment transaction as goal incongruent, then they are more likely to experience unpleasant emotions, such as anxiety. Those unpleasant emotions may increase the desire to escape, which in turn may effect the transaction. With such a defensive coping, they may develop depersonalized attitudes towards students. If this changed person-environment transaction is reappraised, it may lead to a change of intensity or quality in emotions. Given this reappraisal, this teacher may experience stronger stress, which in turn eventually results in dropping out of the profession.

Better understanding of teacher emotions and emotional labor may not only assist in interpreting and preventing teacher dropout phenomena, but also has the potential for improving preservice teacher education programs. Preservice teacher education may be able to assist in reducing teacher burnout through the training to build resilience from unpleasant emotional experiences (Martinez, 2004). In addition, the transition between preservice teacher preparations to in-service professional development can be improved through deeper understanding of their identities, goals, and beliefs about emotions in the classroom.

Teachers must also be provided with opportunities to generate appropriate and realistic beliefs about the nature of teaching and the learning environment. This, coupled with the knowledge of the emotional nature of schooling, would better equip teachers with the tools to handle their own emotional experiences and emotional events within the classroom. For example, school environments are continuously changing. These changes range from policy mandates involving curriculum reform to minor district concerns, such as the issue of mandatory uniforms. This constant change tends to have the most direct impact on those intimately involved in the learning process, namely teachers and students. For teachers, these initiatives often involve transforming current perceptions or beliefs about curriculum and pedagogy. As such, implementing new curriculum mandates and pedagogical standards becomes an effort in belief change, which is not only long and slow but also can be an emotionally laborious process. Having further insight into the role emotions play in belief modification can help to reduce the negative impact that often accompanies changes in educational policy. In this regard, research in the area of emotions should not be limited to teacher-student transactions, but extended to include their role in the success of school reform.

As educational researchers, we seek to identify and understand dimensions of the educational process that have the potential to influence the effectiveness of different learning environments. It seems clear that emotion is one of those dimensions. Emotional experiences within the classroom have been associated with the beliefs, goals, and identities of those involved in the educational process. These beliefs, goals, and identities influence individuals' personal judgments and their motivation to progress towards their goals. As such, it becomes incumbent on us not only to further investigate the causes or antecedents of these emotional events, but also to come to a better understanding of how these events influence students' and teachers' success in the classroom.

References

Ainley, M. (2007). Being and feeling interested: Transient state, mood, and disposition. In P. A. Schutz and R. Pekrun (Eds.), *Emotion in education*. (pp. 141-158). San Diego: Elsevier Inc.

Ashton, P. T, & Gregoire-Gill, M. (2003). At the heart of teaching: The role of emotion in changing teachers' beliefs. In J. Raths & A. McAninch (Eds.), *Advances in teacher education* (Vol. 6, pp. 99-121). Norwood, NJ: Ablex.

Aultman, L. P., Williams, M. R., Garcia, R. I., & Schutz, P. A. (2006). *Emotions in the classroom: Where is the "line"?* Manuscript submitted for publication.

Boekaerts, M. (2007). Understanding students' affective processes in the classroom. In P. A. Schutz and R. Pekrun (Eds.), *Emotion in education*. (pp. 33-54). San Diego: Elsevier Inc.

Bronfenbrenner, U. (1986). Ecology of the family as a context for human development: Research perspectives. *Developmental Psychology, 22*, 723-742.

Calderhead, J. (1996). Teachers: Beliefs and knowledge. In D. C. Berliner & R. C. Calfee (Eds.), *Handbook of educational psychology* (pp. 709-725). New York: Macmillan/American Psychological Association.

Carlye, D. & Woods, P. (2002). *Emotions of teacher stress*. Stoke on Trent, UK: Trentham Books.

Carver, S. C., & Scheier, M. F. (2000). On the structure of behavioral self-regulation. In M. Boekaerts, P. R. Pintrich, & M. Zeidner (Eds.), *Handbook of self-regulation* (pp. 41-84). San Diego: Academic Press.

Clandinin, J.D & Connelly, M.F. (1987). Teachers' personal knowledge: What counts as personal in studies of the personal. *Journal of Curriculum Studies, 19*(6), 487-500.

Clore, G. L., & Gasper, K., (2000). Feeling is believing: Some affective influences on belief. In N. Frijda, A. Manstead & S. Bem (Eds.), *Emotions and beliefs: How feelings influence thoughts* (pp. 10-44). Cambridge: Cambridge? University Press.

Danielewicz, J. (2001). *Teaching selves: Identity, pedagogy, and teacher education*. Albany: State University of New York Press.

Ekman, P. (1973). Cross-cultural studies of facial expression. In P. Ekman (Ed.), *Darwin and facial expression: A century of research in review* (pp. 169-222). New York: Academic Press.

Folkman, S., & Lazarus, R. S. (1988a). Coping as a mediator of emotion. *Journal of Personality and Social Psychology, 54*(3), 466-475.

Folkman, S., & Lazarus, R. S. (1988b). The relationship between coping and emotion: Implications for theory and research. *Social Science & Medicine, 26*(3), 309-317.

Ford, M. E. (1992). *Motivating humans: Goals, emotions and personal agency beliefs*. Newbury Park, CA: Sage.

Frijda, N. H. (1993). The place of appraisal in emotion. *Cognition and Emotion, 7*, 357-387.

Frijda, N. H., Manstead, A., & Bem, S. (2000). The influence of emotions on beliefs. In N. Frijda, A. Manstead & S. Bem (Eds.), *Emotions and beliefs: How feelings influence thoughts* (pp. 1-9). Cambridge: The University Press.

Frijda, H. H., & Mesquita, B. (1994). The social roles and functions of emotions. In S. Kitayama and H. R. Marcus (Eds.), *Emotion and culture: Empirical studies of mutual influence* (pp. 51-87). Washington, D.C.: American Psychological Association

Frijda, N., & Mesquita, B. (2000). Beliefs through emotions. In N. Frijda, A. Manstead & S. Bem (Eds.), *Emotions and beliefs: How feelings influence thoughts* (pp. 45-77). Cambridge: University Press.

Gaziel H. H. (1995). Sabbatical leave, job burnout and turnover intentions among teachers. *International Journal of Lifelong Education, 14*(4), 331-338.

Grandey, A. A. (2000). Emotional regulation in the work place: A new way to conceptualize emotional labor. *Journal of Occupational Health Psychology, 5*(1), 95-110.

Green, T. (1998). *Teaching and the formation of beliefs*. Troy, NY: Educator's International Press Inc.

Gross, J. J. (1998). The emerging field of emotion regulation: An integrative review. *Review of General Psychology, 2*(3), 271-299.

Hochschild, A. (1979). Emotion work, feeling rules and social structure. *American Journal of Sociology, 85*, 551-575.

Hochschild, A. R. (1990). Ideology and emotion management: A perspective and path for future research. In T. D. Kemper (Ed.), *Research agendas in the sociology of emotions* (pp. 117-142). Albany, NY: State University of New York Press.

Hughes, E. (2001). Deciding to leave but staying: teacher burnout, precursors and turnover. *International Journal of Human Resource Management, 12*(2), 288-298.

Hume, D. (1739/1969). *A treatise of human nature*. (Ed. E. C. Mossner) Harmondsworth: Penguin.

Jackson, S. E. Schwab, R. I., & Schuler, R. S. (1986). Toward an understanding of the burnout phenomenon. *Journal of Applied Psychology, 71*, 630-640.

Kemper, T. D. (1987). How many emotions are there? Wedding the social and the autonomic components. *American Journal of Sociology, 93*(2), 263-289.

Kemper, T. D. (1993). Sociological models in the explanation of emotions. In M. Lewis and J. Haviland (Eds.), *Handbook of emotions* (pp. 41-51). New York and London: The Guilford Press.

Knowles, G. J. (1992). Models for understanding pre-service and beginning teachers' biographies: Illustrations from case studies. In I. F. Goodson (Ed.), *Studying teachers' lives* (pp. 99-152). London: Routledge

Lazarus, R. S. (1991). *Emotion and adaptation*. New York: Oxford University Press.

Lazarus, R. S. (1999). *Stress and emotions: A new synthesis*. New York: Springer.

Lazarus, R., & Smith, C. (1988). Knowledge and appraisal in the cognition-emotion relationship. *Cognition and Emotion, 2*, 281-300.

LeCompte. M. D., & Preissle, J. (1993). *Ethnography and qualitative design in educational research*. San Diego: Academic Press.

Markus, H. R., & Kitayama, S. (1994). The cultural construction of self and emotion: Implications for social behavior. In S. Kitayama & H. R. Markus (Eds.), *Emotion and culture: Empirical studies of mutual influence* (pp. 89-130). Washington, D.C.: APA.

Markus, H., & Nurius, P. (1986). Possible selves. *American Psychologist, 41*, 954-969.

Martinez, K. (2004). Mentoring new teachers: promise and problems in times of teacher shortage. *Australian Journal of Education, 48*(1), 95-108.

Maslach, C. (1982). *Burnout: The cost of caring*. Englewood Cliffs, NJ: Prentice Hall.

Massey, D., & Chamberlin, D. (1990). Perspective, evangelism, and reflection in teacher education. In Day, C., Pope, M. L., & Denicolo, P. (Eds). *Insights into teachers' thinking and practice*. London; New York: Falmer Press.

Morris, J. A., & Feldman, D. C. (1996). The dimensions, antecedents, and consequences of emotional labor. *Academy of Management Review, 21*(4), 986-1010.

Nias, J. (1996). Thinking about feeling: The emotions in teaching. *Cambridge Journal of Education, 26*(3), 293-306.

Op 't Eynde, P., De Corte, E., & Verschaffel, L. (2007). Students' emotions: A key-component of self-regulated learning? In P. A. Schutz and R. Pekrun (Eds.), *Emotion in education*. (pp. 179-198). San Diego: Elsevier Inc.

Pajares, F. (1992). Teachers' beliefs and educational research: Cleaning up a messy construct. *Review of Educational Research, 62*, 307-332.

Patton, M. (2002). *Qualitative research and evaluation methods*. Thousand Oaks, CA: Sage Publications.

Pekrun, R., Frenzel, A. C., Goetz, T., & Perry R. P. (2007). The control-value theory of achievement emotions: An integrative approach to emotions in education. In P. A. Schutz and R. Pekrun (Eds.), *Emotion in education*. (pp. 9-32). San Diego: Elsevier Inc.

Pekrun, R., & Schutz, P. A. (2007). Where do we go from here? Implications and further directions for inquiry on emotions in education. In P. A. Schutz & R. Pekrun (Eds.), *Emotion in Education* (pp. 303-321). San Diego: Elsevier Inc.

Piaget, J. (1974). *To understand is to invent. The future of education*. New York: Viking Press.

Powers, W. T. (1971). *Behavior: The control of perception*. Chicago: Aldine.

Ratner, C. (2000). A cultural-psychological analysis of emotions. *Culture and Psychology, 6*, 5-39. *Review of Educational Research, 62*(3), 307-332.

Ratner, (2007). A macro cultural-psychological theory of emotions. In P. A. Schutz and R. Pekrun (Eds.), *Emotion in education*. (pp. 85-100). San Diego: Elsevier Inc.

Rokeach, M. (1968). *Beliefs, attitudes and values: A theory of organization and change*. San Francisco: Jossey-Bass.

Schaubroeck, J., & Jones, J. R. (2000). Antecedents of workplace emotional labor dimensions and moderators of their effects on physical symptoms. *Journal of Organizational Behavior 21*, 163-183.

Schutz, P. A., & Davis, H. A. (2000). Emotions and self-regulation during test taking. *Educational Psychologist, 35*, 243-255.

Schutz, P. A., Crowder, K. C., & White, V. E. (2001). The development of a goal to become a teacher. *Journal of Educational Psychology, 93*, 299-308.

Schutz, P. A., & DeCuir, J. T. (2002). Inquiry on emotions in education. *Educational Psychologist, 37*, 125-134.

Schutz, P. A., Chambless, C. B., & DeCuir, J. T. (2004). Multimethods research. In K. B. deMarrais and S. D. Lapan (Eds.), *Foundations for research: Methods of inquiry in education and the social sciences* (pp. 267-282). Hillsdale, NJ: Lawrence Erlbaum.

Schutz, P. A., Hong J. Y., Cross, D. I., & Osbon, J. N. (in press). Reflections on investigating emotions among social-historical contexts. *Educational Psychology Review.*

Shavelson, R. J., & Stern, P. (1981). Research on teachers' pedagogical thoughts, judgments, decisions, and behavior. *Review of Educational Research, 51*(4), 455-498.

Smith, C. A. (1991). The self, appraisal and coping. In C. R. Snyder & D. R. Forsyth (Eds.), *Handbook of social and clinical psychology: The health perspective* (pp. 116-137). Elmsford, NY: Pergamon.

Smith, C. A., & Ellsworth, P. C. (1987). Patterns of appraisal and emotions related to taking exams. *Journal of Personality and Social Psychology, 52*, 475-488.

Solomon, R. C. (1976). *The passions: Myth and nature of human emotion*. Notre Dame, IN: University of Notre Dame Press.

Stein, N. L., Liwag, M.D., & Wade, E. (1996). A goal-based approach to memory for emotional events: Implications for theories of understanding and socialization. In R.D. Kavanaugh, B. Zimmerberg, & S. Fein (Eds.), *Emotion: Interdisciplinary perspectives* (pp. 91-118). Mahwah, NJ: Lawrence Erlbaum Associates Inc.

Stein, N. L., Trabasso, T., & Liwag, M. (1993). The representation and organization of emotional experience: Unfolding the emotion episode. In M. Lewis & J.M. Haviland (Eds.), *Handbook of emotions* (pp. 279-300). New York: Guilford Press.

Sutton, R. E. (2007) Teachers' anger, frustration, and self-regulation. In P. A. Schutz and R. Pekrun (Eds.), *Emotion in education*. (pp. 251-266). San Diego: Elsevier Inc.

Tharp, R. G., & Gallimore, R. (1988). *Rousing minds to life: Teaching, learning, and schooling in social context*. Cambridge, England: Cambridge University Press.

Thompson, C. J., Locander, W. B., & Pollio, H. R. (1989). Putting consumer experience back into consumer research: The philosophy and method of existential-phenomenology. *Journal of Consumer Research, 16*, 133-146.

Turner, J. E. & Waugh, R. M. (2007). A dynamical systems perspective regarding students' learning processes: Shame reactions and emergent self-organizations. In P. A. Schutz and R. Pekrun (Eds.), *Emotion in education*. (pp. 119-139). San Diego: Elsevier Inc.

Van den Berg, R. (2002). Teachers' meanings regarding educational practice. *Review of Educational Research, 72*(4), 577-625.

Van Fleet, A. (1979). Learning to teach: The cultural transmission analogy. *Journal of Thought, 14*, 218-290.

Vandenberghe, R., & Huberman, A. M. (Eds.). (1999). *Understanding and preventing teacher burnout: A sourcebook of international research and practice*. Cambridge, England: Cambridge University Press.

Virta, A. (2002). Becoming a history teacher: Observations on the beliefs and growth of student teachers. *Teaching and Teacher Education, 18*(6), 687-698.

Vygotsky, L. S. (1986). *Thought and language*. Cambridge, MA: MIT Press.

Weiner, B. (1994). Integrating social and personal theories of achievement striving. *Review of Educational Research, 64*, 557-573.

Weiner, B. (2007). Examining emotional diversity in the classroom: An attribution theorist considers the moral emotions. In P. A. Schutz and R. Pekrun (Eds.), *Emotion in education*. (71-84). San Diego: Elsevier Inc.

Wilhelm, K., Dewhurst-Savellis, J., & Parker, G. (2000) Teacher stress? An analysis of why teachers leave and why they stay. *Teachers and Teaching, 6*(3), 291-304.

Williams, M. R., Cross, D. I., Hong, J. Y., Aultman, L. P., Osbon, J. N., & Schutz, P. A. (2006). *"There are no emotions in math": How teachers approach emotions in the classroom*. Manuscript submitted for publication.

Wisniewski, L. & Gargiulo, R. (1997). Occupational stress and burnout among special educators: A review of the literature. *Journal of Special Education, 31*(3), 325-346.

Woolfolk-Hoy A., Davis, H., Pape, S. (2006) Teacher knowledge and beliefs. In P. Alexander and P. Winne (Eds.), *Handbook of educational psychology* (pp. 715-735). New York: Simon & Schuster/ Macmillan.

Zeichner, K., & Gore, J. (1990). Teacher socialization. In W. Robert Houston (Ed.), *Handbook of research on teacher education* (pp. 329-348). New York: Macmillan.

Zembylas, M. (2003). Emotions and teacher Identity: A poststructural perspective. *Teachers & Teaching, 9*(3), 213.

Zembylas, M. (2005). Discursive practices, genealogies, and emotional rules: A poststructuralist view on emotion and identity in teaching. *Teaching and Teacher Education, 21*(8), 935-948.

Zembylas, M. (2007). The power and politics of emotions in teaching. In P. A. Schutz and R. Pekrun (Eds.), *Emotion in education*. (pp. 285-301). San Diego: Elsevier Inc.

CHAPTER

14

Scaffolding Emotions in Classrooms

DEBRA K. MEYER
Elmhurst College

JULIANNE C. TURNER
University of Notre Dame

Emotions help define classroom experiences, providing powerful rationales for engaging in and avoiding, even abandoning, teaching and learning opportunities. Teachers' interpersonal skills at identifying and supporting students' emotions as well as their intrapersonal understanding and expression of their own emotions are integral to effective teaching (Goleman, 1995; Hargreaves, 2001). Unfortunately, in spite of their ubiquity and power, emotions have mostly been ignored in educational research, professional practice, and teacher education (Linston & Garrison, 2003; Ria, Sève, Saury, Theureau, & Durand, 2003; Rosiek, 2003). Yet, over the past decade a growing interest in emotions in classrooms can be found in the proliferation of books on the topic, such as *emotional intelligence* (Goleman, 1995), *passion* (Fried, 2001; Wink & Wink, 2004), *caring* (Noddings, 1992), *love* (Goldstein, 1997; Linston & Garrison, 2003), and *power* (Boler, 1999). A common theme across these volumes is that emotion has always been essential for understanding classrooms, but was assigned a low priority, as Linston and Garrison (2003) clearly expressed:

> For too long we have left emotions in the ontological basement of educational scholarship, to be dragged up and out only when a particular topic necessitated it (e.g., classroom management, student motivation, or teacher 'burnout'). That seems ill advised, and it is time to rebuild our academic house. When we teach, we teach with ideas and feelings. When we interact with students, we react and they respond with thoughts and emotions. When we inquire into our natural and social worlds, we do so with desire and yearning. (p. 5)

Given the increasingly difficult and complex demands placed on teachers and students in our rapidly changing societies (e.g., single test accountability systems, expanding curriculums, increasing student diversity), now more than ever, we need to further develop current understandings of emotion in classrooms (Hargreaves, 2001). The emerging body of literature offers a starting point for our discussions, but more theoretical and empirical work for exploring these new conceptualizations is also needed.

SCAFFOLDING EMOTIONS IN CLASSROOMS: A THEORETICAL FRAMEWORK

In this chapter we use the theoretical metaphor of *scaffolding* to discuss some of the critical ways that teachers use emotions to support student learning and development. Expanding on traditional views of scaffolding, we define *emotional scaffolding* as temporary but reliable teacher-initiated interactions that support students' positive emotional experiences to achieve a variety of classroom goals. For example, emotional scaffolding may sustain and enhance students' understanding, motivation, collaboration, participation, and emotional well being. Emotional scaffolding differs from other forms of positive teacher-student interactions in that the support has a specific goal of increasing student achievement and autonomy in a particular developmental competency, and often in several areas simultaneously.

The idea of scaffolding emotions is not new, although the focus on supporting emotions and using emotions to support other classroom goals is novel. In 1976, Wood, Bruner, and Ross originated the metaphor by delineating six functions of scaffolded instruction: *recruitment, reduction in degrees of freedom, direction maintenance, marking critical features, frustration control,* and *demonstration*. Within instructional interactions, scaffolding has been identified as temporary teacher support to achieve two interrelated goals: (1) to provide support *only as necessary*, and (2) to move from a position of shared responsibility to one in which the student takes ownership (Hogan & Pressley, 1997; Meyer, 1993; Palincsar, 1986; Rogoff, 1990; Wood, et al., 1976). It is this balance between teacher support and student autonomy that has linked the scaffolding metaphor to Vygotsky's (1978) theoretical construct of "zone of proximal development" (ZPD), defined as an interpersonal space within which a teacher provides support as needed while negotiating the gradual transfer of responsibility to students.

To our knowledge only a few researchers have explicitly examined and identified emotional scaffolding in classrooms. Rosiek (2003) defined emotional scaffolding as "teacher pedagogical use of analogies, metaphors, and narratives to influence students' emotional response to specific aspects of subject matter in a way that promotes student learning" (p. 402). Rosiek's research provides rich descriptions of how teachers enhanced student understanding by evoking student emotions and illustrates how scaffolding emotions is a dimension of pedagogical content knowledge. Schuster (2000) and Henderson, Many, Wellborn, and Ward (2002) also distinguished emotional scaffolding in early childhood classrooms. Identifying children as "emotional compatriots," Schuster (2000) explored how four scaffolding strategies supported children's learning and socio-emotional development: setting a positive emotional tone, building shared understanding, extending understanding, and supporting empathy and mutual respect. Similarly, Henderson et al. (2002) reported that children's development of literacy repertoires was supported through scaffolding with intellectual, academic, and emotional foci. They found that an integral part of helping young children understand how language communicates emotions was assisting them in understanding their own and others' emotions. Important across all of these studies is the notion that emotional scaffolding is distinct and can be achieved at the classroom level as well as in one-on-one or small-group interactions.

Achieving the multiple goals of scaffolding during whole-class instruction, providing assistance only as needed and increasing independence, means inviting students to take risks publicly. Furthermore, classroom-level scaffolding requires a broad context of trust and support for every student's well-being. As Yowell and Smylie (1999) highlighted, one of the greatest challenges for teachers, yet a necessary prerequisite for effective scaffolding, is creating *intersubjectivity*, or "a shared understanding between teacher and student of a problem-solving task within the ZPD" (p. 474). Perhaps more easily established in one-on-one interactions, the foundation for intersubjectivity must also be established at a classroom level because "[t]he creation of intersubjectivity requires mutual trust, respect, and the communication skills necessary to bridge the distance between expert and novice" (Yowell & Smylie, 1999, p. 474). Therefore, to explore the potential for using emotional scaffolding as a framework for studying emotions in classrooms, in the next section we describe empirical examples of how teachers support the development of positive teacher-student relationships and classroom goal structures.

EMOTIONAL SCAFFOLDING: EMPIRICAL EXAMPLES

Establishing positive student-teacher relationships and negotiating learning goals are only two among several categories of emotional scaffolding that have evolved from our classroom research. In our analyses of classroom discourse to discover the instructional characteristics that promote high levels of student involvement in learning, we have found emotional scaffolding to be critical in sustaining students' understanding of challenging concepts, students' demonstration of their competencies and autonomy, students' involvement and persistence, and students' emotional or personal experiences (cf. Henderson et al., 2002; Rosiek, 2003; Shuster, 2000). Emotions have become increasingly important across our classroom research program (Meyer & Turner, 2002) and were critical in the development of our discourse coding. For example, we initially coded instructional discourse for supportive and nonsupportive teacher statements regarding negotiating understanding, transferring responsibility, and encouraging intrinsic motivation, but we soon became aware that the most effective interactions evoked positive emotions and sustained relationships while accomplishing these three outcomes. The reliability of our transcript coding actually improved when we allowed for multiple simultaneous codes (e.g., cognitive support with emotional support). Our findings are not unique across the motivational literature in classroom research. Emotions have been identified as powerful influences on teaching-learning relationships, and some of these studies we use as examples in the following sections. However, only a few studies, (e.g., Henderson, Many, Wellborn, & Ward, 2002; Rosiek, 2003; and Shuster, 2000), provide rich descriptions of what teachers do and say to scaffold emotions. Thus we begin each section with excerpts from our research and then discuss the implications for scaffolding emotion from the broader motivational research literature.

Emotional Scaffolding of Positive Student-Teacher Relationships

The following excerpt is from the first day of school in a sixth-grade classroom. It illustrates how a teacher uptakes a spontaneous student interaction to begin to scaffold trust by explicitly stating her expectations for how students should treat each other and what they can expect from her—a message we observed being repeated throughout the first two days of school. The teacher's response occurred after she finished calling names from the attendance roster in the first hour of the new school year. During this roll call she had heard students making comments when other students' names were called:

> When I called some kids' names . . . I heard someone go "yewww." PLEASE, don't do that any more in my room. Is that clear? As my mother used to say . . . 'If you can't say anything good, then don't say anything AT ALL.' Keep your mouth closed! If you can't give somebody a compliment, if you can't say something nice, like 'I JUST LOVE THAT HAIR CUT!' ' Look at those beautiful eyes there!' 'How much money you got?!' (moving around to address different students in the room in the role play as students giggle) . . . But you don't talk about anybody in my room and you don't put anybody down. All the years I've been teaching, I've never stood it once and I am not changing now. I don't want to hear the 'he said, she said.' Calling her on the phone, 'Do you like him?' . . . all of that mess, leave it out of my room! I don't play that. Is that clear? Like I said before, I am here for you 100%! And my goal is to be fair and to have fun! And I want to have your input. What do you think we should do to make this project better? Or how

can I make this grade better for you as a whole? What can I do to help you? And what can you do help me? Okay?

Not only did the teacher explicitly uptake the "yewww" as an opportunity to clearly state her expectations, but she provided examples of what could be said and stated her expectation that she and the students would be a community. As she continued to take attendance she made eye contact with each student and said something personal (e.g., recalling personal events or qualities of each child), welcoming new students to the school. Using lots of humor and a little drama, she stated that she was glad to have each child in her class. She commented on students' strengths. As she modeled respect for each and every child, she began to establish the foundation for intersubjectivity by building trust among the students and between the students and teacher.

When we returned in late fall and early spring to observe different units of mathematics instruction, we witnessed the continuation of this teacher's emotional scaffolding and her students emerging independence in taking responsibility for supporting each other. In the following excerpt a student who had difficulty with the homework on factoring is working a problem on the overhead projector. The teacher stands near her and has scaffolded her through the problem. The teacher also has invited the class to participate as their peer struggles at the overhead. As the student, Jade, continues to factor she becomes confused again and turns to her classmates for help. It is important to note that Jade has commented to the teacher earlier that she feels stupid about the difficulty she had with the homework assignment.

Teacher: Um hmm. Is she right so far? (to class) Is she doing a good job? (Class: YES) Is she stupid? (Class: NO)

Jade continues on the overhead.

Teacher: UH! ((indicating a wrong move))

Jade: It's wrong, ain't it? Tell me the truth, people.

Jade stands and looks at the class, raises her hands as if pleading, and says in a trusting tone: "Help me out, then."

Teacher: (to the class) Raise your hand. (to Jade) And answer, say "what did I do wrong?" "Did I do anything wrong?" Well, call on somebody.

Jade calls on Kyra.

Kyra: You need to break down 4. (class response)

Jade: Should I bring the 3 down too?

Teacher: I don't know! Listen, should she bring the 3 down? (to class)

Jade: Yes!

One student says: "There you go." (there are sounds of students coaching her–sounds supportive)

The class breaks into applause. Jade finishes. The class applauds again.

When we first observed and coded this interaction we were surprised by the routine in this class that students who needed the most help were called to demonstrate their thinking, not students who could readily work the problem or had volunteered. The teacher and students had established a classroom norm that everyone could learn with the help of others. Jade frequently struggled in mathematics, but the normative belief that she could do math like her peers and expect that they would support her efforts promoted her public success. Students seemed to know that they could trust their teacher and classmates if they were the ones at the overhead or needed assistance. We believe that the teacher began scaffolding these relationships on the first days of school, then consistently provided emotional scaffolding to create and sustain a supportive emotional environment of trust that students themselves then initiated and supported.

Classroom Research on Teacher-Student Relationships and Emotions

Establishing and maintaining positive teacher-student relationships is essential to developing the trust needed for scaffolding positive classroom environments that support student competence and autonomy through relationships. Teachers' relationships with students have been found to be associated with students' academic achievement and school adjustment. Relationships of sufficient intensity and duration (i.e., adult-child, teacher-student) also are related to students' beliefs, expectations, and feelings in their subsequent relationships, the prior experiences serving as guides for current behaviors and interpretations of others' behaviors (Stuhlman & Pianta, 2001; Valeski & Stipek, 2001). Similarly, Wentzel (1997) has reported that middle schoolers' perceptions of caring teachers included common adult-child forms of support, such as democratic interaction styles that promote the development of competence and autonomy.

Emotional scaffolding can help to establish and sustain positive relationships and classroom climate that support student engagement, learning, and perceptions of competence. For example, classroom features have been found to be major influences on the supportive student-teacher relationships, and students' positive emotions about these teacher-student relationships and school in general are related to their engagement (Connell & Wellborn, 1991; Valeski & Stipek, 2001). We believe that it is through emotional scaffolding that these relationships are negotiated with students, providing the opportunity for students to develop competence and demonstrate autonomy. Skinner and Belmont (1993) reported a significant correlation between the quality of teachers' reports of personal relationships with students and the students' perceptions of feeling socially connected in class. Stipek and Daniels (1988) and Harter (1992) specifically described significant correlations between students' perceptions of competence and positive emotions. Thus students' feelings of relatedness appear to derive from a caring supportive teacher, which would be a primary goal of emotional scaffolding.

Although we have observed several teachers who scaffolded consistently and equitably, we have also observed differential scaffolding. Research suggests that teachers may experience more positive emotions toward students who are warm toward them, and these students are more likely to become the recipients of emotional scaffolding. For example, Valeski and Stipek (2001) found that first grade students and teachers agreed on the quality of their relationships and the first graders' emotions about their relationship with their teacher corresponded to their perceptions of their academic competence. Kindergartners and first graders who reported that they liked their teacher also reported that their teacher cared about them and expressed positive attitudes about school. As Valeski and Stipek (2001) suggested, "children's perceptions of these relationships early in their school careers may become the lenses through which they gauge their school experiences" (p. 1211).

In terms of building the foundation for positive student relationships, a teacher's emotional scaffolding must convey consistently positive support of all students to firmly establish a context of trust. In classroom research with our colleagues, Helen Patrick, and the late Carol Midgley, we found that teachers who consistently scaffolded socio-emotional competencies, like the teacher in the previous excerpt, were viewed as significantly more supportive by their students (Patrick, Turner, Meyer, & Midgley, 2003). Furthermore, teachers who offered support inconsistently ("ambiguous environments") were perceived by students to be unsupportive, not different from teachers who did not scaffold and were explicitly negative on the first days of school. Our analyses of ambiguous and negative classroom socio-emotional environments were related to student reports of more avoidance behaviors, whereas in supportive classrooms, students reported significantly lower incidences of avoidance behaviors. Consistent, positive emotional scaffolding appears to have helped establish the necessary foundation of trust needed for taking the risks and accepting the responsibility so essential to learning in classrooms.

Similarly, other researchers have reported that teachers' discipline, instruction, and emotional support are organizing dimensions of teacher-child interactions that also predict children's social and emotional behaviors. For example, in Stuhlman and Pianta's (2001) study of first grade and full-day kindergarten, teachers' narratives were collected in a semistructured interview format and coded for: (1) control and compliance, (2) student achievement, and (3) view of self as a secure base for supporting the student. The researchers also observed and coded classroom interactions, correlating the narratives with the observed behaviors. The major dimension they found to link teacher narratives, teacher behavior, and child behavior was teachers' negative emotions. Students observed as behaving more negatively had teachers who talked about them more negatively. Teachers also were observed to have more interactions with and to interact more negatively with students for whom they had expressed more negativity in their narratives. These findings suggested to the authors that, "when teachers talk about children, it is their emotional responses to the children that are most closely related to behavior in the classroom, and that internally represented affect can be successfully elicited through a brief interview" (p. 160). Similarly, Hamre and Pianta (2001) reported that kindergarten teachers' reports of relational negativity were predictors of students' social and academic outcomes through fourth grade. This body of research is important because it validates that teachers scaffold positive emotions and suggests that the absence of positive scaffolds may result in negative interactions with students—influences that are just as powerful, but ineffective.

In sum, emotions, as reflected in emotional scaffolding, are essential to achieving multiple goals through classroom relationships. Regardless of which emotions may be "appropriate" for teachers to express (see Boler, 1999, for this debate), emotional scaffolding at the classroom-level assists in building relationships with and among students and teaches multiple simultaneous lessons. These "lessons" begin on the first days of school and continue throughout the academic year as well as begin early in students' classroom experiences and persist throughout their academic careers.

Scaffolding Emotions to Promote Learning Goals

An important feature of classrooms that has been associated with positive affect (i.e., emotions and moods) is an emphasis on learning goals in contrast to performance goals. Learning goals promote effort and developing competence, whereas an emphasis on performance goals favors demonstrating competence in comparison to others (Ames, 1992). Teachers initiate and sustain learning goals by scaffolding "positive affective classroom climates" through instructional interactions. Researchers have found that instructional practices and discourse are important indicators of classroom values, beliefs, and norms, which highlight the contextual meanings of motivation and learning (Stipek, Salmon, Givvin, Kazemi, Saxe, & MacGyvers, 1998; Turner, Meyer, Cox, Logan, DiCintio, & Thomas 1998; Vermunt & Verloop, 1999), demonstrating how emotions are integral to understanding student motivation and learning. As Ford (1992) stated, "[E]motions are not simply motivational "add-ons" or "afterthoughts"—they are major influences in the initiation and shaping of goals and personal agency belief patterns that may seem relatively ephemeral or labile at the level of specific behavior episodes, but that in fact may be every bit as influential as cognitive processes in terms of enduring motivational patterns" (p. 147).

In the following excerpts from two seventh-grade mathematics classrooms, we illustrate how the *same* teacher, who taught both mathematics and pre-algebra classes, scaffolded learning goals. She scaffolds by supporting students cognitively, emotionally, and motivationally in ways that emphasize the importance of challenge to learning. Therefore in supporting students to seek and persist during challenging activities, she responds directly to students' emotional needs. In addition, the emphasis on learning goals reflects an interest in and a value for learning content. The teacher scaffolds student engagement both by modeling her own enthusiasm (i.e., emotion) and by noting interesting features of student contributions. Her instructional discourse, while focused on mathematics, is laced with a strong and personal message that she and the students are partners in the important and engaging enterprise of learning mathematics. Moreover, the teacher supports these norms by showing her respect for the importance of learning; by characterizing mathematics and problem solutions as valuable, interesting, and exciting; by providing structures to help students learn; and by supporting students during effortful learning.

The teacher began a class in December by explaining to her students that she had been away at a math in-service session related to their new math curriculum. She demonstrated her commitment to their teaching-learning partnership by explaining to the students that she had attended the session because she wanted to help them, both as learners and as people.

> I apologize for being gone. Once again, I was at another math conference where we were learning how to teach out of the next couple books . . . and it has been very beneficial. So I am learning a lot about how I . . . can help you to be the best learner, best person.

As students finished individual activities that day, the teacher distributed a holiday-related problem-solving exercise. She introduced it as something that students would want to complete, assuming that it was interesting enough for students to work on voluntarily over the holidays. At the same time, she alluded to its challenge, stating that students could succeed with hard work. By mentioning interest and challenge in successive sentences, the teacher seemed to convey the message that learning and effort were worthwhile and that students shared her values.

> The problem solving activity for the holidays is for anyone and you can take it home and even do it over the holidays. It is umhhhhh challenging, but it is nothing you can't do with a lot of hard work. It is called 'Santa's Pack Held 30 Toys.'

After students had been working on the problem for a while, the teacher introduced the day's lesson. She scaffolded positive emotion about the problem by comparing it to a treat for which one must delay gratification, making it all the more desirable. She almost apologized for taking the challenge away, sympathizing that "it is such a neat thing."

> What I would like you to do right now, I know that you are very excited about the problem solving and I really hate. . . . .it is kind of like showing someone a piece of candy, then taking it away and saying you can't have it until later. You get so excited about the problem solving and it is such a neat thing, but I am going to ask you to set it aside . . .

While the class worked on a problem solving activity, the teacher complimented students on their thinking and varying approaches, specifically on how the student responses demonstrated their competence. As she did, she conferred expertise on the student and reinforced the value of learning as "interesting."

> You know that was really interesting Tommy, it was really interesting that you added using the decimal form. We can sometimes use the fraction form, sometimes we use the decimal and sometimes one form is easier than the other....

When the teacher ended the class, she again emphasized the importance of learning and her desire to support students' understanding. Noting that the homework had been very challenging and implying that she did not think the students had necessarily understood it, she designed another opportunity for them to master it. By giving students a second chance before she graded their work, she once again conveyed that learning was *more* important than evaluation and she expressed confidence that they could help each other.

> ...you had homework, which is investigation 2 and investigation 3. I understand that is very challenging homework and I looked at your sheets. I intend to grade it, but this is what I would like to do before I grade it. Sometimes the words get in the way and it helps if we work in groups to try and determine what the questions mean. So I am going to have you work in groups and look at your investigation 2 and 3 and talk about the answers and how you got the answers...

One way that the teacher communicated the importance of learning was by helping students take intellectual risks and to learn from their mistakes. Because of the inherent risks to students' competence, and possible negative emotions and threats to motivation, the teacher consistently provided cognitive, motivational, and emotional supports for students. When she did this publicly, as in the following examples, she communicated to the whole class that she cared about their learning and that she would not let others denigrate them. By modeling respect and engagement, she helped establish such supports among the members of the class for each other.

During a mathematics class in March, students were learning ratio and proportion. Deon disagreed that that the ratio of 3 (agree) to 2 (disagree) is not the same as 60% of a population agreeing.

Teacher: Okay, Deon, what do you think?
Deon: Because they said 60% and 3 out of 2 is not 60%.
Teacher ((to class)): Okay...do you agree with what Deon just said? Why do you disagree?
(The class laughs at the answer and the teacher scolds the class.)
Teacher: Now, now...don't be laughing. First of all, we are making comparisons and we are learning. So just listen. I liked what Deon said and I am sure that there are other people who agree with him. Go ahead, what do you think, Jerrold?

Similarly, in her prealgebra class, the teacher tried to help students take risks while supporting them emotionally. The teacher had welcomed Manuel, who had attended another school briefly, back to their class two days before this lesson in April, so he was a newcomer to the activity. During a challenging activity in which students used red and black chips to learn subtraction of integers, the teacher posed the problem, (-3)–(-7). Although a student volunteered, "I know this one," the teacher offered the problem to Manuel.

Teacher: Manuel, you want to try it or you want to pass?
Teacher: Come on Manuel you can do it.
Student: If I can he can.....
Teacher: Yeah, but he is new, I don't want to scare him to death.
Student: Don't be scared.... it is your birthday...
(Manuel says to add 4 pairs of zeros)
Teacher: He said add 4 pairs of zero, now what?... And what is your answer?
Manuel: 4
Teacher: (writes, "$-3 + 7 = +4$").... give Manuel a round of applause! Good job.

In sum, this teacher directed both her instruction and her interactions with students explicitly toward establishing the worth and enjoyment of learning. To invite students to adopt positive beliefs and to encourage their effortful engagement, she characterized activities as interesting, and students as interested in them. She used encouragement, group work, and peer interaction to support students as they took risks in the classroom. Her discourse was at once instructional and personal, acknowledging that they were a community of learners who cared about learning and cared about each other's learning.

Classroom Research on Motivation to Learn

The importance of emotion to student learning and motivation has emerged from several other studies. For example, in researching the optimal learning experiences of talented teenagers, Csikszentmihalyi, Rathunde, and Whalen (1993) asked teens to describe teachers who engaged them in challenging learning. The students described teachers who balanced teacher support with student autonomy. These teachers frequently showed their personal enthusiasm for subjects, which often seemed boring in other teachers' classes. Not only did the teachers demonstrate personal intrinsic motivation, but they also downplayed extrinsic pressures on students, using mistakes as occasions for learning.

Similarly, Stipek et al. (1998), in a study of elementary mathematics learning and motivation, described teacher practices related to students' learning and affect. These researchers found that affective climate was the best predictor of student motivation and that positive affect was associated with a mastery orientation. Teachers who scored high on positive affect were "sensitive and kind (without being artificially sweet)" (p. 478). They described them this way:

> [Teachers] showed an interest in what students had to say, listened to their ideas, avoided sarcasm or put-downs, and did not allow students to put each other down. The teachers appeared genuinely to like and respect their students. They also appeared to enjoy mathematics, and they made an effort to make mathematics problems interesting. The teachers conveyed that they valued all students' contributions by, for example, calling on students having difficulties... (p. 478)

Throughout these descriptions of teachers are the markers of emotional experiences that appeared to engage and sustain effortful learning. The "high affect" teachers described previously demonstrated similar characteristics to the high-involvement teachers from our research (e.g., Turner et al., 1998), such as their enjoyment of mathematics and respect for the students, and a personal mastery orientation and enthusiasm toward learning mathematics.

Findings from our research also support the critical role for affect in students' reports of intrinsic motivation and have linked the role of emotion to traditional goal theory (Turner et al., 1998; Turner, Thorpe, & Meyer, 1998; Turner et al., 2002; Turner, Meyer, Midgley, & Patrick, 2003). Our research suggests that motivational messages embedded in teacher discourse involves emotional, as well as social supports, which are indicators of strong perceptions of a mastery classroom goal structure and infrequent use of avoidance behaviors. Conversely, teachers' non-supportive discourse is significantly associated with reports of lower classroom mastery goal structure and more frequent use of avoidance behaviors (Turner et al., 2002).

In addition, this body of work has suggested an extension of goal theory's cognitive focus to include the role of affect. Our research (Turner et al., 2003) has demonstrated that an emphasis on learning is necessary, but not sufficient, for positive motivational outcomes. We compared teacher discourse and student reports of academic self-regulation, positive coping, self-handicapping, and negative affect after failure in two classrooms perceived as emphasizing both mastery and performance approach goal structures. Consistent with traditional goal theory, students in both classrooms reported similar levels of self-regulation and positive coping with learning difficulties. However, students reported different levels of negative affect and self-handicapping in the two classrooms. In the motivationally and emotionally supportive classroom, students reported low levels of negative emotions and avoidance. In the classroom with non-supportive discourse, students reported higher negative affect and self-handicapping. Thus, although there was an emphasis on learning in both classrooms, the affective climates differentiated the positive outcomes in one classroom and negative outcomes in the other. Our data suggested that mastery classroom goal structures are most effective when they reflect an emphasis on learning and autonomy *in conjunction with motivational, affective, and social supports*. We interpret these findings to mean that, at least for upper-elementary and middle-school students, teachers' emotional responses, both supportive and nonsupportive, are closely interconnected with student motivation.

TEACHERS' PEDAGOGIES AND PROFESSIONALISM OF EMOTIONS: LOOKING FORWARD

> In order to name, imagine and materialize a better world, we need an account of how Western discourses of emotion shape our scholarly work, as well as pedagogical recognition of how emotions shape our classroom interactions. (Boler, 1999, p. xv)

We have selected examples and research that support the notions that emotional scaffolding shapes classroom interactions, and highly effective teachers engage in it to the benefit of their students. We also acknowledge that teachers who seek to create caring classroom communities need to take the time, gain the experience, and develop the patience to perfect their pedagogy (Oakes & Lipton, 2003). Thus a central question remains unanswered: *How do teachers develop the values and pedagogies needed for emotional scaffolding?* Teacher educators and education researchers need to prioritize the search for answers to: (1) *why* understanding emotion is essential to effective pedagogy and learning, and (2) *how* emotions are defined and experienced within classrooms and the broader profession.

To move forward, teachers and researchers need theories of emotion that explain the multiple ways emotions are experienced and evolve as part of the social and historical contexts of classrooms. Researchers need to develop methods for capturing emotions in valid and reliable ways *during* learning activities and for interpreting their data within classroom and educational contexts. Researchers also need a historical perspective (i.e., longitudinal studies) to understand how pedagogy and relationships develop and change over time. Furthermore, emerging theories and new research findings will need to be translated to practice in ways that will help teachers understand the roles of emotions in their teaching and their influence on students' learning and development. Teachers need to be educated in how emotions may be scaffolded—the importance of consistency, equity, and issues of power.

Just as scaffolding has been used in the service of mainly cognitive goals, teachers also will need to develop an appreciation of emotions in teaching and learning to confront professional norms that may be barriers to effective practice. Thus, to move forward, theorists, researchers, and teachers need to understand, evaluate, and act upon the professional norms surrounding emotions in education. Many of the teachers we have found to be exemplars of emotional scaffolding are often viewed as exceptions to the rule of practice—they are funny, passionate, and emotionally engaged. In comparison, their colleagues have appeared "affectless" or negative and controlling. As Zembylas, 2003, concluded, "Teachers learn to internalize and enact roles and norms (for example, emotional rules) assigned to them by the school culture through what are considered 'appropriate' expressions and silences" (p. 119). Therefore, we need to help teachers acknowledge and change these "emotional rules" (Zembylas, 2003), which have become prescriptions for how to act or which emotions are appropriate to express (e.g., showing too much affection or too much anger). To provide useful implications for practice, our theories and research must address emotions in their professional contexts—their expression and their silences. Emotions are powerful pedagogical tools, but a culture that denies emotions to teachers also denies them to students (Oakes & Lipton, 2003). Thus the proposition that teachers' emotional scaffolding is influenced by professional context is in itself an exciting and important new area of research.

In conclusion, teachers' emotions are instrumental in their choices to become teachers (Schutz, Crowder, & White, 2001), to remain in the profession, as well as to leave it (Wilhelm, Dewhurst-Savellis, & Parker, 2000). Teachers and students are infused with emotions throughout their daily interactions and they attempt to understand these experiences to make sense of themselves as well as each other (Hargreaves, 2001). Embracing emotion as a central process in teaching and learning requires us to challenge our own intellectual and social histories regarding the definitions of emotions and their appropriateness in classrooms as well as to debate the need to control them. Within our scholarly explorations and discussions, the scaffolding of emotions presents a theoretical construct for exploring how to create positive classroom environments that engage teachers and students as life-long, passionate learners.

References

Boler, M. (1999). *Feeling power: Emotions and education*. New York: Routledge.

Connell, J., & Wellborn, J. (1991). Competence, autonomy, and relatedness: A motivational analysis of self-system processes. In M. Gunner & A. Sroufe (Eds.) *Self-processes and development* (pp. 43-77). Hillsdale, NJ: Erlbaum.

Csikszentmihalyi, M., Rathunde, K., & Whalen, S. (1993). *Talented teenagers: The roots of success and failure*. Cambridge, England: Cambridge University Press.

Ford, M. E. (1992). *Motivating humans: Goals, emotions, and personal agency beliefs*. London: Sage Publications.

Fried, R. L. (2001). *The passionate teacher: A practical guide*. Boston: Beacon Press.

Goldstein, L. S. (1997). *Teaching with love: A feminist approach to early childhood education*. New York: Peter Lang Publishers.

Goleman, D. (1995). *Emotional intelligence*. New York: Bantam Books.

Henderson, S. D., Many, J. E., Wellborn, H. P., & Ward, J. (2002). How scaffolding nurtures the development of young children's literacy repertoire: Insiders' and outsiders' collaborative understandings. *Reading Research and Instruction, 41*, 309-330.

Hamre, B., & Pianta, R. C. (2001). Early teacher-child relationships and the children's social and academic outcomes through eighth grade. *Child Development, 72*, 625-638.

Harter, S. (1992). The relationship between perceived competence, affect, and motivational orientation within the classroom: Process and patterns of change. In A. Boggiano & T. Pittman (Eds.), *Achievement and motivation: A social-developmental perspective* (pp. 77-114). New York: Cambridge University Press.

Hargreaves, A. (2001). Emotional geographies of teaching. *Teachers College Record, 103*, 1056-1080.

Hogan, K., & M. Pressley (Eds.). (1997). *Scaffolding student learning: Instructional approaches and issues*. Cambridge, MA: Brookline Books.

Linston, D., & Garrison, J. (2003). *Teaching, learning, and loving: Reclaiming passion in educational practice*. New York: Routledge Falmer.

Meyer, D. K. (1993). What is scaffolded instruction? Definitions, distinguishing features, and misnomers. C. J. Kinzer & D. J. Leu (Eds.), *Forty-second yearbook of the National Reading Conference* (pp. 41-53). Chicago, IL: National Reading Conference.

Meyer, D. K., & Turner, J. C. (2002). Discovering emotion in classroom motivation research. *Educational Psychologist, 37*, 107-114.

Noddings, N. (1992). *The challenge to care in schools: An alternative approach to education*. New York: Teachers College Press.

Oakes, J., & Lipton, M. (2003). *Teaching to change the world (2nd ed.)*. Boston: McGraw-Hill.

Palincsar, A. S. (1986). The role of dialogue in providing scaffolded instruction. *Educational Psychologist, 21*(1 & 2), 73-98.

Patrick, J., Turner, J. C., Meyer, D. K., & Midgley, C. (2003). How teachers establish psychological environments during the first days of school: Associations with avoidance in mathematics. *Teachers College Record, 105*, 1521-1558.

Ria, L., Sève, C., Saury, J., Theureau, J., Durand, M. (2003). Beginning teachers' situated emotions: A study of first classroom experiences. *Journal of Education for Teaching, 29*, 219-233.

Rogoff, B. (1990). *Apprenticeship in thinking: Cognitive development in social context*. New York: Oxford University Press.

Rosiek, J. (2003). Emotional scaffolding: An exploration of the teacher knowledge at the intersection of student emotion and the subject matter. *Journal of Teacher Education, 54*, 399-412.

Schuster, C. (2000). Emotions count: Scaffolding children's representation of themselves and their feelings to develop emotional intelligence. (electronic version) Issues in Early Childhood Education: Curriculum, Teacher Education, & Dissemination of Information of the Proceedings of the Lilian Katz Symposium (Champaign, IL, November 5-7, 2000). Retrieved March 30, 2006.

Schutz, P. A., Crowder, K. C., & White, V. E. (2001). The development of a goal to become a teacher. *Journal of Educational Psychology, 93*, 299-308.

Skinner, E. A., & Belmont, M. J. (1993). Motivation in the classroom: Reciprocal effects of teacher behavior and student engagement across the school year. *Journal of Educational Psychology, 85*, 571-581.

Stipek, D., & Daniels, D. (1988). Declining perceptions of competence: A consequence of changes in the child or the educational environment? *Journal of Educational Psychology, 80*, 352-356.

Stipek, D., Salmon, J. M., Givvin, K. B., Kazemi, E, Saxe, G., & MacGyvers, V. L. (1998). The value (and convergence) of practices suggested by motivation research and promoted by mathematics education reformers. *Journal for Research in Mathematics Education, 29*, 465-488.

Stuhlman, M. W., & Pianta, R. C. (2001). Teachers' narratives about their relationships with children: Associations with behavior in classrooms. *School Psychology Review, 31*, 148-163.

Turner, J. C., Meyer, D. K., Cox, K. C., Logan, C., DiCintio, M., & Thomas, C. T. (1998). Creating contexts for involvement in mathematics. *Journal of Educational Psychology, 90*, 730-745.

Turner, J. C., Meyer, D. K., Midgley, C., & Patrick, H. (2003). Teacher discourse and students' affect and achievement-related behaviors in two high mastery/high performance classrooms. *Elementary School Journal, 103*, 357-382.

Turner, J. C., Thorpe, P., & Meyer, D. K. (1998). Students' reports of motivation and negative affect: A theoretical and empirical analysis. *Journal of Educational Psychology, 90*, 758-771.

Turner, J.C., Midgley, C., Meyer, D. K., Gheen, M., Anderman, E. M., Kang, Y., & Patrick, H. (2002). The classroom environment and students' reports of avoidance strategies in mathematics: A multimethod study. *Journal of Educational Psychology, 94*, 88-106.

Valeski, T. N., & Stipek, D. J. (2001). Young children's feelings about school. *Child Development, 72*, 1198-1213.

Vermunt, J. D., & Verloop, N. (1999). Congruence and friction between learning and teaching. *Learning and Instruction, 9*, 257-280.

Vygotsky, L. S. (1978). *Mind in society: The development of higher psychological processes*. (M. Cole, V. John-Steiner, S. Scribner, E. Souberman, Eds.). Cambridge, MA: Harvard University Press.

Wentzel, K. R. (1997). Student motivation in middle school: The role of perceived pedagogical caring. *Journal of Educational Psychology, 89*, 411-419.

Wilhelm, K., Dewhurst-Savellis, J., & Parker, G. (2000). Teacher stress? An analysis of why teachers leave and why they stay. *Teachers & Teaching, 6*.

Wink, J., & Wink, D. (2004). *Teaching passionately. What's love got to do with it?* Boston: Allyn & Bacon.

Wood, D., Bruner, J. S., & Ross, G. (1976). The role of tutoring in problem-solving. *Journal of Child Psychology and Psychiatry, 17*, 89-100.

Yowell, C. M., & Smylie, M. A. (1999). Self-regulation in democratic communities. *Elementary School Journal, 99*, 469-490.

Zembylas, M. (2003). Interrogating teacher identity: Emotion, resistance, and self-formation. *Educational Theory, 53*, 107-127.

CHAPTER

15

Teachers' Anger, Frustration, and Self-Regulation

ROSEMARY E. SUTTON
Cleveland State University

> A first year teacher with a class of 13- to 14-year-old students sternly says, "Jessica, a reminder, I want to see you after school." Violet (Jessica's best friend), then says, "Miss you shouldn't punish Jessica, you both just lost your tempers."

This incident from my first year teaching is still vivid and contains the elements of a line of research it took me more than 20 years to begin. As a beginning teacher from a family that stressed the importance of managing intense negative emotions, I was embarrassed that a 14-year-old girl believed that I had "lost my temper," even though I had decided not to punish Jessica before hearing Violet's advice. For many years, I believed that my experiences with anger and frustration in the classroom were idiosyncratic and that other teachers did not experience them. However, after 20 years of talking to K-16 teachers and then reading *Losing Control: How and Why People Fail at Self-Regulation* (Baumeister, Heatherton & Tice, 1994), I learned that the overwhelming majority of American adults report regulating their emotions consciously and that emotional regulation is so common that we typically only notice its absence (Gross, 1998).

In this chapter I summarize what my colleagues and I have learned about teachers' anger, frustration, and attempts to regulate these emotions since we began this work in 1998. I begin by providing an overview of my current theoretical perspective of emotions and then summarize our work on teachers' anger and frustration. The next section focuses on our studies of teachers' emotion regulation, and implications for teachers are considered in the final section.

THEORETICAL PERSPECTIVE OF EMOTIONS

The dominant current social psychology theoretical perspective is that emotions are complex processes that evolved to enhance adaptation to environmental challenges. The processes consist of multiple components including *appraisal, subjective experience, physiological change, emotional expression*, and *action tendencies* (Lazarus, 2001). According to this perspective, the emotion process begins with an individual making a judgment or *appraisal* of the situation (Roseman & Smith, 2001). Appraisals are often made instantaneously and unconsciously, even in complex circumstances. However, slower conscious appraisals also occur, including reappraisals that correct the initial evaluations because additional information permits more thorough processing (Scherer, 2001). For example, a teacher who becomes angry when a student steps on her foot then reappraises the event as accidental when she sees the untied shoelace that caused the student to trip, thus reducing her feelings of anger.

Lazarus (1991) distinguished between primary and secondary appraisal. Primary appraisal involves the assessment that the situation is relevant and congruent (for positive emotions) or incongruent (for negative emotions) with one's goals. During secondary appraisal judgments about possible blame, coping potential and future expectations are made. Both frustration and anger involve the primary appraisal that the situation is relevant and incongruent with one's goals, and current research suggests that anger typically involves the secondary appraisal that someone is to blame for a blocked goal or an arrogant entitlement or unfairness (Kuppens, Van Mechelen, Smits, & De Boeck, 2003; Smith & Kirby, 2004). However, there is disagreement whether frustration involves any secondary appraisal and therefore is not an emotion (e.g., Lazarus, 1991) or involves the secondary appraisal that circumstances rather than individuals are to blame (e.g., Roseman, 2001). Teachers' frustration may be particularly important to examine in contexts that emphasize educational accountability, since frustration may be specific to cultures that emphasize goals, plans, and achievements (Wierzbicka, 1999).

Appraisals are dependent on each individual's goals, personal resources, and experiences so there are individual and temporal differences in appraisals of the "same" interaction. For example, one teacher may become angry at a student who often disrupts class, and blame that student for deliberately creating problems, whereas another teacher may become frustrated, believing that the student's behavior is caused by the current educational system that requires students to sit for long periods of time. A third teacher may become sad at the same situation, perceiving the disruptions as an indication of neglectful parents and an irreversible loss in the child's learning and development. In addition, one teacher may appraise the "same" interaction differently on various occasions and so could experience anger, frustration, and sadness.

The *subjective experience* of emotion is a distinct type of private mental state. Thus, anger does not feel like sadness or joy. Teachers who feel much anger and little joy have a different experience of teaching than those who feel little anger and much joy. The experience of emotions has been studied through the metaphors people use, and a central metaphor for anger is a boiling liquid. One teacher said, "... I was very angry, and if steam could come out of my nose and ears, it probably could [*sic*]" (Sutton & Wheatley, 2003, p. 330).

The emotion process also involves *physiological changes*, affecting body temperature, heart rate, blood pressure, and oxygen saturation of the blood. Anger is associated with physiological arousal, including increases in cardiac contractility and heart rate (Herrald & Tomake, 2002), diastolic blood pressure, and finger temperature (Cacioppo, Bernston, Larsen, Poehlmann, & Ito, 2000). *Emotional expression* includes facial expressions, and the angry face is associated with a lowered brow, tensed eyelids, penetrating or hard stare, and a mouth with lips pressed close together or parted in a square shape (Ekman & Friesen, 1975).

Some physiological changes and nonverbal expressions of emotion are consciously felt by teachers and are observed by their students. More than 20 years ago, Violet not only observed my anger but also gave me some good advice. Teachers report that students are good observers; one middle school teacher, when asked how successfully she masked her emotions, said:

> Not very ... they know ... when my body language is not saying what my mouth is saying or what my other behaviors are saying ... Sometimes they will come and ask me, are you sure you are ok? (Sutton & Wheatley, 2003, p. 331)

Emotions also involve *action tendencies*. A teacher may want to laugh out loud at a student's joke or yell at a disrespectful student. The action tendencies typically associated with anger are attacking or retaliating (Lazarus, 1991), and with frustration, overcoming an obstacle or increasing the vigor of the behavior (Roseman, 2004). However, action tendencies for frustration and anger overlap: participants in one study were as likely to say they felt like hitting someone or yelling when they were frustrated as compared to when they were angry (Roseman, Wiest, & Swartz, 1994). Action tendencies can be regulated. A teacher with the urge to yell at a student may just quietly say, "Please see me after class." Emotion regulation is important for teachers as they talk about holding in anger, keeping themselves in check, and looking at their own tone (Sutton, 2004).

The multicomponential perspective of emotions assumes response coordination among the various components during the emotion episode. For example, the subjective experience of anger is typically associated with specific facial expressions, action tendencies to strike out, and the appraisal of blaming others (Roseman, 2004). However, current evidence supporting coherence is mixed, as the size and direction of the correlations among components varies by the individual, situation, and type and intensity of the emotion (Mauss, Levenson, McCarter, Wilhelm, & Gross, 2005). A recent study reported coherence among the self-reported feelings of anger, facial expressions of anger, and the appraisal theme of injustice when recently bereaved individuals talked freely (Bonanno & Keltner, 2004). However, this study also indicated the complexity of emotional experiences, as instances of Duchenne laughter (genuinely felt laughter) were more likely to occur during appraisals of injustice than happiness or pride. The authors argue that laughter is used as a dissociation that accompanies reductions in anger, indicating that the distinction between the initial experience of the emotion and the self-regulation of that emotion can be difficult to make. I discuss teachers' emotion regulation later in this chapter.

TEACHERS' ANGER AND FRUSTRATION

A person who teaches usually acquires a temper worthy of remark, or, if he is already blessed with such a characteristic, he learns how to use it more effectively. (Waller, 1932, p. 205)

As an educational sociologist Waller (1932) was particularly interested in the use of anger as a means of establishing power and control in the classroom. At a sociological level, anger serves to uphold accepted standards of conduct (Averill, 1982), whereas from a psychological perspective anger typically arises because someone is to blame for a blocked goal (Kuppens et al., 2003) or commits a "demeaning offense against me and mine" (Lazarus, 1991, p. 222). Most of the limited research on teachers' anger prior to 2001 was embedded in broad sociological studies of beginning and experienced teachers' lives and was stimulated by changes in education resulting from school reform (Sutton & Wheatley, 2003).

These early studies indicated that frustration and anger arise from a number of sources related to thwarted goals including students' misbehavior and violation of rules, factors outside the classroom that make it difficult to teach well, uncooperative colleagues, and parents who do not follow appropriate behavior norms or are perceived as uncaring and irresponsible (Sutton & Wheatley, 2003). Teachers also become angry when they believe that students' poor academic work is due to controllable factors, such as laziness or inattention (Reyna & Weiner, 2001).

Exploring Teachers' Anger and Frustration

My first study on teachers' emotions differed from the earlier work in several ways: the focus was teachers' emotion rather then a broad understanding of teachers' lives and change; the theoretical perspective was social

1. When you think about emotions and classroom teaching what comes to mind?

2. You mentioned the emotion(s)____. Other common emotions are on the list I am giving you. Could you look at the list and tell me which 1 or 2 or 3 seem most relevant to you when teaching? (List: Anger, Fear, Sadness, Joy, Disgust, Surprise, Love/Affection)

3. Do you ever try to control, regulate or mask your emotional experiences in the class room?

4. Is there anything else you would like to say about emotions and teaching?

FIGURE 1

Anchor questions for interviews with teachers.

psychological; and teachers were asked about specific positive and negative emotions,[1] their everyday experiences, and their emotion regulation. The 30 teachers interviewed taught students of ages 10 to 15 years, and had varied experiences in the types of schools in which they taught and the number of years teaching. The majority of the teachers were European American females (Sutton, 2004). The four anchor questions in the semistructured interview are in Figure 1.

Teachers reported experiencing a wide range of positive and negative emotions, but were more likely to describe experiences of their negative emotions as frustration rather than anger: two-thirds of the teachers spontaneously talked about their frustration, but only one-third talked about their anger in response to the first interview question, "When you think about emotions and classroom teaching what comes to mind?" However, when asked, in the second question, if anger was one of the most relevant emotions, three-quarters of the teachers agreed. Some teachers talked about their difficulty with their own anger. One third-year teacher said,

> I don't like to think of anger as much as frustration. Anger is something else. They have a hard time with their own anger; they don't need to deal with the teacher's anger.[i]

No teachers talked about having difficulty with their own frustration.

Teachers reported that they most commonly get angry and frustrated when their academic goals are blocked by the misbehavior, inattention, or lack of motivation of students. For example, a second-year teacher in a suburban school district said,

[1] The terms positive and negative emotions are used by social psychologists to indicate the relationship between emotions and goals, not whether the emotions are "good" or "bad." Positive emotions (e.g., happiness, joy, love) arise from goal congruence whereas negative emotions (e.g., anger, disgust) arise from goal incongruence. According to recent theorists both positive and negative emotions are functional, e.g., anger can communicate that another person has committed a moral offense, and love communicates trustworthiness and strengthens attachment.

> If I'm really into something and everybody's doing really well [learning] and we're having a great day, and a kid just decides to get up and slide across the floor for no reason, just to get attention, that upsets [angers] me.

A teacher in his 10th year in an urban school district had a longer view of academic goals:

> So many kids go from day to day and they don't realize how important education is, not even trying . . . and it is so frustrating when you have a kid, especially for two years in a row, and you know they can do it . . . but you can't [get to] them. . . . Nothing works. . . .

Some of teachers' anger and frustration arises from demeaning offenses of students, which typically include not treating other students well, using inappropriate manners, and using profane language. For example, a fifth-year teacher in a suburban district explained an incident that triggered her anger:

> I had a student that said something to another student that was way off line . . . and I was really angry with that student because he didn't care about anyone's feelings. . . . I mean it wasn't like there was frustration than anger, it was just anger . . . He was so heartless.

The teachers' comments did not support Waller's (1932) belief that it is helpful for teachers to have a temper; rather, teachers in this study regretted losses of control and learned with experience to express less negative emotion in the classroom. The interviews confirmed the psychological research indicating that appraisals for anger and frustration include thwarted goals and suggested that "demeaning offenses against me and mine" were more likely to be associated with anger than frustration. Some teachers were uncomfortable with anger and were much more likely to spontaneously talk about their frustration than their anger. Since these findings suggested that it was important to continue exploring these negative emotions, a colleague (Paul Conway) and I decided to conduct a diary study, adapting the methodology used by Averill (1982) in his classic research comparing everyday anger and annoyance. However, we decided to compare anger and frustration, because the teachers we interviewed talked about their frustration and anger, but not their annoyance.

Focusing on Everyday Anger and Frustration

Emotion diaries are often used when studying everyday emotions and this methodology encourages participants to attend to and record daily experiences, not just those experiences salient enough to recall weeks, months, or even years later. The majority of the teachers were female, European American, and had been teaching for more than five years.

The diary consisted of five daily questionnaires (Monday to Friday), which asked respondents about the frequency and nature of anger and frustration episodes. Respondents were also asked to describe and identify various components of the most intense episode of anger and frustration they experienced each day. The emotion components included: triggers, bodily responses, intrusive thoughts, as well as immediate and longer-term actions.

The diary study added to our understanding of teachers' anger and frustration in several ways. First, the teachers were much more likely to report incidents of frustration than anger, consistent with the findings from the interview study. Detailed descriptions of 87 episodes of anger (median two per week) and 195 episodes of frustration (median seven per week) formed the basis of a series of analyses.

Second, the majority of the teachers reported that most experiences of anger and frustration were not minor, momentary feelings, but were intense, lasted more than one hour, and were associated with noticeable bodily sensations.[ii] In addition, the teachers reported that their anger and frustration led to changes in their classroom behaviors and coping strategies. The teachers said that intrusive thoughts made it difficult for them to concentrate on what they were doing before the emotion episode, that students were the immediate target of the anger and frustration, and that their planning for the next workday changed because of the anger and frustration experiences. This indicates that the classroom impact of these negative emotions is considerably longer than the specific episodes (Sutton, Genovese, & Conway, 2005).

Third, the distinctions between frustration and anger were blurred in several areas. The instructions in the diary assumed that frustration and anger were separate emotions. Teachers were asked to record details about an episode in which they felt frustrated but not angry, and also an episode in which they were angry but not frustrated. However, teachers reported that they experienced anger and frustration simultaneously for 17% of the episodes, and that frustration turned into anger (or anger into frustration) for 13% of the episodes.

Distinctions between episodes of anger and frustration were also blurred in teachers' reports of various emotion components. There were no significant differences between the anger and frustration incidents in the likelihood that respondents reported bodily responses (e.g., tenseness), intrusive thoughts that made it difficult to concentrate, immediate actions (e.g., teacher stare), coping strategies, and changes in planning (Sutton, Genovese, & Conway, 2005).

Our interest in teachers' anger and frustration focuses not on their private experiences, but on the role emotions play in classroom interactions. The findings that there were no reported significant differences in a variety of classroom behaviors, as well as the inability of many teachers to distinguish the experiences of anger and frustration, suggests that in future studies we need to be careful in devising interview questions or survey items.

One important difference in the anger and frustration episodes was in the coded appraisal of the open-ended questions. Respondents were most likely to attribute the cause of the anger and frustration incidents to others (83%), followed by circumstances (19%), and then self (6.5%). Frustration was significantly related to higher levels of circumstances, which is consistent with some views of frustration (e.g., Clore & Centerbar, 2004). However, anger was not related to higher levels of agency to others, although the appraisal that others are to blame underlies many theoretical views of anger (e.g., Roseman, 2004; Smith & Kirby, 2004). Everyday meanings of anger and frustration vary widely (Clore & Centerbar, 2004), but teachers may describe their feelings as frustration rather than anger because they believe it is more socially acceptable.

Although the type of emotion did not predict a variety of responses, the intensity of the emotion did. Teachers reported high levels of intensity: 25% of the episodes were rated 7 or 8, on an 8-point scale. More intensely felt emotions lasted longer and were more likely to be associated with bodily sensations, intrusive thoughts, immediate actions, doing something to cope, and planning different activities. Anger episodes were rated significantly more intensively than frustration episodes, but the effect size was small (Sutton Genovese, & Conway, 2005).

These findings are consistent with those of Sonnemans & Frijda (1995), who asked undergraduate students to recall emotion episodes. "Overall felt intensity" was positively correlated with emotion duration, perceived bodily changes, severity of action tendencies, the influence on long-term behavior, and efforts to regulate the emotion. (Teacher's emotion regulation is discussed in the next section.)

This study has confirmed the importance of understanding teachers' anger and frustration when studying classroom processes, especially since teachers reported that their thoughts and behaviors are influenced by the experiences and intensity of these negative emotions.

EMOTION REGULATION

> Anyone can get angry—that is easy . . . but to do this to the right person, to the right extent, at the right time, with the right motive, and in the right way is no longer something easy that anyone can do. (Aristotle, 1962, p. 50)

When individuals attempt to modify aspects of the emotion experience, the process is called "emotion regulation" (Ochsner & Gross, 2004). Emotion regulation can influence the *intensity* and *duration* of the experience, as well as *how* the emotions are expressed (Larsen & Prizmic, 2004). Americans are more likely to try to down-regulate or reduce the intensity, duration, and expression of their negative emotions than up-regulate or increase the intensity, duration, and expression of their positive emotions (Gross, Richards, & John, 2006).

Current emotion regulation research arises from a variety of theoretical perspectives (e.g., Baumeister & Vohs, 2004), and several of the current models are relevant for teachers (Sutton & Knight, 2006a). Our work has used the process model proposed by Gross and colleagues (e.g., Gross, 1998; Ochsner & Gross, 2004), which focuses on the timing of emotion regulation strategies. Gross assumes that emotions are processes that unfold over time and that emotion regulation can occur at five points between the initiation of an emotion and its expression (Gross, 1998). Four of these points—*selecting situations,*[iii] *modifying situations, attention deployment,* and *cognitive change*—are antecedent-focused or preventive (i.e., what individuals do before the emerging emotion becomes fully activated). The fifth point is response-focused and involves the *modulation of experiential, behavioral,* or *physiological emotion responses.* We have applied this model in our attempts to understand the strategies teachers use to regulate their emotions.

Emotion Regulation Goals and Strategies of Teachers

In the initial interview study described earlier in this chapter, the third anchor question focused directly on emotion regulation (Figure 1), but nearly two-thirds of the respondents spontaneously talked about emotion self-regulation before being asked about it, indicating the salience of this topic. Teachers spontaneously talked about holding in anger, gritting their teeth, stepping back and breathing, and having regrets about their losses of control (Sutton, 2004).

Middle-school teachers said that reducing their negative emotions aides their effectiveness by keeping them focused on their goal of academic learning and helping them nurture relationships with students. Twenty percent of the teachers ($n = 6$) stated that regulating their emotions was part of the teachers' role (e.g., "that was the way I was taught," illustrating an idealized emotion teacher image). The teachers spoke about their teaching with enthusiasm, passion, and humor, but the majority did not want their classrooms too "hot" with anger or too "cold" without affection, because they had learned that a temperate display of emotions was "just right," serving their effectiveness goals and/or their goal of an idealized emotion teacher image.

Teachers reported using a myriad of strategies to help them with their emotion regulation goals, and examples of the preventative categories identified by Gross (1998) were evident. For example, teachers *modified the situation* before school by being extra careful to be well prepared, so there were likely to be fewer problems; or they told the students that they were not feeling well, which was intended to influence students' behavior and make it less likely that the teachers' negative emotions would be triggered. Teachers modified the situation at the time of the specific emotion cue by the management and discipline techniques they used, including asking the students to do something quiet at their desks or telling targeted students, "See me after school as I'm not ready to deal with you now." Teachers said using these strategies not only improved the situation, but also helped them regulate their own emotions and did not interrupt their academic goals.

Examples of the *attention deployment strategy* were frequently used before school and included talking to peers or reading positive thoughts each morning. Several experienced teachers also said they had learned to ignore students' minor behavioral infractions, since this helped them regulate their emotions in the immediate situation. Teachers who talked to themselves about staying calm or reflected on their prior experiences when they had failed to regulate their emotions used the *cognitive change* strategy at the specific emotional cue. Teachers also said they learned with time to take students' behavior and comments less personally.

Teachers used a variety of *responsive strategies* to modify the emotions experienced in the classroom. These included physically moving away, taking deep breaths, controlling facial features, and "thinking of a serene place." The most common strategy used by teachers after school was talking to colleagues and friends. Other strategies included either reducing physical activity (e.g., sitting quietly) or increasing it (e.g., working out). These responsive strategies are not specific to the teaching activities and can be used by nonteachers or teachers in their nonwork lives, whereas the preventative strategies such as modifying the classroom situation seem to require experience in the classroom. Research is needed to determine if teachers learn with experience to use different emotion regulation strategies and if some strategies are more effective than others. Gross and John (2002, 2003) have shown that the college students' emotion regulation strategies vary in effectiveness, with reappraisal, an antecedent form of cognitive change, more effective than suppression, a responsive strategy. Speculating that these findings may also apply to teachers, Cathy Knight and I (Sutton & Knight, (2006a) explored the relationship between these two strategies and perceived teaching effectiveness (i.e., efficacy).

Emotion Regulation Strategies and Teacher Efficacy

People using reappraisal to modify their emotions try to change the way they think about a situation early in the generation of an emotion. Teachers using reappraisal to regulate their anger and frustration talk about stopping and thinking as well as reminding themselves that they are teaching kids (Sutton, 2004). In contrast, suppression as a form of emotion regulation typically occurs late in the generation of the emotion response and involves trying to control facial expressions and utterances. The teachers who use suppression to modify their anger and frustration report "just doing it," or forcing themselves to be calm (Sutton, 2004).

A series of studies with college students in the United States found that self-reported reappraisal use is positively correlated with positive mood, positive emotion expression, and sharing of emotions, but it is negatively correlated with neuroticism. In contrast, the use of suppression as an emotion regulation strategy is associated with feelings of inauthenticity and depression. There are also cognitive consequences for these two types of emotion regulation (Gross & John, 2002). Suppression of emotions requires continuous self-monitoring and self-corrective actions for as long as the emotion process lasts, and therefore reduces cognitive resources for other activities. In contrast, reappraisal occurs early in the emotion process and thus does not require continuous self-monitoring, which frees cognitive resources for other activities. In an experimental study, respondents instructed to use emotion suppression in a conversation were more distracted and less responsive than a control group not so instructed (Butler, Egloff, Wilhelm, Smith, Erickson, & Gross, 2003). These findings suggest that teachers who use reappraisal compared to those who use suppression should have more cognitive resources to monitor classroom activities and sustain better relationships with students.

Based on these findings and teachers' beliefs that appropriate emotion regulation makes them more effective teachers, we predicted there would be positive relationships between reappraisal and forms of teacher efficacy, and negative relationships between suppression and forms of teacher efficacy (Sutton & Knight, 2006a). Questionnaires developed by Gross and John (1995, 2003) were adapted for teaching situations and administered to 413 early childhood, middle-school, and high-school classroom teachers. Nearly 80% of the teachers were female; over half were in their twenties.

Four items assessed the suppression strategy (e.g., "I control my emotions by not expressing them in the classroom."), and six items assessed the reappraisal strategy (e.g., "When I want to feel less negative emotion [such as sadness or anger] when teaching, I change what I'm thinking about."). Emotion intensity was assessed with six items (e.g., "When I am teaching I experience intense emotions."). Three aspects of teacher efficacy were assessed: efficacy for instructional strategies, classroom management, and student engagement (Tschannen-Moran & Woolfolk Hoy, 2001). We also asked teachers if they believed they were more or less effective when they showed their positive and negative emotions in the classroom (Sutton & Knight, 2006a).

The prediction that there was a positive relationship between reappraisal and efficacy was supported with early childhood and middle-school teachers reporting higher levels of reappraisal and also higher levels of efficacy for student engagement and classroom management. These findings are consistent with the findings from my first interview study with middle-school teachers that constructive interactions with students and successful classroom management are the reasons teachers regulate their anger and frustration. The data also indicated that teacher efficacy was related to the perceived effectiveness of showing positive emotions. The prediction concerning the negative relationship between suppression and efficacy was not supported. Teachers who reported high levels of emotion intensity had an intriguing pattern of relationships. They were more likely to endorse the effectiveness of showing negative emotions, less likely to endorse suppression use, but no more or less likely to endorse reappraisal. "High" intensity teachers reported lower levels of efficacy for instructional strategies and classroom management. It may be that these "high" intensity teachers show their emotions more often in the classroom than "low" intensity teachers but feel less confident in their skills to manage or teach effectively. These findings have confirmed our earlier conclusion that exploring the self-reported emotion intensity of teachers' anger and frustration is important. Such exploration should consider both stable individual differences in emotion intensity as well as deviations from baseline within individuals (Chow, Ram, Boker, Fujita, & Clore, 2005).

Measurement Issues

We have encountered a number of measurement issues in our studies of teachers' anger and frustration. First, we have relied on self-report measures, which assess subjective components of emotions, even though emotions are multifaceted. Social psychology research on emotions is often experimental and uses multiple measures including self-report, physiological indices, and observation of facial expressions. However, obtaining facial expressions and physiological indices of teachers in classrooms is very complex and beyond our current resources. The self-report measures we used in the self-regulation survey need modification. In that study, we adapted existing measures used in social psychology with limited success, and future work should more accurately reflect teachers' reported strategies. Second, we need to develop scales that separately measure teachers' emotion regulation goals, strategies, beliefs in the effectiveness of the regulation of emotions (outcome expectancies), and confidence in regulating emotions (efficacy) that are grounded in teacher's lives. As part of this scale development, regulation of positive and negative emotions should be assessed separately, as there is much less variation in teachers' beliefs about the effectiveness of regulating positive emotions than negative emotions (Sutton & Knight, 2006a). Recent pilot data (Sutton & Knight, 2006b) indicates that the standard deviations are much higher and the curves much flatter for the negative emotions subscales than the positive emotions subscales, which are highly skewed. Developing reliable and valid questionnaires is important for future quantitative research.

IMPLICATIONS FOR PRACTICE

A teacher in his 7th year said,

> I've gotten much better at masking my emotions in the classroom. I do like to have some emotion in there. I don't want to appear like a robot; I want the students to be interested . . . to trust me and have faith in what I say. I want them to know when I'm happy and when I am not, but going too far one way or the other—I learned just by mistake, by actually doing it—it's not a good thing to do one way or the other. (Sutton, 2005, p. 232)

Although many experienced teachers say they learn to manage their emotions, especially their negative emotions when teaching, they discover how to do this on their own or with the help of colleagues, but with little or no assistance from their teacher education programs. It is important for preservice teachers to learn that their own and their students' emotions will permeate the classroom and influence their goals, motivation, and teaching strategies (Hargreaves, 2000). Nevertheless, educational psychology textbooks for preservice teachers contain chapters on learning, assessment, and motivation, but not on emotions (e.g., Ormrod, 2002).

One important implication of research on teachers' emotions is for preservice and in-service teachers to understand the current psychological view that emotions are multicomponential, an essential part of a productive adult life, and are important in understanding the goals we attain, rather than primitive and irrational. It is also important that teachers know that their peers experience intense emotions when teaching, and that emotional swings are particularly powerful for beginning teachers (Erb, 2002). It may help beginning teachers cope with their first-year teaching if they understand that novice as well as expert teachers often experience intense anger and frustration.

Beginning teachers should also know that most experienced teachers seek to regulate their negative emotions, believing that this makes them more productive. Effective teachers learn how to manage a classroom effectively and have a productive learning environment, using humor and the expression of positive emotions rather than a predominance of negative emotions (Turner, Meyer, Midgley, & Patrick, 2003). The amount of emotion regulation that is desirable is situated, varying by grade level, subject matter, and sociohistorical context.

Teachers can also learn the strategies for emotion regulation that other teachers use. These strategies include those that help to prevent the situation, help teachers keep their cool in the heat of the moment, as well as those that help them cope after the emotion episode (Sutton, 2005). Helping teachers understand the role of appraisal and reappraisal in emotions should also help them cope with the day-to-day aggravations inherent in classroom teaching.

In this chapter, I have provided an overview of my colleagues and my work on teachers' anger and frustration and regulation of these emotions. Our work is new and much more research is needed to understand the role of anger and frustration in classroom interactions. However, these initial studies already suggest important implications for teachers, and hence the growing field of emotions in education will continue to provide insights and applications for educational practice.

References

Aristotle (1962). *The nicomachean ethics*. (M. Ostwald, Trans.). New York: Bobbs-Merrill Co.

Averill, J. A. (1982). *Anger and aggression: An essay on emotion*. New York: Springer.

Baumeister, R. F., Heatherton, T. F., & Tice, D. M. (1994). *Losing control: How and why people fail at self-regulation*. San Diego: Academic Press.

Baumeister, R. F., & Vohs, K. D. (Eds.). (2004). *Handbook of self-regulation*, New York: Guilford Press.

Bonanno, G. A., & Keltner, D. (2004). The coherence of emotion systems: Comparing "on-line" measures of appraisal and facial expressions, and self report. *Cognition and Emotion, 18*, 431-444.

Butler, E. A., Egloff, B., Wilhelm, F. H., Smith, N. C., Erickson, E. A., & Gross, J. J. (2003). The social consequences of expressive suppression. *Emotion, 3*, 48-67.

Cacioppo, J. T., Bernston, G. G., Larsen, J. T., Poehlmann, K. M., & Ito, T. A. (2000). The psychophysiology of emotion. In M. Lewis and J. M. Haviland-Jones (Eds.), *Handbook of emotions* (2nd ed.). (pp. 173-191). New York: Guilford Press.

Chow, S-M., Ram, N., Boker, S. M., Fujita, F., & Clore, G. (2005). Emotion as a thermostat: Representing emotion regulation using a damped oscillator model. *Emotion, 5*, 208-225.

Clore, C. L., & Centerbar, D. B. (2004). Analyzing anger: How to make people mad. *Emotion, 4*, 139-144.

Ekman, P., & Friesen, W. V. (1975). *Unmasking the face: A guide to recognizing emotions from facial clues*. Englewood Cliffs, NJ: Prentice Hall.

Erb, C. S. (2002, May). *The emotional whirlpool of beginning teachers' work*. Paper presented at the annual meeting of the Canadian Society of Studies in Education, Toronto, Canada.

Gross, J. J. (1998). The emerging field of emotion regulation: An integrative review. *Review of General Psychology, 2*, 271-299.

Gross, J. J., & John, O. P. (1995). Facets of emotional expressivity: Three self-report factors and their correlates. *Personality and Individual Differences, 19*, 555-568.

Gross, J. J., & John, O. P. (2002). Wise emotion regulation. In F. F. Barrett & P. Salovey (Eds.), *The wisdom in feeling: Psychological processes in emotion intelligence* (pp. 297-318). New York: Guilford Press.

Gross, J. J., & John, O. P. (2003). Individual differences in two emotion regulation processes: Implications for affect, relationships, and well-being. *Journal of Personality and Social Psychology, 85*, 348-362.

Gross, J. J., Richards, J. M., & John, O. P. (2006). Emotion regulation in everyday life. In D. K. Snyder, J. A. Simpson, & J. N. Hughes (Eds.). *Emotion regulation in families: Pathways to dysfunction and health* (pp. 13-35). Washington, D.C.: American Psychological Association.

Hargreaves, A. (2000). Mixed emotions: Teachers' perceptions of their interactions with students. *Teaching and Teacher Education, 16*, 811-826.

Herrald, M. H., & Tomake, J. (2002). Patterns of emotion-specific appraisal, coping, and cardiovascular reactivity during an ongoing emotional episode. *Journal of Personality and Social Psychology, 83*, 434-450.

Kuppens, P., Van Mechelen, I., Smits, D. J. M., & De Boeck P. (2003). The appraisal basis of anger: Specificity, necessity, and sufficiency of components. *Emotion, 3*, 254-269.

Larsen, R. J., & Prizmic, Z. (2004). Affect regulation. In K. D. Vohs & R. F. Baumeister (Eds.). *Handbook of self-regulation: Research, theory, and applications* (pp. 40-61). New York: Guilford Press.

Lazarus, R. S. (1991). *Emotion and adaptation*. New York: Oxford University Press.

Lazarus, R. S. (2001). Relational meaning and discrete emotions. In K. S. Scherer, A. Schorr, T. Johnstone (Eds.), *Appraisal processes in emotion: Theory, methods, research* (pp. 37-67). New York: Oxford University Press.

Levine, L. J., & Safer, M. A. (2002). Sources of bias in memory for emotions. *Current Directions in Psychological Science, 11*, 169-173.

Mauss, I. B., Levenson, R. W., McCarter, L., Wilhelm, F. H., & Gross, J. J. (2005). The tie that binds? Coherence among emotion experience, behavior, and physiology. *Emotion, 5*, 175-190.

Ochsner, K. N. & Gross, J. J. (2004). Thinking makes it so: A social cognitive neuroscience approach to emotion regulation. In K. D. Vohs & R. F. Baumeister (Eds.), *Handbook of self-regulation: Research, theory, and applications* (pp. 229-255). New York: Guilford Press.

Ormrod, J E. (2002). *Educational psychology: Developing learners* (4th ed.). Upper Saddle River, NJ: Merrill/Prentice Hall.

Reyna, C., & Weiner B. (2001). Justice and utility in the classroom: An attributional analysis of the goals of teachers' punishment and intervention strategies. *Journal of Educational Psychology, 93,* 309-319.

Roseman, I. R. (2001). A model of appraisal in the emotion system. In K. R. Scherer, A. Schorr, & T. Johnson (Eds.), *Appraisal processes in emotion: Theory, methods, research* (pp. 68-91). New York: Oxford University Press.

Roseman, I. R. (2004). Appraisals, rather than unpleasantness or muscle movements, are the primary determinants of specific emotions. *Emotion, 4,* 145-150.

Roseman, I. J., & Smith, C. A. (2001). Appraisal theory: Overview, assumptions, varieties, controversies. In K. R. Scherer, A. Schorr, & T. Johnson (Eds.), *Appraisal processes in emotion: Theory, methods, research* (pp. 3-19). New York: Oxford University Press.

Roseman, I. R., Wiest, C., Swartz, T. S. (1994). Phenomenology, behaviors, and goals differentiate discrete emotions. *Journal of Personality and Social Psychology, 67,* 206-221.

Scherer, K. R. (2001). Appraisal considered as a process of multilevel sequential checking. In K. S. Scherer, A. Schorr, & T. Johnstone (Eds.), *Appraisal processes in emotion: Theory, methods, research* (pp. 92-120). New York: Oxford University Press.

Smith, C. A., & Kirby, L. D. (2004). Appraisal as a pervasive determinant of anger. *Emotion, 4,* 133-138.

Sonnemans, J., & Frijda, N. H. (1995). The determinants of subjective emotional intensity. *Cognition and Emotion, 9,* 483-506.

Sutton, R. E. (2004). Emotional regulation goals and strategies of teachers. *Social Psychology of Education, 7*(4), 379-398.

Sutton, R. E. (2005). Teachers' emotions and classroom effectiveness: Implications from recent research, *The Clearing House, 78,* 239-234.

Sutton. R. E., Genovese. J., & Conway. P. F. (2005 April). *Anger and frustration episodes of teachers: Different emotions or different intensity?* Paper presented at the American Educational Research Association Meeting, Montreal, Canada.

Sutton, R. E., & Knight, C. C. (2006a—in press). Teachers' emotion regulation. In A. V. Mitel (Ed.). *Trends in educational psychology* (pp., Hauppauge). NY: Nova Publishers.

Sutton, R. E., & Knight, C. C. (2006b, April). *Assessing teachers' emotion regulation.* Paper presented at the American Educational Research Association Meeting, San Francisco, CA.

Sutton, R. E., & Wheatley, K. (2003). Teachers' emotions and teaching: A review of the literature and directions for future research, *Educational Psychology Review, 15,* 327-358.

Tschannen-Moran, M., & Woolfolk Hoy, A. (2001) Teacher efficacy: Capturing an elusive construct. *Teaching and Teacher Education, 17,* 783-806.

Turner, J. C., Meyer, D. K., Midgley, C., & Patrick, H. (2003). Teacher discourse and sixth graders' reported affect and achievement behaviors in two high mastery/high performance mathematics classrooms. *Elementary School Journal, 103,* 357-382.

Waller, W.W. (1932). *Sociology of teaching.* New York: John Wiley & Sons.

Wierzbicka, A. (1999). Emotions across languages and cultures: Diversity and universals. Cambridge, UK: Cambridge University Press.

Notes

i The quotations from teachers in the article come from the interview study with 30 teachers. If the quotations have been included in previously published work, the appropriate citation is included.

ii Memories for emotions are influenced by a variety of factors including current emotion, appraisals, and intensity of the emotion (Levine & Safer, 2002). This influence appears to have less importance in diary studies where information is recorded each day as opposed to those questionnaire or interview studies, which ask respondents to recall much older memories. However the respondents in this study may have been more likely to remember emotion episodes that were more intense.

iii I do not discuss the strategy "selecting situations" further, because my focus was on the strategies teachers used when teaching middle-grade students, not how and why they had decided to teach in the middle grades.

CHAPTER

16

"There's No Place for Feeling Like This in the Workplace": Women Teachers' Anger in School Settings

ANNA LILJESTROM, KATHRYN ROULSTON, & KATHLEEN DEMARRAIS
University of Georgia

> *I think it was that I lost control. That surprised me of myself because that's just not something that I do. I experience anger. I experience disappointment but I don't typically lose control over whatever those feelings are. I can't tell you why I did it in this moment because it wasn't this huge infraction. There are much worse things that can happen in a high school classroom. And I don't know why I would have done it with this particular student.* (Julia, high school teacher)

This unsettling moment in a young high school teacher's professional life centers around the confusion she experiences as she, for a moment, experienced a loss of emotional control. Julia's words focus our attention not to the pleasant feelings involved in her work as teacher but rather to the difficult emotions, and the struggle to control these, which is a topic less visible in the educational literature. In this chapter we explore women teachers' descriptions of their experiences with anger, drawing on supporting data from interviews with teachers conducted in multiple studies led by Kathleen deMarrais. These studies have examined women teachers' anger through phenomenological interviews over a number of years.

We first review the literature on emotions, focusing on anger as it relates to the emotional lives of teachers. Second, we discuss methodological issues relevant to the study of the difficult emotion of anger. Third, we focus on a concept that pervades our examinations of teachers' narratives of anger, that of "teachers' moral purposes." Fourth, we turn to teachers' descriptions of longer-term solutions to situations that evoke highly emotive responses in teachers, including opting out of school systems altogether. We conclude with a discussion of the implications for preservice teacher education, and recommendations for further research.

LITERATURE REVIEW

Emotions as a topic of research have been widely discussed from a number of theoretical positions by researchers within psychology and sociology (cf. Boler, 1999; Denzin, 1984; Hochschild, 1983; Kemper, 1990, 2000; Williams, 2001). In *On Understanding Emotion,* Denzin (1984, p. 1) argued that people "are their emotions. To understand who a person is, it is necessary to understand emotion." He argued that emotions must be studied as lived experiences (Denzin, 1990). From a similar position, Williams (2001, p. 54) argues that this field would benefit from

> a more phenomenologically informed exploration of emotions, which enables us to bring the lived body 'back in' to our understanding of intersubjective, intercorporeal social life, including the circuits of selfhood upon which it rests and the micro-macro linkages this provides.

Drawing on social psychology and sociology, we have explored teachers' professional lives with the intent of situating their emotions within the context of schools. Specifically, we employ Denzin's (1984, p. 3) definitions:

> *Emotions* are self-feelings. *Emotionality,* the process of being emotional, locates the person in the world of social interaction. . . . All emotions refer back to the person who feels, defines, and experiences them.

Thus we locate emotions within the individual; although like Kemper (2000) and Hochschild (1983), who have proposed that, depending on social status, people have access to varying kinds of emotions, we recognize that individuals' emotional responses are embedded within sociocultural contexts and are inseparable from the contexts in which they occur.

Hochschild (1983) emphasized the appropriateness of emotions as functions of cultural expectations and theorized that "feeling rules," situated within social structures such as race, class, and gender, guide our experiences and expressions of emotions. A key concept within the sociology of emotions literature is that of "emotional labor," a term coined by Hochschild, and adopted within the field of education, particularly in the work of Andy Hargreaves and his colleagues (Beatty, 2000; Hargreaves 1998a, 1998b, 2000, 2001, 2004; Lasky, 2000). From this perspective, teachers (like social workers or day-care providers) "supervise their own emotional labor by considering informal professional norms and client expectations" (Hochschild, 1983,

p. 153). This is particularly pertinent when we consider teachers' work, which involves an ability and expectation to manage on an everyday basis a multitude of complex social interactions with others—including students, colleagues, administrators, parents, and community members.

Emotional labor has specific implications for women teachers. Traditional gender expectations in the home as well as in the workforce require women to perform a substantially larger portion of emotional labor than men (Bellas, 2000). Blackmore (1998) argued that women are often portrayed in our society as possessing innate caring and nurturing qualities that draw on common sense assumptions of stereotyped characteristics of men and women. This contributes to the "invisibility" of this aspect of the teaching profession and the important skills and effort involved in the doing of emotional work.

Although he does not focus on women teachers as a group, Hargreaves employs the idea of emotional labor (1998a, 1998b) in a conceptual framework he has developed to understand how emotions are located and represented in teachers' work and professional development. Like Tickle (1991), Golby (1996), Nias (1996), and Noddings (1996), he has argued that teachers' emotions are crucial for bonding with the students and parents, and must, therefore, be taken seriously by those involved in teacher development. Hargreaves' (1998a) theoretical framework for the "emotional politics of teaching" is based on seven important assumptions that embed individual experiences of emotions within the sociocultural contexts of schooling (p. 319):

(1) Teaching is an *emotional practice*;
(2) Teaching and learning involve *emotional understanding*;
(3) Teaching is a form of *emotional labor*;
(4) Teachers' emotions are inseparable from their *moral purposes* and their ability to achieve those purposes;
(5) Teachers' emotions are rooted in and affect their *selves*, identities, and relationships with others;
(6) Teachers' emotions are shaped by experiences of *power* and powerlessness; and
(7) Teachers' emotions vary with *culture and context* (italics in original).

Hargreaves (1998a, pp. 330–333) has emphasized that the particular structures and work conditions of teachers have a profound impact on their emotional state; and suggested that school environments, often complicated by internal issues, such as overcrowding, and external political mandates, can be contexts in which teachers are likely to experience emotions that may negatively impact interactions with others.

Literature concerning difficult emotions, such as anger, in school contexts is typically not research-based, but includes recommendations for teachers to deal with angry students or parents, as well as ways to prevent violence and aggression in classrooms and schools (e.g., Marion, 1997). A few empirical studies, however, have focused on teachers' emotional lives, although not explicitly on anger. For example, Golby's study (1996) includes teachers' descriptions of how they managed negative emotions in their work, in particular annoyance, irritation, and anger. Golby found that teachers learned to control their anger, viewing professionalism as controlling one's emotions to effectively work within the institution. Golby suggested that the strong need to be in control might result in a restricted set of emotional responses in the workplace. Tickle (1991), in a study of beginning teachers, found that teachers' emotions were often the focus of their discussions as they met for their weekly debriefings. With findings similar to Golby, he theorized that these new teachers perceived emotional management to be a pertinent aspect of growing into their professional role. Dorney (2000) has examined women's expressions of anger in stories told at two retreats, undertaken as part of action research projects to intervene in girls' and boys' education. In these narratives, Dorney (2000, p. 4) found anger to be "a theme" identified as an "issue in personal relationships" as well as "very present in [the women's] work."

While there is a growing body of literature detailing the emotional aspects of teachers' work and teachers' lives, there has been little work that has dealt specifically with the emotion of anger as experienced by teachers (see Sutton & Wheatley, 2003). Over a number of years, Kathleen deMarrais has led a number of investigations of women teachers' lived experiences of anger in school settings. In these studies, 49 women teachers ranging in age from their early 20s to late 50s in two states within the United States have been interviewed. Eleven of the women were in their first four years of teaching, eighteen were experienced teachers with between five and fifteen years of experience, and twenty of the teachers had between fifteen and thirty years in the profession. The teachers interviewed taught in all levels of K-12 schooling, in different disciplines, and in different contexts (rural, suburban, and urban). Participants were selected for maximum variation in age, ethnicity, and experience. These interview studies have used a phenomenological perspective with the purpose of examining "human experience in close, detailed ways" (deMarrais, 2004, p. 56). In the next section, we discuss methodological issues relevant to the study of difficult emotions, such as anger.

METHODOLOGICAL ISSUES

We have used phenomenological interviews to study teachers' anger (deMarrais, 2004; Kvale, 1996), since this type of interview generates in-depth, first-person descriptions of particular human experiences, such as anger (see also van Manen, 1990; Moustakas, 1994). From a phenomenological perspective, the researcher examines "everyday human experiences in close, detailed ways" (deMarrais, 2004, p. 56). In the phenomenological interview, "the researcher assumes the role of learner in that the participant is the one who has had the experience, is considered the expert on his or her experience, and can share it with the researcher" (deMarrais, 2004, p. 57).

To focus on the topic of "anger" in a research interview, the researcher asks participants to describe specific school- and classroom-related situations in which they have experienced anger. The primary interview question for our studies of teachers' anger is: "Think of a time when you experienced anger in a school or classroom setting and tell me about it." This question has been an effective means of generating narratives of teachers' experiences of anger in school contexts. With a focus on difficult emotions, however, this kind of interview may result in emotional responses on the part of both the researcher and participant (deMarrais & Tisdale, 2002); and for some participants, describing their experiences has resulted in reliving them. Yet, with an ethical and respectful approach, interviewers can focus the discussion through follow-up questions or probes that use participants' own words to elicit specific details about the experience. For example, after an initial description of one participant's experience of anger, deMarrais asked Julia the following question, eliciting an in-depth description. Julia's response incorporates a description of the physical experience of anger and her purposes and expectations as a teacher. As well, the response shows her reluctance to label her response as anger, substituting the term "disappointed."

KD: And you mentioned that you were vibrating. I wonder if you could kind of get back into that moment and tell me what that was like for you when you were experiencing that anger.

J.: I don't get extremely angry very often, but I know when it's happening. I have high expectations of people that I'm close to, including my students, and I'm typically disappointed when they don't meet those and it takes me a while to say, "Well, they're kids." It takes me a while to step back from that expectation. So anyway, physically, when I get extremely angry what it feels like to me is that the cells all over my body are just shaking. That's just a very strange sensation. I also don't often lose control in situations like that.... but I can remember that same sensation. When I'm disappointed in something that's happened, it's a physical sensation of my whole body just vibrating.

Phenomenological interviews are often lengthy, and it may take from one to four hours for participants to provide in-depth descriptions of two or three incidents of anger. Sometimes, follow-up interviews are necessary to clarify details provided in the first interview. Prior to analysis, interviews are audio-recorded and transcribed verbatim.

One approach to the analysis of interview transcripts is that of analysis via a research group, as we have done. As members of a research group, we meet regularly to collectively analyze successive rounds of interview data. Our research groups have been composed of three to five members, and have included experienced and beginning researchers. Each group member is responsible for initial analysis of a portion of the data set. Meeting on a weekly basis, group members compare analyses, discuss findings and emerging patterns, as well as search for discrepant cases in the data. After sharing understandings of categories and assertions, group members come to a consensus on their interpretations of the data.

Conventional definitions of validity call for researchers to provide statements concerning how data analyses accurately represent reality. With respect to teachers' recollections of anger incidents in interview settings, we can make no assertions as to what *actually* happened in each of the events described. What we can claim, however, is that for these teachers, their recollections speak to their personal responses and perspectives of these events. Reliability as applied to qualitative research methods refers to the consistency with which different researchers would come to similar interpretations of the data. In our examinations, this was achieved via a subjective and iterative approach to data analysis, in which members of each research group supported assertions with excerpts from the data sets, searched the data set for negative instances, and achieved consensus on interpretations. For example, one pattern apparent across our data set was that teachers' descriptions of larger incidents often included many smaller anger episodes that complicated their experiences. We found that data analysis shifted from an examination of the interview as a whole to a close examination of each small episode.

This recursive approach to the analysis of teachers' narratives revealed that teachers' descriptions of their anger experiences were intimately connected to teachers' expressed beliefs and assumptions about teaching, schooling, education, and students' welfare. Hargreaves' (1998a) assertion proved to be worthwhile examining in further detail: "Teachers' emotions are inseparable from their *moral purposes* and their ability to achieve those purposes" (p. 319). Below, we discuss in detail the ways in which teachers' moral purposes are tied to their experiences of anger.

TEACHERS' MORAL PURPOSES

Moral Purposes and the Role of the Teacher

What makes teachers angry? In our analyses, teachers' moral purposes were inextricably intertwined with their descriptions of anger experiences. Teachers were clear in what they proposed to achieve in their work, and many times expressed anger when they perceived that they were impeded from attaining their purposes. Obstacles came from many directions—colleagues, administrators, parents, students, or society at large.

Teachers expressed strong views about purposes in regard to their teaching role. All the participants described "good" teachers as dedicated and hardworking. For example, Cleo, a high school teacher, talked of teaching as more than an "eight-to-five job"; it was a life style in which you "live education" and "cry, laugh, and be with students, be a part of the community." For many of the teachers, the act of teaching was about the "rapport you can develop with your students" and the relationships that develop. New teachers, in particular, spoke of their dedication and love for students, and how they wanted to create change. For example, Zoë, a high school teacher, reported beginning her teaching career with the intent of change in a difficult school with many social problems. In discussing her attitude towards teaching, she reported a lack of solidarity from fellow teachers and administrators to her outspoken views during her first year of teaching. Ignored by colleagues, she retreated to her classroom to work with her students.

> I was focusing just on my students but it took away that desire in me to be an agent of change, which is what I really wanted to be. You know, I wanted to be revolutionary. I wanted to change things and especially the way I saw things going in this particular school.

Julia, a high-school science teacher, also described her intense involvement with her students. In her self-description, Julia was explicit about her 'stand-in' role for the parents of her students.

> That's part of my conception of my role as a teacher. Now if you were to ask my parents, they would say I was too involved. I mean, I would make it a point to go to as many football, baseball, basketball games, tennis, plays, as I could. I honestly felt that, in many of the cases, their parents weren't going to be there sitting in the stands. Maybe they were working, maybe they were just too busy. Maybe they just didn't care. But I thought if they could see consistently a face that they recognized and know that the person actually cared about what they were doing outside of class as well, that that would matter.

Although not all the teachers compared their work to that of parenting as Julie, many of the teachers spoke of their responsibility for students' holistic development, describing themselves as protectors of the students, and, in many cases their confidantes. Krystal, a high school teacher, described her work as providing students with important "life-skills," which would make a difference in their lives.

In our examinations of teachers' narratives, we have found that when teachers feel themselves to be obstructed in the pursuit of their goals, whether it is developing relationships, caring for the "whole" child, enacting change for social justice, or giving students life skills, anger, albeit expressed in multiple ways, commonly results. For example, Kathy, an elementary school teacher, expressed the following: "I am truly an advocate of quality education and it makes me angry when I see students not getting the education that they should."

As mentioned, anger resulted when teachers sensed that they were impeded in carrying out their moral purposes. Yet, the experience of anger itself often caused further moral qualms. Anger was a complex and problematic topic for many of the teachers and some teachers spoke explicitly about not expressing anger at their students or in their presence. For teachers, exhibiting anger in the classroom provides an unhealthy model to students. Kathy, for example, commented:

> I just didn't go into the classroom with a lot of anger because I felt like if the students saw you angry then often times you are modeling part of those students and often they would just emulate your behavior.

Carrie reserved her anger for adults, not children. She expressed some wonderment at certain parents' actions: "How in the world could a parent who has a darling child like this let that child come to school, no food, dirty, what have you—that made me very, very angry and upset."

In instances where students, and particularly young children, do not have access to adequate services, some teachers reported becoming especially frustrated and angry. For example, Lauren, an elementary teacher, described one experience of anger where one of the more difficult students in her class did not have access to adequate health care. Her anger was not directed at the student, who she perceived to be "suffering," but the welfare agencies to which she had referred his case.

In their descriptions, many teachers in our studies portrayed themselves as champions for their young charges, and expressed frustration when adequate resources were not available to help them in their care and support for the "whole" child. Such accounts indicate that teachers are concerned not merely for the curriculum content delivered in the classroom setting, but also for the health and well-being of children and young people entrusted to their care.

Owning Anger, Others' Anger

Although teachers frequently described a reluctance to express any anger in front of or at students, students were expected to be "team players" and to respect and learn from their teachers. This was most notable in the case of high school students. Krystal, a high school teacher, outlined her classroom management style as one where her students were expected to follow the rules. When they chose not to, they were excluded from class. In some instances when students expressed anger and resentment, she responded by ignoring them. In one example she talked about, an excluded student became furious and walked out. Krystal said:

> Then what got me angry is that some of the kids in the class were angry because I had done that. One little girl said, "Ms. Krystal, that wasn't such a big deal, his talking. Why couldn't you just let him off?" and I said, "He knows my rules and you know my rules and if it wasn't such a big deal, why didn't he back off? Why is it that you expect me as a teacher, the one who has to put up with thirty-five kids in the class, why is it that you expect me to back off, don't you think that's his place, if he felt that I was being unfair my door is always open, you can always come back to me after school, but I will not argue with him, I will not argue with you during class time." I felt better when I said that, especially when the rest of

> the kids in the class started clapping, they agreed with me then, they knew that he was out of place.

In this description, Krystal's authority as teacher is affirmed and applauded by other students who "agreed" with her actions. The example provided here by Krystal shows how teachers not only confront their own experiences of anger, but those of others.

While Krystal provides an example of how a teacher might deal with infractions of classroom rules, and confront challenges from students, other teachers excused students' infractions under certain circumstances. For example, students who came from difficult home situations, had been abused, or were hungry, dirty, and uncared for were excused by their teacher for "off task" behavior. With the identification of these physical and social characteristics of students, some teachers perceived it to be the teacher's responsibility, or moral duty, to assist and help the student and their family as part of their professional roles. A high school teacher, Christie, for example, reported that *any* misbehavior on the part students was her responsibility, taking this approach to an extreme. Describing an experience where she publicly reprimanded a student for calling out correct answers during a test, she took full responsibility for what she described as an inappropriate and angry response on her part.

> I was angry with him at that instance and then I have been incredibly angry with myself and ashamed that's what I am. . . . Two years later I am still crying about it.

Other teachers speak of modeling appropriate emotion management strategies to handle their students. For example, Zoë noted that dealing with a group of problematic high school students, she "would talk about my feelings with them, you know, and say I want to tell you where I am so we're going to do things this way." Susan also noted that she modeled appropriate emotion management. She said that "some children have not learned to deal with anger in a rational way—we can find a way to do it together."

Other teachers admitted to expressing their anger to students on occasion. Cleo, for example, spoke of dealing with one student's anger quite publicly. In this incident, an angry student defied her authority in class. She said,

> One time, I guess she had a fight with her boyfriend. I wasn't sure why this girl was so angry. Bad attitude. . . . She blurted, "Yeah I've got a complaint. I think you and your class are stupid."

Cleo handled the incident by publicly sparring with the student, explicitly punishing what she perceived to be unacceptable in terms of emotional behavior in her classroom to set an example in front of the rest of the class:

> I said, "You think I'm stupid? I'm on this power trip? You're so informative about me, why don't you tell us about yourself? Are you stupid?" She said, "Of course not." I asked the rest of the class, "Do you think she's stupid?" They all went, "Yeah, we think she's real stupid." By then, it was almost like a gang up on her.

Similarly to the excerpt from Krystal above, in her narrative, Cleo highlighted the preservation of authority and respect associated for the teacher role. Krystal repeatedly noted that she took no responsibility for the consequences of her students' actions, and in several examples made no attempt to control students' expression of emotions. The position that Krystal articulated in terms of her students was clear: teachers, administrators, and students all have certain responsibilities. Krystal did not speak of excusing students from responsibilities for their actions and emotions regardless of their personal and social circumstances, unlike many other teachers whose accounts we have examined.

In contrast to Krystal, many teachers reported using numerous strategies to excuse the behaviors of others, especially students. The strategies ranged from prophylactic strategies of "moving, walking," or leaving the classroom to regain control, and were often aimed at helping teachers avoid their own feelings of anger. Teachers experienced discomfort and qualms when their anger was triggered, and considerable energy appeared to be expended at avoiding expressing or even feeling anger. Anger was, in these teachers' view, incompatible with being a good teacher, and the teachers often reported assuming full responsibility for handling their own and the emotions of others as part of their role as teacher.

For the women we have interviewed, their expectations of professional roles as teachers were closely intertwined with their moral purposes. The context of these teachers' work, and the power structures inherent in schools, often posed serious threats to teachers' objectives; however, in interview data these factors are explicitly linked to teachers' expressions of intense emotional distress and anger. The structural constraints mentioned by the teachers were connected to their position in the school and community and often entailed the length of experience and time spent in the context, but their positions were also complicated by race and gender. Learning to negotiate the possibilities for carrying out their teaching purposes appeared to be an important part of these teachers' professional lives. This process of learning to navigate a system that was not always compatible with their own moral purposes and expectations was painful for many teachers. Caught in a double-bind, women weighed the pursuit of their moral purposes as teachers, while still keeping students' best interests at the center. Teachers expressed different ways of dealing with these constraints, which we explore in the next section.

LONG-TERM SOLUTIONS

Some teachers made explicit references to how they came to compromises or resisted impediments to their professional role. Lauren, for example, explicitly talked about how she tried to change a system, which she believed short-changed her moral purposes. For example, she bombarded the central office with letters about students' problems and took her complaints to the county central administration office as well as constantly reporting. Lauren indicated that she refused to let her teaching purposes be impeded and continued to pursue what she perceived to be unfair treatment of students, although the efforts were not always appreciated by administrators. However, Lauren was the only teacher we have found who reported taking this very active approach to complain about the injustices and problems she perceived.

Margaret, an elementary special education teacher, reported a different approach. Margaret was promised a class size of four hearing-impaired students, but then, unexpectedly, had yet another student added to the class at the same time she lost her experienced teaching assistant. Margaret described intense anger at this breach of agreement as she felt that the quality of instruction would be severely damaged and, in the extension, the students' educational rights impeded. Margaret did not mention ever contemplating openly expressing her anger to the administrator regarding the new arrangements. Instead, Margaret chose a more subtle, psychological resistance to school duties when meeting with her administrator. She said,

> I've not been as cooperative as I normally am, I don't volunteer for little extra assignments like I used to do . . . instead of being the concerned teacher as I normally would have been, saying, "These students are on this level, these three are on his level . . . "; I said, "I want my original four—the two high ones, the two low ones, not the two new ones." She [the administrator] looked at me real funny, because that is not my normal way.

Here, we see Margaret's attempt at resistance towards the administrator whom she had identified as the source of the traumatic changes in her working conditions. But, at the same time, she also admitted the futility of this approach in the long term, commenting:

> No, I'll go back. I don't have any way to protest, and this is about the only way that I can make a protest without being ugly, or writing the letters . . . I will probably go back to my old ways of volunteering, I'll do it. I'll assume my responsibility.

Margaret's perceived moral responsibility towards her students put her in a double bind. First, this care of her students' well-being was implicated in her initial anger, since she perceived that they were the ones who would suffer from the administration's reallocation of resources. Second, caring for students was also the reason that she could not continue her protest. In effect, the actions Margaret described above were merely tokens for the moment rather than factors for real change.

Other teachers, after assessing their place in the system, expressed that their moral purposes were doomed to failure. To avoid further pain they decided to "opt out" in a more concrete sense. For Alice, this opting out meant a decision to leave regular classroom teaching to pursue a career in special education. She stated,

> I didn't like the climate. The parents controlled the things you said to the kids and the things you could write on a piece of paper . . . a lot of pandering towards parents. Making concessions about kids when you know in your heart that you are doing the wrong thing. I couldn't do that. Special Ed has a lot more flexibility.

Zoë, similarly, felt devalued and disrespected by the administrators with whom she worked. She perceived that accomplishing her moral purposes in the teaching profession to be impossible. In addition, she talked of feeling "pushed out" of classroom teaching by an administrative team that did not value her pedagogy or respect her teaching purposes. In a particular instance as the leader of the student congress, she described her work as "babysitting" a very privileged group of students who refused to do the assigned work—and with the principal's blessing. The ensuing anger she experienced, together with her sense of powerlessness in contributing to positive change, proved unbearable to Zoë. She said,

> It made me even angrier because that particular role [facilitator of student congress] took me away from my teaching . . . I mean I had planned on coming back to graduate school, that this is good—I'll stay in teaching for a couple more years, three more years before I go back. But because of that experience and because I saw them pushing me towards administration, I got out. They were putting me in administrative roles when I wanted to teach.

Zoë decided that she had to opt out of the teaching profession all together as she could see no room for the pursuit of her goals within the structures of the school. Zoë went into teaching with a strong desire to make a difference via critical pedagogy, and specifically chose to work within a less privileged school. She felt confused and sad when it appeared that the administrators had different objectives and did not support her practices.

Although not every teacher takes such drastic measures as to leave the profession, the teachers in our studies described different survival strategies including distancing themselves from the physical context and/or psychologically distancing themselves from their work. Although some teachers indicated that this initial withdrawal from their usual dedicated teacher selves was a temporary solution to deal with immediate anger, we can only speculate what these perceived injustices' long-term effects might be in their professional lives.

CONCLUSIONS

In this chapter we have discussed how women teachers frequently describe anger in relation to unfulfilled moral purposes as professionals. As reported by teachers in a series of interview studies of women teachers' anger, being professional involves both emotional management and avoidance of the experience of anger. In cases where it was not possible to avoid expressions of anger, anger was nevertheless perceived to be incompatible with the professional role of teacher. This emotional management amounts to considerable work, or "emotional labor" (Hochschild, 1983). These women teachers spoke of adopting particular pedagogical practices to avoid classroom situations that might anger them, for example, by being well-prepared and also by deepening their understanding of students' backgrounds and other circumstances. Unlike Hargreaves (2000), we found that the high school teachers in our study did not report that they were less likely to express anger to students and displayed no greater emotional "distance" than elementary teachers (p. 823). As we have shown, high school teachers sometimes expressed anger to students as they viewed students able to assume responsibility for their actions. Teachers, in addition, often took it upon themselves to manage their students' emotions through a variety of strategies. When facing the wrath of students, some teachers reported modeling appropriate emotional management; others spoke of undertaking a type of "pedagogical diagnosis"—in effect, taking the "temperature" of the student and/or the classroom to find the underlying causes of problems.

Our examinations of teachers' anger further supports earlier suggestions by Hargreaves (1998a; 1998b; 2000) in that the particular structures and work conditions of teachers have a profound impact on their emotional state. Teachers often mentioned the school contexts as restricting possibilities for emotional expression as well as affecting their emotional state. In addition, teachers saw these contexts as being complicated further by race and gender. New, young teachers, in particular, reported that they found themselves at the bottom of a hierarchical structure, and felt restricted as they were learning appropriate ways of expressing their emotions.

Some teachers in our interview data spoke of guilt arising from their own anger reaction to students' inappropriate behavior. It appears that anger is a particularly complicated emotion to deal with, since it *always* involves interactions with others. Furthermore, anger is intense and complex, as well as personal, and some teachers found it difficult to talk about their anger. Interestingly enough, we saw a "clustering" of emotions—disappointment, frustration, fury, rage, surprise, shock, anger, guilt, sadness, shame, and fear—in these teachers' descriptions of anger experiences. Teachers appeared to be unable to separate these difficult emotions and they often occurred at the same time. Fear, as a topic in these descriptions, was often of the very expression of anger itself, since teachers often talked of the fear of being out of control.

Teachers dealt differently with the challenges to their moral purposes, and it appeared that these emotional experiences had important consequences for their professional lives. In fact, some teachers reported making major career decisions based on a traumatic anger experience, such as altering their level of professional involvement, or changing schools or area of instruction. For teachers who came to believe that their moral purposes could no longer be achieved within the constraints of school systems, the decision to leave teaching appears to be inevitable. Nine of the 23 teachers who provided narratives of anger experiences that had occurred in their first three years of teaching spoke of decisions to leave schools, leave teaching altogether, or change the focus of their work as a direct result of the incidents described in interviews. This finding is pertinent in light of estimates that as many as 50% of beginning teachers in the United States leave the profession in the first five years of teaching (Smith and Ingersoll, 2004, p. 682).

IMPLICATIONS

Implications from our examinations of teachers' anger fall into two areas: (1) practice, and (2) further research. First, within preservice education, there has been little systematic treatment of the emotional nature of teaching and learning. Our data suggest that teaching and learning are intensely emotional experiences, encompassing the full gamut from joy to rage. Given the data presented in this chapter, emotions have a central role in teachers' work, since emotional experiences were, at times, described by the teachers as having serious repercussions. We urge teacher educators to deal in explicit and proactive ways with issues relating to emotions, preparing teachers for the possibility of experiencing difficult emotions. The teachers in this study talked about how they were completely surprised and even felt frightened by their own emotional reactions. This in turn led to feelings of loss of control. We believe, therefore, that an awareness of emotional reactions may help teachers in making decisions that are beneficial to themselves and others. Examining and identifying their specific moral purposes of teaching within the sociopolitical context of schools, and learning to expect that difficult emotions will arise around deeply held beliefs when these are challenged, may help new teachers prepare for their induction into the profession.

Second, administrators, and particularly principals, were cast as central figures in many of these teachers' descriptions of anger experiences. Teachers reported that it was of utmost importance that they had support in difficult situations with students and parents, as they often felt isolated and alone on "the battle field." It seems, then, that it is important for administrators to be aware of teachers' emotional lives to support them in their work and the fulfillment of teaching purposes. This point echoes Hargreaves (1998a, 1998b), in that teachers' purposes, as expressed in this study, often were connected to having sufficient time and material resources as a measure of quality instruction. In addition, we suggest that teachers need to feel safe to express their purposes and emotions in school contexts (see also Dorney, 2000).

Finally, we believe that teachers' emotional responses when dealing with students' emotions need to be further explored, discussed, and theorized. We suggest that two areas for research deserve more attention: teachers' emotional lives and teachers' moral purposes and how these areas intersect. Research needs to be conducted into (1) teacher attrition and anger; and (2) beginning teachers' stories of emotional experiences. Given the feminist discourses concerning "caring and sharing which refer only [to] the upsides—trust, service, dedication, patience and love" (Blackmore, 1996, p. 348), it would seem timely that those emotions of teaching deemed to be negative (anger being but one), be explored further. In our data, there are suggestions that teachers who feel consistently thwarted in their moral purposes of teaching and experience negative emotions may be at risk of leaving the profession. In light of current pressing concerns in the United States and elsewhere regarding teacher attrition, we believe that further attention to these issues is of utmost importance. The teachers in our study complicate traditional assumptions that when supported, care and high "emotional intensity" of classrooms necessarily benefit teachers as well as students (Hargreaves, 2000, p. 823). What is missing in this assumption is a clearer separation and examination of different emotions and teachers' experiences of these. Our findings included many descriptions of the anxiety and guilt that complicated teachers' possibilities for expressing anger and other difficult emotions, constraints that were perceived to be contextual as well as personal. Researchers may have to question whether encouraging "emotional intensity" will help teachers improve the quality of instruction as well as their own job satisfaction. Like Acker (1992), we recognize the "irony" in teachers' lives that they cannot afford *not* to believe they have the power to "shape their own destinies," but yet cannot risk a "full-blown feminist analysis" in the current society (p.159). We suggest that further examination of how the structure of schools plays out in teachers' emotional lives is needed, because the teachers in this study reported that race and gender often posed additional challenges when carrying out their work

We believe teachers' work to be difficult. Passionate teachers are necessary, but perhaps not sufficient, to do this work. We also believe that effective teaching and learning in even the most difficult situations are possible when adequate support in terms of material resources, leadership, and professional training and development is provided. Working with and for teachers to improve our understanding of the emotional labor and the moral purposes involved in the act and art of teaching is one step towards this possibility.

References

Acker, S. (1992). Creating careers: Women teachers at work. *The Curriculum Inquiry*. *22*(2), 841-863.

Beatty, B. R. (2000). The emotions of educational leadership: Breaking the silence. *International Journal of Leadership in Education, 3*(4), 331-357.

Bellas, M. (2000). The gendered nature of emotional labor in the workplace. In D. Vannoy (Ed.), *Gender mosaics: Social perspectives*. Los Angeles: Roxbury.

Blackmore, J. (1996). Doing 'emotional labour' in the education market place: Stories from the field of women in management. *Discourse: Studies in the cultural politics of education, 17*(3), 337-349.

Blackmore, J. (1998). The politics of gender and educational change: Managing gender or changing gender relations? In A. Hargreaves, A. Lieberman, M. Fullan & D. Hopkins (Eds.), *International handbook of educational change* (Vol. 5, pp. 460-481). Dordrecht: Kluwer Academic Publishers.

Boler, M. (1999). *Feeling power: Emotions and education*. New York: Routledge.

deMarrais, K. (April, 1994). *Angry voices in the workplace: Teachers tell their stories*. American Educational Research Association Meetings, New Orleans, LA.

deMarrais, K. (2004). Qualitative interview studies: Learning through experience. In K. deMarrais & S. D. Lapan (Eds.), *Foundations for research: Methods of inquiry in education and the social sciences* (pp. 51-68). Mahwah, NJ: Lawrence Erlbaum Associates.

deMarrais, K. & Tisdale, K. (2002). What happens when researchers inquire into difficult emotions?: Reflection on studying women's anger through qualitative interviews. *Educational Psychologist, 37*(2), 115-124.

Denzin, N. K. (1984). *On understanding emotion*. San Francisco: Jossey-Bass.

Denzin, N. K. (1990). On understanding emotion. The interpretive-cultural agenda. In Kemper, T. D. (ed.), *Research agendas in the sociology of emotions* (pp. 85-116). Albany, NY: State University of New York Press.

Dorney, J. (2000). "Turning anger into knowledge": Exploring anger and advocacy with women educators. (ERIC Document Reproduction Service No. ED 444 983).

Golby, M. (1996). Teachers' emotions: An illustrated discussion. *Cambridge Journal of Education, 26*(3), 423-434.

Hargreaves, A. (1998a). The emotional politics of teaching and teacher development: with implications for educational leadership. *International Journal of Leadership in Education, 1*(4), 315-336.

Hargreaves, A. (1998b) The emotional practice of teaching. *Teaching and Teacher Education, 14*(8), 835-854.

Hargreaves, A. (2000). Mixed emotions: Teachers' perceptions of their interactions with students. *Teaching and Teacher Education. 16*(8), 811-826.

Hargreaves, A. (2001). Emotional geographies of teaching. *Teachers College Record, 103*(6), 1056-1080.

Hargreaves, A. (2004). Distinction and disgust: The emotional politics of school failure. *International Journal of Leadership in Education, 7*(1), 27-41.

Hochschild, A. R. (1983). *The managed heart: Commercialization of human feeling*. Berkeley: University of California Press.

Kemper, T. D. (Ed.). (1990). *Research agendas in the sociology of emotions*. Albany: State University of New York Press.

Kemper, T. D. (2000). Social models in the explanation of emotions. In M. Lewis & J. M. Haviland-Jones (Eds.). *Handbook of emotions* (2nd ed.) (pp. 45–58). New York: Guilford Press.

Kvale, S. (1996). *Interviews: An introduction to qualitative research interviewing*. Thousand Oaks, CA: Sage.

Lasky, S. (2000). The cultural and emotional politics of teacher-parent interactions. *Teaching and Teacher Education, 16*, 843-860.

Marion, M. (1997). Guiding young children's understanding and management of anger. *Young Children, 52*(7), 62-67.

Moustakas, C. (1994). *Phenomenological research methods*. Thousand Oaks, CA: Sage Publications.

Nias, J. (1996). Thinking about feeling: the emotions in teaching. *Cambridge Journal of Education, 26*(3), 293-306.

Noddings, N. (1996). Stories of affect in teacher education. *Cambridge Journal of Education, 26*(3), 435-447.

Smith, T. M., & Ingersoll, R. M. (2004). What are the effects of induction and mentoring on beginning teacher turnover? *American Educational Research Journal, 41*(3), 681-714.

Sutton, R. E., & K. F. Wheatley, K. F. (2003). Teachers' emotions and teaching: A review of the literature and directions for future research. *Educational Psychology Review, 15*, 327-358

Tickle, L. (1991). New teachers and the emotions of learning teaching. *Cambridge Journal of Education, 21*(3), 319-329.

van Manen, M. (1990). *Research lived experience: Human science for an action sensitive pedagogy*. London, Ontario: State University of New York Press.

Williams, S. J. (2001). *Emotion and social theory: Corporeal reflections on the (ir)rational*. London: Sage.

CHAPTER

17

The Power and Politics of Emotions in Teaching

MICHALINOS ZEMBYLAS
Open University of Cyprus

Emotions are increasingly the focus of analysis across a number of disciplines and professional settings, including education. To date, emotions have largely been investigated through a range of scientific, biomedical, and psychological discourses that consider emotional phenomena primarily as "individual" and "private" (Lupton, 1998). The assumption that emotions are individual and private phenomena (e.g., Ekman & Rosenberg, 1997; Plutchik, 1980) is considered problematic because it draws heavily on presupposed psychological and biological categories; the social and cultural aspects are hardly ever present in studies that focus on emotion as an intrapersonal characteristic. However, since the early 1990s, this position has been heavily criticized and alternative views have been presented branching out into the sphere of culture and politics (e.g., Cornelius, 1996; Lutz & Abu-Lughod, 1990).

Nowadays, emotions are increasingly being recognized as part of everyday social, cultural, and political life. Emotions in the classroom, for example, are not only a private matter but also a *political* space in which students and teachers interact with implications in larger political and cultural struggles (Albrecht-Crane & Slack, 2003; Zembylas, 2005a). The notion of "politics" here refers to "a process of determining who must repress as illegitimate, who must foreground as valuable, the feelings and desires that comes up for them in given contexts and relationships" (Reddy, 1997, p. 335). That is to say, power is located in emotional expression (Campbell, 1997)—in who gets to express and who must repress various emotions. The *politics of emotions*, then, is the analysis that challenges the cultural and historical emotion norms with respect to what emotions are, how they are expressed, who gets to express them and under what circumstances. It is in this sense that it may be argued that there is always something political in which teachers and students are caught up as they relate emotionally to one another across classroom spaces, because power relations are essentially unavoidable; there are always emotion norms influencing emotion discourses and emotional expressions (Albrecht-Crane, 2003; Zembylas, 2005a).

But how are emotions implicated in power relations, resistance, and transformation in teaching? How do emotions create *affective connections* in the classroom, and what are the (political) implications of these connections? In this chapter, I argue that emotions play a crucial role in the ways that individuals come together and constitute collective "bodies" in the classroom—that is, I show how affective connections are not individualized but work to bind together a whole community. Such an argument clearly challenges the assumption that emotions are "individual" and "private" phenomena and supports the position that emotions are crucial to politics, in the sense that power is an inextricable aspect of how individuals come together, move, and dwell in the classroom (see Ahmed, 2004; Butler, 1997). In other words, emotions are not only considered private reactive responses to events, but are also socially organized and managed through "social conventions, community scrutiny, legal norms, familial obligations and religious injunctions" (Rose, 1990, p. 1). Therefore, power—that is, who exercises power, through which mechanisms, and with what implications—as well as resistance are at the center of exploring emotions in education.

Consequently, such a perspective acknowledges the constitutive effects of emotions as "discursive practices" (Abu-Lughod & Lutz, 1990). By that it is meant that the words used to describe emotions are not simply names for "emotion entities," preexisting situations with coherent characteristics; rather, these words are themselves "actions or ideological practices" that serve specific purposes in the process of creating and negotiating reality (Lutz, 1988). Therefore this approach emphasizes the roles that language, body, and culture play in constituting the experience of emotion. Emotions function as discursive practices in which emotional expressions are *productive*—in other words, emotions make individuals into socially and culturally specific persons engaged in complex webs of power relations. In this sense, emotions are particular forms of social practices and performances that involve processes of production, embodiment, and interpretation of meanings, which are based on particular social conventions. Unavoidably, then, *power* is an integral part of discursive practices of emotions because "power relations determine what can, cannot, or must be said about self and emotion, what is taken to be true or false about them, and what only some individuals can say about them. . . . The real innovation is in showing how emotion discourses establish, assert, challenge, or reinforce power or status differences" (Abu-Lughod & Lutz, 1990, p. 14). Abu-Lughod and Lutz argue that such an understanding of emotion "leads us to a more complex view of the multiple, shifting, and contested meanings possible in emotional utterances and interchanges, and from there to a less monolithic concept of emotion" (1990, p. 2).

In this chapter, I explore the role of emotions in the classroom using the theoretical concepts of power and politics to show how affective connections are formed and transformed over time through resistance and transformation. I begin by clarifying the theoretical approach that guides my investigations and continue by analyzing two examples from my research program on emotions in education. These examples critically analyze the discursive practices that underlie the flow of emotions in the ways that a teacher and her students' identities come to be constituted and assigned to them through classroom discourses, practices, and performances. As a body of literature, this research program emphasizes that emotions play an important political role in enabling *resistance* and *transformation*, two aspects that are currently missing from many accounts of emotions in education. The chapter ends by sketching some implications for teacher practice. By understanding how emotion norms and affective connections are historically contingent, I argue that teachers and students can begin to deconstruct the power relations that normalize their lives in the classroom. After all, if emotion norms and affective connections are politically constructed, they can also be politically deconstructed.

THEORETICAL PERSPECTIVES ON RESEARCHING EMOTIONS IN EDUCATION

For some time the view of emotion as a social and cultural phenomenon was not particularly accepted among educational researchers, who tended to emphasize teaching practice as primarily a cognitive activity (Boler, 1999). In the context of social sciences in general, emotions were predominantly seen as originating within the individual, confined to interior aspects such as brain functioning and personality (Lupton, 1998). As a result, many educational researchers viewed the study of emotion as the province of the "psy" disciplines, particularly cognitive (and social) psychology. For example, cognitive psychologists would be interested in the interrelationship among bodily response, context, and the individual's recognition of an emotion (Lupton, 1998).

However, over the last decade or so there has been an increased interest among educators for the role of emotions in teaching from a variety of perspectives, as part of a broad reaction against the oversimplified views that prevailed in older conceptions of teaching as mainly a cognitive activity (e.g., Hargreaves, 1998, 2000, 2001; Nias, 1996; Schutz & DeCuir, 2002; Van Veen & Lasky, 2005). Other researchers in education have begun constructing accounts about emotions in teaching using feminist, anthropological, and poststructuralist ideas (e.g., Boler, 1999; Evans, 2002; Zembylas, 2005a).

Feminist and poststructuralist views are the critical frame that primarily guides my view on emotion. These views examine the role of culture, power, and ideology in creating discourses and how teachers and students participate in this process by adopting or resisting dominant discourses. The interest of feminist and poststructuralist views focuses on language and social practices and moves away from privileging individualism and self-consciousness. My own interest has been on problematizing the "control" of emotions in education and the process with which *emotional rules* are constructed (i.e., norms that shape the expression of emotions by permitting teachers and students to feel some emotions and by prohibiting them to experience or express other emotions).

My conceptualization of emotion rests on four important assumptions: (1) Emotions are not private or universal and are not uncontrollable innate impulses that just happen to passive sufferers (the Aristotelian view). Instead, emotions are constituted via language and are about social life. This view challenges the divisions of "private" (the existentialist and the psychoanalytic concern) and "public" (the structuralist concern). (2) Power relations are inherent in "emotion talk" and shape certain emotional rules and the expression of emotions by permitting us to feel some emotions and by prohibiting other emotions—for example, through moral norms, and explicit social values, such as efficiency, objectivity, and neutrality. (3) When using emotions, one can create sites of social and political resistances. Feminist poststructuralist criticism, for example, exposes contradictions within discourses of emotions—what may be called "counterbalancing discourses" or "disrupting discourses" (Walkerdine, 1990). (4) Finally, it is important to recognize the role of the body in emotional experience, a view that is not related to any notion of emotion as "inherent" but emphasizes how embodiment is integral to the constitution of affective connections. If the corporeality of emotion is also acknowledged—and this is precisely the advantage of theorizing emotion as a discursive practice—then it is shown how individuals come to be aligned with communities, and the notion of an individualized psychological self is completely subverted.

Ahmed's (2004) notion of *affective economies* is another way of arguing that emotions do not reside in a subject but rather circulate involving relations of difference, whereby what is "moved" and what "moves" is the effect of affective intensities and energies. In other words, Ahmed acknowledges the significance of corporeality in the constitution of emotional attachments and meanings; that is, how emotions become attached to objects, bodies, and signs—a process that is crucial in the constitution of each individual. What characterizes emotionality, according to Ahmed, is that it functions as an *economy*; it separates us from others as well as connects us to others. For example, she points to an economic understanding of hate and explains that hate does not reside within an individual, but is circulated and draws other bodies together making them members of a group united by their hatred to other groups. This economy of hate works to differentiate some bodies from other bodies, a differentiation that is never over.

While Ahmed's argument draws on psychoanalysis and focuses on how emotions circulate, the French philosopher Michel Foucault's (1980, 1983a, 1983b, 1990a, 1990b, 1990c) analyses of power and politics are helpful in examining how emotions are constituted and managed—that is, how and under what circumstances emotions are considered appropriate or inappropriate, and how they function with and in power relations. Although Foucault did not engage directly with emotions, by examining the ways in which emotions function simultaneously as a sign of power and as one of its effects, we can use Foucauldian ideas to describe emotions as historically and culturally constituted. Because emotions may be described both as a way of knowing and as a distinct realm in which meaning is constructed, we are urged to put our discursive practices about emotions and our discursive practices about power and politics into dialogue with one another (Jaggar, 1989).

For Foucault, discourses do not simply reflect or describe reality, knowledge, experience, self, social relations, social institutions, and practices; rather, they play an integral role in constituting (and being constituted by) them. In other words, discursive practices establish what can be felt. In and through these discursive practices we ascribe to ourselves bodily feelings, emotions, intentions, and all the other psychological attributes that have for so long been attributed to a unified self. In this sense, subjects *do* their emotions; emotions do not just happen to them. This is where an examination of how and why subjects are constituted as such opens the possibility of creating new forms of *subjectivity*. Foucault prefers to use the term "subjectivity" instead of "self-identity" to describe the manifold ways in which individuals are historically constituted (Rose, 1990). The concept of subjectivity implies that what is called self-identity, like society and culture, is fractured, multiple, contradictory, contextual, and regulated by social norms.

However, it is important to emphasize that power in Foucault's work is not repressive. Power, according to Foucault, is dispersed, manifest in discursive practices, and exercised; it is not a possession, but it is unstable and localized. Discourses produce power, which in turn continuously produce and constitute the self (Foucault, 1977, 1983a). Foucault developed two major ways of describing the formation of subjectivity: one based on discipline and normalization—*technologies of power*—and the other based on care of the self and the uses of pleasure—*technologies of the self*. The notion of technology refers to "any assembly structured by a practical rationality governed by a more or less conscious goal" (Rose, 1998, p. 26), such as a school, a prison, and an asylum. Technologies operate in terms of a detailed structuring of space, time, and relations among individuals, through hierarchical observation, and normalizing judgments that the individual utilizes to conduct his or her own conduct (Foucault, 1977).

Technologies of power focus on the relationship between discourses and regimes of power/knowledge and are internalized and self-regulating. Technologies of the self "permit individuals to effect by their own means or with the help of others a certain number of operations on their own bodies and souls, thoughts, conduct, and way of being" (Foucault, 1988, p. 18). Thus individuals play a large part in their own control and take the form of techniques for the conduct of one's relation with oneself. These techniques are embodied such as in confession, diary writing, group discussion, and so on, and they are practiced under the actual or imagined authority of some regime of truth. The technologies of the self are part of an assemblage of power relations, knowledges, and practices.

This discursive production of the self is both constraining and liberating; thus subjectivity is understood through resistance and domination. Foucault (1990a) maintained that "where there is power there is resistance" (p. 95), suggesting that power and resistance together define agency; the notion that self-identity or agency exists prior to the interplay between power and resistance is problematic. Attending to the local manifestations of power allows one to track resistances, to be critical, and to develop strategies for (re)constituting one's power relations. "The problem is not changing people's consciousness—or what's in their heads—but the political, economic, institutional regime of the production of truth" (Foucault, 1990a, p. 135). In other words, people choose among various discourses that are available to them or act to resist those discourses. From a Foucauldian perspective, no discourse is inherently liberating or oppressive (Johannesson, 1998).

The preceding analysis of emotions and power relations can contribute to a critical understanding of emotionality in education. For example, technologies of power can serve as a conceptual tool to investigate the role of emotions in the constitution of power relations in the classroom, how emotion discourses are formed and mobilized, and what their political implications are. The focus on the notion of power informs a study of emotions in education because it turns our attention toward examining dominant and/or resistant discourses and teaching/learning practices and their effect on teacher-student relationships. Thus an analysis of the reciprocal relationship between emotion and politics identifies the possibilities for affective connections and reminds us of the potential to subvert normalizing practices in the classroom.

To study emotions in education within the previous set of assumptions allows the exploration of spaces that move beyond theories that psychologize emotions and treat them as internalized (e.g., psychoanalysis) or structural theories that emphasize how "structures" shape the individual (e.g., Marxism). In this sense, emotions are neither private nor merely effects of outside structures. Emotions are not simply language-laden, but they are also "embodied" and "performative" within particular communities; the ways in which teachers and students understand, experience, perform, and talk about emotions are highly related to their sense of body. The role of power relations in how emotions are constructed directs attention to an exploration of emotion discourses and the mechanisms with which emotions are "disciplined" and certain norms are imposed and internalized as "normal." This kind of theorization allows teachers and students first to identify such discourses and then to destabilize and denaturalize the regimes that demands certain emotions to be expressed and others to be disciplined. The place of emotion in self-formation in the classroom becomes an element in circuits of power that constitute some selves while denying others, thus performing an act of discipline and domination. But, as I will show through my research examples, an analysis of power that attends to what resistances mean is useful in articulating the possibility of new forms of subject-formation in the classroom.

Finally, it is important to acknowledge that theoretical assumptions have always had important methodological implications in studying emotions, even when these assumptions are not explicitly acknowledged (Savage, 2004; Schutz & DeCuir, 2002). Attentive or not, researchers start from a certain perspective, and in taking that perspective much depends on their understanding of emotion and its relation to other aspects of individual and social life. In dealing with emotions, many fields (e.g. psychology, philosophy, anthropology, sociology, cultural studies) are involved in ongoing debates about the extent to which emotions involve the mind or the body, meaning or feeling, or whether these dichotomous poles can be transcended (Lupton, 1998; Williams, 2001). For example, if emotions are perceived as primarily private, bodily phenomena, they are likely to be studied within a universally law-bound psychodynamic framework. On the other hand, if emotions are seen as culturally relevant, public performances, reflecting power relations and mediating between subjective experiences and social practices, they are likely to be studied in specific sociocultural contexts. Furthermore, the assumptions held by researchers will undoubtedly influence the research design, data collection, and interpretation.

Based on the preceding theoretical assumptions, my own approach is situated between those methodologies focusing on emotions as private (psychodynamic approaches) or emotions as sociocultural phenomena (social constructionist approaches) and aims at bridging their differences by following an interactionist approach (Savage, 2004). This approach suggests a closer look to the sociocultural context emphasizing that within the repertoire of emotions different peoples experience different ranges of emotions (Beatty, 2005). For example, while people everywhere feel something like anger, some societies cultivate it while others avoid it. Thus "the salience or absence of certain emotions will be linked to a range of customary usages, mythology, and socialization practices" (Beatty, 2005, p. 22). Consequently, methodological practices that embrace ethnographic and narrative research open possibilities for rich descriptions of the salience of emotions in everyday life. The convergence between sociocultural approaches and psychodynamic ones takes place around the notion that "socialized human bodies, bodies that normally exist as groups and in interaction rather than as isolated entities, have their being in recurrent situations that call forth the meaning/feeling responses we recognize as emotions" (Leavitt, 1996, pp. 524–525). This perspective, argues Leavitt, seems truer to our common, daily life experience of emotion than a vision of emotion as either bodily or simply sociocultural.

RESEARCH EXAMPLES

In my research, I have been primarily interested in the connection between emotion discourses and power relations in everyday classroom contexts. In this section, I provide two examples that describe: (1) how emotional rules are constituted in the classroom and what implications they have for teachers and students; and (2) how the affective economies that are created in the classroom have a subversive potential. Both of these examples show the dialectics of politics and emotions in the classroom, and in particular, how emotion discourses construct specific notions of emotionality and connectivity that have political implications in terms of teacher and student normalization, resistance, and transformation. These examples are drawn from a series of ethnographic studies I conducted on emotions in education over the last six years (see Zembylas, 2005a, 2005b, 2005c). The focus of these studies was to explore how social conventions, community scrutiny, and school policies and obligations in a teacher's (Catherine) classroom constituted how she felt about teaching and learning. Catherine (not her real name) was an experienced early childhood and elementary educator. She had been teaching for 25 years and had worked with children from the kindergarten through fifth grade. She usually taught multiage classes of kindergarten and first grade, or first and second grade at a multiethnic elementary school of 400 students. The school was located in a medium-size university city in Illinois. The subject matter that constituted the initial focus of this research program was science (Zembylas, 2004a, 2004b), but I expanded the focus in later studies (Zembylas, 2005b, 2005c) because Catherine uses an interdisciplinary approach in her teaching.

The Constitution of Emotional Rules and Their Implications

First, it needs to be emphasized that although we searched carefully over the years, Catherine and I could not find any explicit rules (e.g., written school memos or protocols) that laid out norms of emotional display and expression for teachers. Also, while conducting this research, teachers in Catherine's school did not participate in any workshops organized to discuss how to exercise emotion-management practices. However, as Catherine often pointed out, there was always talk among teachers about "appropriate" and "inappropriate" emotions. For example, she described that for many years—in particular at the beginning of her long teaching career that extended over

25 years—she had to strive for "neutrality." As she explained, there was a perception among teachers that showing strong emotions in school was not considered a "professional" thing to do. "For a long time," Catherine highlighted, "I prevented myself from expressing what I *really* felt, because it was not considered *professional* to do that." It was not appropriate to display one's emotions as one felt them at all times, nor was it desirable to maintain too tight a hold over them. This power over Catherine's expressions of emotions had real effects because she often had to "manage" and "control" how she felt.

Emotional control and regulation ranged from issues of teaching pedagogies to feelings of well-being and functioned to produce a particular kind of teacher self; that is, a teacher who was able to control how she felt at all times. For example, the continuous self-observation and monitoring practices of the administration and fellow teachers constituted the notion of the "normal" teacher self, against which all teachers were urged to measure themselves. Catherine happened to be enthusiastic about pedagogies that were deviant from the "norm"—for example, teaching science in a "progressive" manner instead of emphasizing "teaching to the test" like everybody else—thus she was systematically told by several of her colleagues to achieve "normality." This is evident in what one of Catherine's colleagues urged her once: "*Do* whatever everyone is doing. Why do you want to be so *different*? Why don't you just teach science the way it's supposed to be taught?" Catherine's prevailing feeling was a sense of powerlessness and personal inadequacy, precisely what Bartky (1990) defined as *shame*. It is worthwhile to further analyze this feeling and examine some of the consequences of this particular emotion discourse in the process of Catherine's subjectification as a "different" teacher.

Shame was a profound affective attunement in Catherine's career, because she was constantly being exposed as having some kinds of flaws in her teaching philosophy. As Bartky (1990) says: "Shame can be characterized as a species of psychic distress occasioned by a self or a state of the self apprehended as inferior, defective, or in some way diminished" (p. 85). A major reason for this feeling of shame was Catherine's use of progressive pedagogies that were not "approved" by her colleagues and the administration; therefore, she was identified as someone "different." A more implicit, unstated issue underlying this "difference" was that Catherine was a deviant teacher who should have been managed through measures that prescribed the appropriate teaching (e.g., memos, discussions in the staff room, speeches about the importance of state testing, etc.). That such strategies had a role in Catherine's teaching practices is not controversial. Rather, my argument is that this emotion discourse masked disciplining agents by employing teacher classifications about what it meant to be a professional teacher. Disciplinary strategies aimed not only at disciplining Catherine's pedagogy, but also at prescribing her emotions; thus emotions of excitement and passion about pedagogies other than "teaching to the test" were perceived in a suspicious manner.

Consequently, emotion management was legitimated through social networks of "professional" knowledge reinforcing the "rational" process of emotion management. For instance, setting up meetings to discuss the importance of state testing or to interpret the results from the previous year assisted in reinforcing the rational management of emotions that did not express approval of pedagogical practices guided by teaching to the test. Emotion management, in other words, became a technique of power that depended on an emotion discourse of "normal" and "standard." Rather than regarding this discourse as something to be challenged, emotion management became a performative "truth."

Furthermore, Catherine's feeling of shame manifested a profound mode of disclosure, both of herself and of the setting. Shame can be a devastating experience characterized by many negative evaluations of one's self and by a sense of worthlessness, dismissability, and powerlessness. Campbell (1997) analyzes the consequences of shame arguing that: "[W]hen our feelings are trivialized, ignored, systematically criticized, or extremely constrained by the poverty of our expressive resources, this situation can lead to a very serious kind of dismissal—the dismissal of the significance to a person of his or her own life, in a way that reaches down deeply into what the significance of a life can be to the person whose life it is" (p. 188). Catherine's feeling of shame was the realization that her educational aims were not worthy. The normative expectation was that Catherine should assimilate into predetermined roles and expectations and thus manage "outlaw" emotions accordingly. Silence and isolation was often a direct outcome of this feeling of shame for Catherine, but at another level she was constituted as a "failure" because she became unsure of her teaching philosophy: was she doing the "right" thing to teach science using inquiry, emphasizing passion and love for it, and making connections to other subject matters when most teachers accused her of depriving her students the opportunity to get a good score in the state test? Not to mention, of course, that the way this discourse was constituted implied problematically, in my view, that teaching to the test and teaching using progressive pedagogies were contradictory; that is, the use of the one necessarily excluded the use of the other.

The above research example politicizes issues of teacher and student subjectification and the normalization of affective connections in the classroom or school. The emotional rules developed and legitimated in schools through the exercise of power are used to "govern" teachers and students' performances by putting limits on their emotional expressions. These limits "normalize" teachers and students and thus turn appropriate behavior into a set of skills, desirable outcomes, and dispositions that can be used to examine and evaluate them. More than a decade ago, Bartky (1990) called for "a political phenomenology of the emotions—an examination of the role of emotion, most particularly of the emotions of self-assessment both in the constitution of subjectivity and in the perpetuation of subjection" (p. 98).

A political phenomenology of emotions in education emphasizes the importance of exploring the political roots and implications of emotions in teaching and learning. This research example urges us to *historicize* emotions in education; that is, to contextualize them in and across specific social and cultural spaces within power relations and resistances.

The Subversive Power of Affective Economies in the Classroom

More recently, in a follow-up study conducted three years after the previous one, Catherine provided new narratives and practices that added new insights and enriched our understanding about how teachers and students can subvert classroom and school emotion discourses. In particular, the example described here indicates the ways in which creating new affective connections in the classroom based on compassion and understanding for others constituted a subversive affective community in the months after September 11, 2001. Evidence from Catherine's narrations suggests how that tragic day initiated a shared "communal" visceral response that marked the emotional culture in the classroom for many months. This is an excerpt of Catherine's provocative narrative two years after 9/11.

> Looking back on what happened that day and how it affected us as a classroom community in the following months, there was a miraculous transformation that began in the middle of all those painful feelings.... The next day we came together as a class, we had a class meeting... Everybody had a chance to talk about it.... The only thing that I did was when they had totally glaring inaccuracies I would say, "Well actually...," but I let kids talk about it. And then letting them know that it's important to talk about how you feel....
>
> We did that for several days...we came together as a group and shared how we felt.... The feelings we expressed were mainly sorrow for the devastation and sympathy for the victims' relatives and the suffering that this incident caused.... We expressed those feelings through a variety of activities: painting, journaling, discussing, researching the Internet for more information about how people felt around the country, and so on.... And you could see almost right away that in their interactions kids began to treat each other differently! I mean it was amazing! Everyone became more kind and self-conscious not to hurt each other's feelings in their everyday encounters... and kids became more eager to find out how other children felt in different parts of the world in which there was pain, suffering, and devastation. We began investigating how poor children felt in Africa and other areas in which there was conflict and devastation....this became a project that went on for many weeks. And I have to tell you that in all the years I'm teaching I haven't felt this powerful sense of community in the classroom.... everyone caring about each other, feeling so close to each other, so passionate about learning how others felt around the world.... And it seems to me that what brought us together was how we all felt about what had happened on 9/11 and then all the things we did together to explore those feelings in relation to the rest of the world. Before this tragic event, we may have

talked about children in other parts of the world, but I don't think there was *this* kind of emotional connection. (Interview, April 23, 2003)

This narrative, of course, is far from extraordinary. In fact, it shows a situation in which the teacher and her students can come together through the mobilization of the power of emotions. The emotions of sorrow and sympathy worked initially to bring Catherine and her students together. It is important to clarify that students did not "inhabit" the "skin" of the classroom community; rather, the skin of the classroom community was an effect of the affective economy that was gradually being created in the classroom (see Ahmed, 2004). Thus, the response of compassion for other children around the world became an effect of the affective encounters that were developed in the classroom. It was through moving towards others who were also suffering in some ways that students created affective connections and became aligned with others. Through this corporeal affectivity that brought them together, Catherine and her students reformulated the social and affective landscape of the classroom. Emotions of compassion circulated and were distributed across a social as well as a psychic landscape.

To theorize this finding in a wider context, I want to turn our attention to the dynamic character of children's emotional responses and the resulting transformation in the classroom. In particular, I want to argue that this transformation has to do with the power of *empathetic understanding* to subvert things that get stuck in habitual tendencies and normalized situations. Empathetic understanding refers to two things: first, becoming aware of how the other feels and moving into rhythm with these feelings—in a sense, "feeling with"; and second, developing a passionate affection for the "object" one studies—that is, being caring and passionate about what one explores. In short, empathetic understanding has to do with the creation of affective connections that allow for empathetic communication across possible gaps.

Far from being a mere cultural construct or discourse in Catherine's classroom, empathetic understanding in the above example was an essential part of the affective connections that were created. From this perspective, Catherine's pedagogic actions and children's learning practices took on a fascinating new significance. Empathetic understanding became a valuable source of emotional knowledge—of a personal, practical, and pedagogical nature—that strengthened Catherine and her students' affective connections. This affective economy made possible a sense of empowerment to creatively subvert previous negative feelings such as sorrow or despair. Again, the concepts of power and politics allow educators to regard affective connections in the classroom as powerful tools for shaping emotional material that either stands in tension with the prevalent emotional regime at the school (the first research example) or subverts the negative emotional climate in the classroom and creates an emotional landscape of solidarity and empathy (the second research example).

As both of the above research examples show, while teachers or students are forced to engage in processes of emotion management, they can always create new affective economies within particular localities and subvert prevailing economies. As some ideologies gain acceptance and others decline, certain emotional rules rise and fall. Constructing an emotional landscape that subverts the prevailing emotional rules, as a means of questioning these rules and their assumed ideologies and truths, invokes vulnerability as well as resistance. As Boler (1999) reminds us, vulnerability often provides the turbulent ground on which to negotiate new affective connections that are necessary for transformation to take place. Resistance functions both as a defense against vulnerability and as an assertion of power in the face of impositions. Thus the mechanisms of empathetic understanding and affective connections in the classroom may constitute powerful forms of resistance.

IMPLICATIONS FOR TEACHER PRACTICE

The kind of theory and research described here makes possible a theorization of the success of some emotional regimes in schools, the failure of others, and therefore it helps us to understand how emotional rules come and go. In making these observations, it seems that teachers' emotional development is profoundly influenced by their participation in particular forms of discursive and embodied practices at school. By making this point, I wish to avoid a suggestion that subordinates the individual to the social and loses sight of the reciprocal relation between the two. As I have shown, there is a great deal at stake in the emotional regimes that govern the lives of teachers and students. However, professional and classroom communities are able to constitute affective spaces that have the potential to subvert disciplinary mechanisms and practices.

In view of the demands for conformity and homogeneity within schools, it is not surprising that emotional suffering is induced—that is, the pressure to control one's emotions. Emotional vulnerability, resistance, and transformation of teachers and students emphasize the importance of being attentive to the project of ungrounding and unmaking of normative rules. To analyze and challenge these rules means to reveal their historicity and contingency that have come to define the limits and possibilities of teachers and students' understandings of themselves, individually and collectively. By doing so, it is to disturb, destabilize, and subvert these rules, to identify some of the weak points and lines of fracture where new affective connections (as counter-hegemonic) might make a difference.

In particular, the theorization of empathetic understanding and affective connections gives political meaning back to research on emotions in education and allows educators to discern the successes and failures of particular emotional regimes within a school culture. In light of this contention, it should be apparent that I view the process of constantly negotiating classroom or school emotional rules as one in which teachers and students coconstruct the activity system that constrains or encourages their individual actions. Teachers and students can create "emotional refuges," like Catherine and her students did, that constitute the greatest possible spaces for emotional freedom; that is, espousing affective connections that minimize emotional suffering (Reddy, 2001). Armed with this recognition, it would be possible to write ethnographies of emotions in teaching that take into account the political significance of emotionality. There are already efforts that combine narrative-biographical and micropolitical approaches to explore how pedagogical actions are influenced by professional and other interests (see Kelchtermans & Ballet, 2002). The dialectics of emotions and power relations in teaching focuses on often neglected political implications of the emotional aspects in education. Thus the ways in which empathetic understanding and affective connections are used to subvert prevailing emotional rules in the schools reveal interesting micropolitical mechanisms in teaching.

The ultimate aim of such mechanisms is to reshape and expand the terms of emotion discourses in education, enabling different questions to be asked, modifying the relations between teachers and students and the rules in the name of which they govern or are governed. The contribution of the political implications of emotions in education amounts to an intervention in a much larger debate about subjectivities in the classroom, in which concepts of affective elements of consciousness and relationships, community, and reform are slowly being reexamined. This sociopolitical character of emotion in education creates the difference between possible and real transformation, and it is this difference that constitutes the power of the above mechanisms as tools to subvert existing conditions. The need for a deeper conceptualization of this sociopolitical character can guide ethnographic work on emotions in education in whatever locality, work informed by a genuine search to understand the power and the limitations of the political merits or demerits of any emotional regime at a school or in a classroom.

Consequently, the task of progressive pedagogies is to identify the practices, strategies, and spaces where affective empowerment might be possible. Things that matter affectively can be taken up as sites of ideological struggle (Grossberg, 1992). Political positions can be claimed through and constituted by modes of felt experiences—such as solidarity as a response to poverty, exploitation, and injustice. These felt experiences indicate that "positive" and "negative" emotions are not attributes of emotions or of individuals who feel those emotions, but they represent provisional readings and judgments in community settings and have profound political implications. The acknowledgement of the politics of emotions in education clarifies the kind of work that "decolonization pedagogies" (Tejeda, Espinoza, & Gutierrez, 2003) require and suggests that the intervention needed to initiate personal and collective transformations can be nothing less than a radical subversion of the existing regime of truth. Such decolonization must work at the level of emotions by identifying our emotional attachments and their effects.

Examining emotions in the manner suggested in this chapter creates possibilities for establishing and continuing an "affective struggle" (Grossberg, 1988) within and against systems of oppression. Listening to students and teachers' affects about injustices, for example, can be constructive, and thus affect can constitute a central part of challenges to any form of domination (see Holmes, 2004). Affective connections may call attention to a group's demand for respect and recognition, but also highlight inequalities more generally. The manifestation of this affective struggle, however, remains to be fully recognized. The task at hand is to find new ways to enable productive affective connections between teachers and students that "can form a strong, wondrous sense of vitality, potentiality, and creation.... Affect operates precisely as a limit-experience, as a force that exceeds and escapes the subject. Affect allows for (communal) connections at a level within and yet outside of social organization" (Albrecht-Crane, 2003, pp. 587–588). An analysis that accounts for teachers' and students' affective relations and empathetic understanding in the classroom provides an opening for community-building and the critical tools for responding to the political implications of emotions in education. The critical task of engaging pedagogies affectively implies inventing strategies of subverting the disempowering ways that affect might function in schools. According to Harrison (1985), the question is not, "What do I feel," but rather, "What do I do with what I feel?" (p. 14).

My argument in this chapter has been that emotions are crucial to the way that teachers and students relate to one another in educational environments. Unraveling the different aspects of using feminist and poststructuralist thought—both as analytic tools and as points of departures for conducting research on emotions in education—creates possibilities for enriching our perspectives about the dynamics of affective relations in the political landscape of the classroom. While my argument questions any politics based on notions of liberation as a source of personal/political transformations, neither should it be interpreted as utopian politics. We need pedagogies that account for the intersections of affective connections, power relations, and ideology. A theory and practice that account for affective relations in our pedagogies allow us to grasp the processes of affective struggles in community-building. As an educator, I welcome such an opening.

References

Abu-Lughod, L., & Lutz, C. A. (1990). Introduction: Emotion, discourse, and the politics of everyday life. In C. Lutz and L. Abu-Lughod (Eds.), *Language and the politics of emotion* (pp. 1-23). Cambridge: Cambridge University Press.

Ahmed, S. (2004). *The cultural politics of emotion*. Edinburgh, UK: Edinburgh University Press.

Albrecht-Crane, C. (2003). An affirmative theory of desire. *JAC: Journal of Composition Theory, 23*, 563-598.

Albrecht-Crane, C., & Slack, J. (2003). Toward a pedagogy of affect. In J. D. Slack (Ed.), *Animations (of Deleuze and Guattari)* (pp. 191-216). New York: Peter Lang.

Bartky, S. (1990). *Femininity and domination*. New York: Routledge.

Beatty, A. (2005). Emotions in the field: what are we talking about? *Journal of the Royal Anthropological Institute, 11*, 17-38.

Boler, M. (1999). *Feeling power: Emotions and education*. New York: Routledge.

Butler, J. (1997). *The psychic life of power: Theories in subjection*. Stanford, CA: Stanford University Press.

Campbell, S. (1997). *Interpreting the personal: Expression and the formation of feelings*. Ithaca, NY: Cornell University Press.

Cornelius, R. R. (1996). *The science of emotion: Research and tradition in the psychology of emotion*. Upper Saddle River, NJ: Prentice Hall.

Ekman, P., & Rosenberg, E.L. (1997) *What the face reveals: Basic and applied studies of spontaneous expression using the facial action coding system*. New York: Oxford University Press.

Evans, K. (2002). *Negotiating the self: Identity, sexuality, and emotion in learning to teach*. New York: Routledge.

Foucault, M. (1977). *Discipline and punish: The birth of the prison* (A. M. Sheridan, Trans.). New York: Pantheon Books.

Foucault, M. (1980). *Power/knowledge: Selected interviews and other writings 1972-1977* (C. Cordon, L. Marshall, J. Mepham, & K. Soper, Trans.). New York: Pantheon.

Foucault, M. (1983a). The subject and power. In H. L. Dreyfus, & P. Rabinow (Eds.), *Michel Foucault: Beyond structuralism and hermeneutics* (pp. 208-226). Chicago: University of Chicago Press.

Foucault, M. (1983b). On the genealogy of ethics: An overview of work in progress. In H. L. Dreyfus, & P. Rabinow (Eds.), *Michel Foucault: Beyond structuralism and hermeneutics* (pp. 229-252). Chicago: The University of Chicago Press.

Foucault, M. (1988). Technologies of the self. In L. H. Martin, H. Gutman, & P. H. Hutton (Eds.), *Technologies of the self* (pp. 16-49). Amherst: University of Massachusetts Press.

Foucault, M. (1990a). *The history of sexuality, volume 1: An introduction*. New York: Vintage Books.

Foucault, M. (1990b). *The history of sexuality, volume 2: The use of pleasure*. New York: Vintage Books.

Foucault, M. (1990c). *The history of sexuality, volume 3: The care of the self*. New York: Vintage Books.

Grossberg, L. (1988). Postmodernity and affect: All dressed up with no place to go. *Communications, 10*, 271-293.

Grossberg, L. (1992). *We gotta get out of this place: Popular conservatism and postmodern culture*. New York: Routledge.

Hargreaves, A. (1998). The emotional practice of teaching. *Teaching and Teacher Education, 14*, 835-854.

Hargreaves, A. (2000). Mixed emotions: Teachers' perceptions of their interactions with students. *Teaching and Teacher Education, 16*, 811-826.

Hargreaves, A. (2001). The emotional geographies of teaching. *Teachers College Record, 103*,1056-1080.

Harrison, B. W. (1985). *Making the connections*. Boston: Beacon Press.

Holmes, M. (2004). Feeling beyond rules: Politicizing the sociology of emotion and anger in feminist politics. *European Journal of Social Theory, 7*, 209-227.

Jaggar, A. (1989). Love and knowledge: Emotion in feminist epistemology. In S. Bordo & A. Jaggar (Eds.), *Gender/body/knowledge: Feminist reconstructions of being and knowing*. New Brunswick, NJ: Rutgers University Press.

Johannesson, I. A. (1998). Genealogy and progressive politics: Reflections on the notion of usefulness. In T. S. Popkewitz & M. Brennan (Eds.), *Foucault's challenge: Discourse, knowledge, and power in education* (pp. 297-315). New York: Teachers College Press.

Kelchtermans, G., & Ballet, K. (2002). The micropolitics of teacher induction: A narrative-biographical study on teacher socialization. *Teaching and Teacher Education, 18*, 105-120.

Leavitt, J. (1996). Meaning and feeling in the anthropology of emotions. *American Ethnologist, 23*, 514-519.

Lutz, C. (1988). *Unnatural emotions: Everyday sentiments on a Micronesian atoll and their challenge to Western theory*. Chicago: University of Chicago Press.
Lutz, C. A., & Abu-Lughod, L. (Eds.). (1990). *Language and the politics of emotion*. Cambridge: Cambridge University Press.
Lupton, D. (1998). *The emotional self: A sociocultural exploration*. London: Sage.
Nias, J. (1996). Thinking about feeling: The emotions in teaching. *Cambridge Journal of Education, 26*, 293-306.
Plutchik, R. (1980). A general psychoevolutionary theory of emotion. In R. Plutchik & H. Kellerman (Eds.), *Emotion: Theory, research, and experience: Vol. 1. Theories of emotion* (pp. 3-33). New York: Academic.
Reddy, W. M. (1997). Against constructionist: The historical ethnography of emotions. *Current Anthropology, 38*, 327-340.
Reddy, W. M. (2001). *The navigation of feeling: A framework for the history of emotions*. Cambridge, UK: Cambridge University Press.
Rose, N. (1990). *Governing the soul: The shaping of the private self*. London: Routledge.
Rose, N. (1998). *Inventing ourselves: Psychology, power and personhood*. Cambridge, UK: Cambridge University Press.
Savage, J. (2004). Researching emotion: The need for coherence between focus, theory and methodology. *Nursing Inquiry, 11*, 25-34.
Schutz, P., & DeCuir, J. T. (2002). Inquiry on emotions in education. *Educational Psychologist, 37*, 125-134.
Tejeda, C., Espinoza, M., & Gutierrez, K. (2003). Toward a decolonizing pedagogy. In P. Tryfonas (Ed.), *Pedagogies of difference: Rethinking education for social change* (pp. 10-40). New York: Routledge Falmer.
Van Veen, K., & Lasky, S. (2005). Emotions as a lens to explore teacher identity and change: Different theoretical approaches. *Teaching and Teacher Education, 21*, 895-898.
Walkerdine, V. (1990). *Schoolgirl fictions*. London: Verso.
Williams, S. (2001). *Emotion and social theory*. London: Sage.
Zembylas, M. (2004a). The emotional characteristics of teaching: An ethnographic study of one teacher. *Teaching and Teacher Education, 20*, 185-201.
Zembylas, M. (2004b). Emotion metaphors and emotional labor in science teaching. *Science Education, 55*, 301-324.
Zembylas, M. (2005a). *Teaching with emotion: A postmodern enactment*. Greenwich, CT: Information Age Publishing.
Zembylas, M. (2005b). Discursive practices, genealogies and emotional rules: A poststructuralist view on emotion and identity in teaching. *Teaching and Teacher Education, 21*, 935-948.
Zembylas, M. (2005c). Beyond teacher cognition and teacher beliefs: The value of the ethnography of emotions in teaching. *International Journal of Qualitative Studies in Education, 18*, 465-487.

PART

V

Implications and Future Directions

CHAPTER

18

Where Do We Go from Here? Implications and Future Directions for Inquiry on Emotions in Education

REINHARD PEKRUN
University of Munich

PAUL A. SCHUTZ
University of Texas at San Antonio (UTSA)

As noted in the introduction to this volume (Schutz & Pekrun, 2007), research on emotions in education has been slow to emerge. While students' test anxiety has been researched since the 1930s (e.g., Brown, 1938; Stengel, 1936), and the attributional antecedents of emotions following success and failure since the 1970s (Weiner, 1985), studies outside of these two research traditions were rare (Pekrun & Frese, 1992; Pekrun, Goetz, Titz, & Perry, 2002). During the past ten years, however, there has been a discernable, steady increase of investigations into the nature of emotions experienced by students and teachers (Efklides & Volet, 2005; Linnenbrink, in press; Schutz & Lanehart, 2002). These investigations have produced new insights, as evidenced in the chapters of this volume. They suggest that multiple emotions in both students and teachers should be considered when reflecting on educational problems; that these emotions are patterned in complex ways within individuals, across individuals, and over time; and that they are influenced by gender, race, individual propensities, classroom interaction, and the sociohistorical context. Also, the findings demonstrate that emotions profoundly affect students' and teachers' engagement, performance, and personality development, implying that they are of critical importance for the agency of educational institutions and of society at large.

At the same time, however, the studies conducted so far seem to pose more new, challenging questions than they can answer. Research on emotions in education is still at an early stage. Theories, strategies, and measures for analyzing emotions in education are yet to be fully developed. Also, to date, studies are too scarce to allow any metaanalytic synthesis based on cumulative evidence, or any firm conclusions informing educational practitioners in validated ways how to deal with emotions, with evidence on test anxiety being an exception (Zeidner, 1998, 2007). The progress made so far is promising, but much more has to be done if educational research on emotions is to evolve over the next years in ways benefiting education and society.

In this concluding chapter, we address three basic questions on where research on emotions in education should go from here (also see Pekrun, 2005; Sansone & Thoman, 2005; Schutz & DeCuir, 2002). First, how should we advance our theoretical thinking about emotions in education? Second, how should we study these emotions empirically? And third, what should be studied? In discussing answers to the first question, we address the need for developing consensus on constructs, as well as more comprehensive theoretical frameworks that work towards integrating multiple theoretical perspectives. Similarly, with regard to the methodology of empirical research, we argue that multimethod approaches are needed to advance this field of research. In addition, we address the need for refining the measurement of emotions. Concerning phenomena to be studied, we suggest that future advances require addressing multiple emotions, studying the organization of these emotions at different levels (individuals, classrooms, institutions, and contexts), as well as examining the dynamics of emotional change within educational activity settings, and over years and historical epochs. Also, we address the need for educational intervention research targeting emotions. In conclusion, we argue that progress in this field requires interdisciplinary perspectives and call for more interdisciplinary collaboration of researchers.

THEORIES ON EMOTIONS IN EDUCATION: BUILDING MORE COMPREHENSIVE FRAMEWORKS

In psychology, education, and sociology alike, different traditions of research on emotions have been working in relative isolation, in spite of often sharing basic assumptions. Generally, whereas more comprehensive theories were developed in these disciplines in the first part of the 20th century (e.g. Allport, 1938; Lewin, 1935), there seems to be a proliferation of small constructs and minitheories in many research fields today, including the study of human emotions (see Lewis & Haviland-Jones, 2000). To make things worse, all too often the wheel seems to be reinvented in each of these disciplines. Authors create "new" constructs and theories that, in fact, have often been proposed previously (see e.g. Skinner, 1996, on psychological constructs of control). This is disguised by inventing new terms and by neglecting to cite those who wrote about the construct previously. As a result, there is a lack of theoretical integration that has to be overcome if cumulative progress is to be made. Specifically, there is a need to find more of a consensus on constructs of emotions, and to integrate theoretical models that share assumptions on the structures, dynamics, and functions of emotions. However, we also need debate and cross-fertilization among researchers pursuing truly divergent approaches, as well as new perspectives that enrich existing theories.

Constructs of Emotions in Education

If researchers define their constructs in idiosyncratic ways that are not shared by others, communication between researchers and an integration of empirical findings become difficult. Fortunately, there seems to be some consensus today as to the basic nature of human emotions. Most researchers in the affective sciences, including educational research on emotions, probably would agree that human emotions are multicomponent systems involving coordinated processes of subsystems of mind and behavior (Kleinginna & Kleinginna, 1981; Pekrun, 1988, in press; Schutz, Hong, Cross, & Osbon, in press; Turner & Waugh, 2007). Important components are affective processes that are physiologically bound to subsystems of the limbic system (e.g., the amygdala; Schafe, Doyere, & LeDoux, 2005), and subjectively experienced as emotional feelings; emotion-specific thoughts that accompany these feelings (e.g., worries in anxiety); peripheral physiological activation (or deactivation) that prepares for action (or inaction); emotion-specific motivational impulses, such as fight, flight, or giving up in anger, anxiety, and hopelessness, respectively; and expressive motor movements, such as facial expressions, communicating the emotion to others.

However, the boundaries of the domain of emotions, the internal structures of this domain, and the universal vs. culture-specific status of many emotions are much less clear to date. As to *boundaries*, while there is consensus that "primary" emotions like joy, anger, or anxiety are members of the domain, this is much less clear for other constructs relating to feelings and affect. Students' interest is a case in point. Some researchers regard interest as an emotion (e.g., Ainley, 2007), whereas others define interest as a more complex construct involving several components (such as the emotion of enjoyment, affectively more neutral subjective values relating to the object of interest, or knowledge structures; Schiefele, 1991). For constructing measures, comparing findings of studies, or designing educational intervention regarding students' interest, it makes a critical difference which of the various definitions of *interest* is used. Similarly, the status of affective constructs like mood (Linnenbrink, in press), curiosity, or metacognitive feelings (Efklides & Petkaki, 2005) is unclear to date.

Concerning *internal structures*, the domain of emotions can be conceptualized by using dimensional circumplex concepts (e.g., Linnenbrink, 2007), or concepts of discrete emotions (e.g., Pekrun, Frenzel, Goetz, & Perry, 2007). Though it seems to be clear, from a conceptual perspective, that these two approaches are complementary rather than mutually exclusive, it is unclear which of the two is more appropriate for describing students' and teachers' affect. Using both of them interchangeably, however, can make it difficult to integrate findings. For example, in studies on achievement goals and students' affect, researchers either used a pleasant (positive) vs. unpleasant (negative) affect approach or a discrete emotions perspective, and the findings from these two approaches are not easily reconciled (Linnenbrink & Pintrich, 2002; Pekrun, Elliot, & Maier, 2006). Making the pros and cons of such divergent approaches explicit, discussing them, and trying to integrate their assumptions at the theoretical or metatheoretical level should prove useful in making progress in our understanding of the structures of emotions.

Finally, as to the *relative universality* of emotion constructs, it seems clear that basic, neuropsychologically defined processes of emotions such as surprise, joy, anger, anxiety, or sadness are universal across cultures and sociohistorical contexts. Many of these basic emotions are universal not only within our species, but beyond our species as well. However, researchers should be aware that the specific contents, expressions, and process parameters (e.g. frequency, intensity, duration) of these emotions can vary widely between cultures and contexts. For example, pride and shame can substantially differ among different cultures. Also, there likely is wide variability in emotions that developed during the course of cultural evolution and are unique to the human race (see Ratner, 2007). One indicator for this variability is the fact that there are emotions for which verbal labels were developed in some languages, but not in others. For example, as noted by Weiner (2007), there is a term for the enjoyment of another person's harm in the German language ("Schadenfreude"), whereas no such term is available in the English language.

Researchers should attempt to reach more of a consensus on the constructs of emotions relevant for education. Consensus on constructs may be a precondition for making it possible to construct consensual measures, integrate the findings of studies, and provide educational practitioners with consistent information. To the extent that different definitions diverge in nonarbitrary ways, however, consensus should not be forced prematurely and at all costs. In theses cases, more dialogue would be helpful to reduce misunderstandings, and to resolve controversial issues, if possible.

Integrating Theories on Emotions

To date, in research on human emotions, many attempts at theory construction ended up in minitheories addressing isolated facets or functions of emotions, or one single emotion only. An example relevant for education is the many theoretical models on the effects of emotions on specific types of cognitive performance that were developed since the 1970s (Lewis & Haviland-Jones, 2000). Many of the theories on the effects of test anxiety are examples for approaches addressing one single emotion, without taking neighboring emotions into account (Zeidner, 1998, 2007).

As with a lack of consensus on constructs, if programs of research are built on one specific theoretical perspective and ignore neighboring perspectives, cumulative progress may be hampered. Specifically, ignoring alternative approaches prevents the detection of commonalities and contradictions. For example, contradictory conclusions seem to follow from different social-psychological theories on the cognitive effects of emotions (Aspinwall, 1998). Some of these theories would imply that students' and teachers' positive affect (such as pleasant mood) is detrimental for any effortful, elaborate processing of information, whereas negative affect (such as unpleasant mood) instigates enhanced effort. A case in point are safety-signal and mood-as-information approaches suggesting that positive mood signals safe conditions and motivates one to relax, whereas negative mood signals unsafe conditions, thus motivating to change the situation by investing task-related effort (cf. Clore, Schwarz, & Conway, 1994). Other theories, however, suggest that positive affect serves beneficial functions for creative problem solving (e.g., Fredrickson's [2001] "broaden-and-build" model of positive emotions), and that negative affect impairs performance (e.g., many models of test anxiety; Zeidner, 2007).

It may be, however, that the contradictions between these approaches are more apparent than real. From an empirical perspective, each of these approaches pertains to specific experimental conditions that served to validate them in the laboratory. In the more complex educational reality outside the psychological laboratory, different conditions are prevalent. The specific processes described by any laboratory-based theory may be at work in specific situations outside the laboratory as well. However, it is very likely that not all of these theories are of equal relevance for the real-life academic achievement of students or occupational achievement of teachers. More comprehensive theories, as well as integrative metatheories, are needed to disentangle commonalities and contradictions, and to make it possible to define the contextual conditions for which different assumptions are valid (see Pekrun, 1992, and Pekrun et al., 2002, for an attempt to integrate assumptions from some of the laboratory-based approaches to affect and performance).

The potential fruitfulness of comprehensive theory building can be seen in research on the appraisal antecedents of emotions. In this research, consensus on some of the dimensions of appraisals seems to be emerging, as evidenced, for example, in Roseman's metatheoretical approach to synthesizing theoretical assumptions (Roseman, Antoniou, & Jose, 1996). Although the exact contents and labeling of appraisal dimensions are still controversial, consensus on basic structures of appraisals can enable researchers to produce cumulative evidence on the functions of these appraisals and to move on to new frontiers of research, once issues of dimensionality are settled.

Research on emotions in education, therefore, should make an attempt to construct more integrative approaches as well. Specifically, frameworks would be needed that integrate perspectives on different levels and contexts of students' and teachers' emotions, including the dynamics of processes within and between levels (e.g., between components of emotions; between emotions, motivation, and cognition; between teachers and students; and between different institutional and sociohistorical contexts).

For creating such dynamic, multilayered, and contextualized frameworks, it seems necessary to integrate theoretical perspectives from a variety of disciplines, including education, psychology, the neurosciences, sociology, and history. Also, an integration of perspectives from different fields within these disciplines is required. For example, education and work share many common features in our societies. A case in point is institution-based competitive vs. cooperative goal structures that are of critical importance both for teachers and students in educational organizations and for employees in the business industry. Nevertheless, educational psychologists and educational researchers more generally, on the one hand, and occupational and work psychologists, on the other, have virtually ignored each other to date. Work on emotions in education is rarely cited by occupational psychologists, and vice versa (e.g., see Grandey, 2000, and Morris & Feldman, 1996, for work on emotional labor that should be relevant for education). The time seems ripe to integrate theoretical perspectives from the various disciplines that contribute to the emerging discipline of research on emotions in education and neighboring fields.

Enlarging Theoretical Perspectives

Beyond integrating existing models, we also see the need for developing fresh perspectives. This pertains not only to the *objects* of theories (see the section on what to study), but also to the *types* of theories that have to be developed. Emotions are organized at multiple levels, change over time, and are situated in sociohistorical contexts (Ratner, 2007; Schutz, Cross, Hong, & Osbon, 2007; Zemblyas, 2007). However, as noted, theories on emotions in education have yet to fully incorporate the multilevel, dynamic, and contextualized nature of students' and teachers' emotions. To do so, it is probably not sufficient to just acknowledge these aspects of emotions. Rather, the type of theorizing used has to be adapted to these targets as well. For example, as to levels of emotions, it would be necessary to construct *multilevel theories* specifying the precise nature of interactions between levels (e.g., in which ways does the composition of the classroom affect students' emotions; see Pekrun, Frenzel, Goetz, & Perry, 2006).

Similarly, concerning the dynamic nature of emotions, *dynamic theories* are needed. In research on motivation and affect, an early approach that has used a dynamic perspective and was based on differential equations modeling of affective processes is Atkinson and Birch's (1970) dynamic theory of action. A second group of approaches are the more recent computational process models of emotions (Wehrle & Scherer, 2001). In contrast to computational black-box models, computational process models specify the types of processes shaping emotions. Research on emotions in education should consider possibilities to adapt dynamic models for its purposes. Assumptions derived from dynamic systems theories probably will be useful for doing so (see Turner & Waugh, 2007).

Finally, concerning cultural and sociohistorical contexts, cross-cultural, sociological, and historical perspectives are needed. *Cross-cultural theories* specifying the nature of differences of emotions in educational systems around the world will have to be developed. Concerning historical aspects, *historical theories* and collaboration with historians of education would be needed. Currently, there are a variety of scholars who have been investigating the history of emotions (Ratner, 2007; Stearns & Stearns, 1985). It would be useful to bring that perspective to the study of the history of emotions in education as well.

MULTIMETHOD APPROACHES TO THE STUDY OF EMOTIONS IN EDUCATION

Different strategies and methods in empirical research on emotions have multiple advantages and disadvantages, and no single strategy or method is suited to answer all questions (Schutz, Chambless, & DeCuir, 2003). Traditionally, psychological research on emotions was based on deductive, quantitative, experimental approaches analyzing emotions in the laboratory by use of nomothetic strategies (Lewis & Haviland-Jones, 2000; Schutz et al., 2003; Schutz & DeCuir 2002). To fully capture the richness of emotions experienced in educational settings, however, exploratory methodology, qualitative data, nonexperimental designs, field studies, and idiographic strategies are needed as well. Therefore, rather than regarding different ways of analyzing emotions as competitive or mutually exclusive, we advocate viewing different approaches as being complementary. In this regard, it also is important to note that there are a number of different empirical approaches. Whereas qualitative research that is exploratory, nonexperimental, and field-based can be contrasted with quantitative, experimental, laboratory-based hypothesis testing, many other combinations of these approaches are possible, and potentially useful. For example, quantitative research can serve exploratory purposes, qualitative research can be laboratory-based, and experiments can be conducted in field settings.

Thus, we believe the study of emotions in education needs a multimethod paradigm that integrates inductive (exploratory) and deductive (hypothesis-testing) strategies, qualitative and quantitative approaches, experimental and nonexperimental designs, laboratory and field studies, as well as idiographic and nomothetic strategies. Finally, as part of a paradigm of studying emotions in education, advances in the measurement of emotions are needed as well.

Exploration and Hypothesis Testing: Integrating Inductive and Deductive Strategies

In the psychology of the 20th century, research that tested theories received high regard, maybe because such research made it possible for researchers to believe that psychology can become part of the world of natural sciences, as defined by such prototypical disciplines as experimental physics. All too often, however, theories were developed without observing reality in the first place, thus deriving assumptions from researcher's mental exercises instead of empirical observations. Such a strategy stands in contrast, for example, to strategies followed in the biological sciences, which often require extended observation of the behavior of members of some species before deriving theoretical assumptions (Lorenz, 1974).

For research on emotions in education, it seems recommendable, from our perspective, to analyze the richness of students' and teachers' affect by using exploratory research strategies. Within programs of research in this field, this may be a useful, if not necessary, first step, as theory is being developed (e.g., Pekrun et al., 2002). Although exploration inevitably is built on theoretical assumptions itself (e.g., on what and where to observe, and which methods to use), these assumptions should be handled in as open-minded a way as possible, such that justice can be done to observations not confirming preconceptions. Later on, observations can be used to create theoretical models that are tested by deductive strategies. Phases of induction and deduction can alternate, such that a program of research uses inductive-deductive loops to create, test, refine, and enlarge theories.

Exploratory strategies are often qualitative and nonexperimental in nature. In fact, however, exploratory strategies can involve any kind of empirical methods, including quantitative methods and experimental designs. For example, exploring the effects of dealing with learning material on various affective processes can use standardized, quantitative methods of sampling affective experiences (see Ainley, 2007).

Integrating Qualitative and Quantitative Approaches

Qualitative research is often needed to explore emotion phenomena and to generate hypotheses. Furthermore, qualitative methods may be best suited to derive in-depth descriptions of these phenomena and to get explanations for unexpected findings (e.g., by asking participants why they responded in unexpected ways to a psychological experiment). Quantitative methodology, on the other hand, is often best suited to test hypotheses on the structures and functions of emotions, and to analyze the generalizability of assumptions across individuals and cultures. In our view, the paradigm battles about whether to prefer one or the other methodology should be disregarded in educational research on emotions in favor of multimethod approaches adapting methods in flexible ways to the research questions that are developed to solve educational problems (Johnson & Onwuegbuzie, 2004; Schutz et al., 2003; Schutz & DeCuir, 2002).

Integrating Experimental and Non-Experimental Designs, and Laboratory and Field Studies

Controlled experiments may provide the best opportunities to study causality. Laboratory experiments have the potential to analyze basic processes of affect as experienced in relation to learning and teaching, such as the activation of different regions of the brain enhancing learning after mood induction, the impact of emotions on basic cognitive processes like task-related attention (e.g. Meinhardt & Pekrun, 2003), or the effects of different types of tasks on students' affective engagement in computer-based collaborative learning. Similarly, field experiments and quasi-experiments can nalyze the effectiveness of educational intervention, and are needed to test the usefulness of intervention programs targeting students' and teachers' emotions.

However, there also are clear limitations to an experimental approach to emotions in education, due to ethical constraints and problems of ecological validity. For example, as argued by Schutz and DeCuir (2002), "what principal or parent would agree to allow a researcher to create a situation in which students could become angry so that researchers could study the experience of anger in education?" (p. 125). Similarly, there are clear limitations to any laboratory-based approaches. Specifically, to the extent that laboratory settings are artificial and do not represent real-life classroom situations, laboratory experiments are in danger of analyzing *potential* causality that can be demonstrated in the laboratory, but possibly never outside the experimental setting. For example, it is currently unclear to what extent the results of the extant laboratory-based studies on mood and learning can be transferred to the real-life, context-bound, and often intense emotions experienced by students in the classroom. Furthermore, many aspects of emotions in education do not lend themselves to laboratory-based investigation. For example, this holds for research questions regarding the impact of educational institutions and sociohistorical contexts on students' and teachers' emotions (see DeCuir & Williams, 2007; Zemblyas, 2007). By implication, there is a clear need to complement laboratory-based strategies with field studies.

Integrating Idiographic and Nomothetic Approaches

Traditionally, many studies on emotions have used sample-based inferential strategies to derive conclusions about the general, nomothetic validity of theoretical assumptions. Typically, in field studies sample-based strategies involve observing distributions of one or more variables across individuals, as in analyzing the variation of mean values across time or groups of individuals, or in analyzing the relation between variables by using their covariances to calculate correlations and structural equations. For example, test anxiety research analyzed relations between test anxiety and students' academic performance by correlating test anxiety and performance measures across students. Similarly, in experimental laboratory investigations, groups of individuals are studied, and the distributions of values of dependent variables are compared across the different experimental groups.

These sample-based strategies, however, have the problem that one cannot infer from the resulting data if conclusions on population parameters are valid for any single individual under study. In the methodological literature, this is a well-known issue, for example, for interindividual correlations between variables. The *interindividual* and the *intraindividual* correlations between variables are statistically independent (Robinson, 1950; Schmitz & Skinner, 1993), implying that any inferences from sample correlations to individual psychological functioning can be invalid. An example cited by Schmitz and Skinner (1993) is the positive interindividual correlation between duration of sleep and frequency of migraine headaches, which seemingly implies that sleeping late can lead to headaches (or vice versa). Such a conclusion, however, would be misleading. These two variables are correlated *negatively* within individuals, implying that headaches occur in combination with *shorter* duration of sleep. Similarly, group differences in experiments can mask differential responses by different participants that are due to different individual causal mechanisms, rather than to one unitary mechanism as suggested by mean value differences of some dependent variable.

By implication, if one wants to study the intraindividual processes of students' and teachers' emotions, one would need to study students and teachers individually, implying an idiographic research strategy. Idiographic analysis based on single cases only, however, is problematic as well. Such an analysis has the complementary disadvantage that the generalizability of findings, beyond single individuals, necessarily remains an open question.

Succinctly stated, this is the dilemma we have to face: If we study *single individuals* only, more general conclusions cannot be drawn, thus reducing the scientific and practical relevance of results. If, on the other hand, we study *samples of individuals*, we cannot be sure whether the resulting findings are valid for any single individual under study. With this strategy as well, it is unclear whether findings are generalizable and useful for reaching scientific and applied goals.

An elegant way out of this dilemma is to combine both strategies by first analyzing reality within individuals, and then testing the generalizability of findings across individuals (Schmitz & Skinner, 1993). For example, Pekrun and Hofmann (1996) used such a combined idiographic-nomothetic strategy in a diary study on students' emotions experienced before and during their final university exams. In this study, the intraindividual relations between emotions and variables of learning over days were examined first (idiographic analysis), before analyzing the generalizability of intraindividual relations across students (nomothetic analysis). One result of this study was that many functions of emotions showed generalizability across students. However, there also were relations that were more specific to individual students (e.g., relations between anxiety and students' motivation to learn; also see Pekrun, in press).

Advancing the Measurement of Emotions

The measurement of students' and teachers' emotions is in its infancy. Tools need to be developed that allow an assessment of different emotions, and components of emotions, in reliable and valid ways (see Benson, 1998; Benson & Clark, 1982; Messick, 1989; Shepard, 1993). Specifically, whereas many measures of test anxiety are available today (Zeidner, 1998, 2007), measures of emotions other than anxiety have to be developed as well (see the *Achievement Emotions Questionnaire*; Pekrun et al., 2002; Pekrun, Goetz, & Perry, 2005). Similarly, measures of students' regulation of and coping with their emotions need to be constructed (see the *Emotional Regulation During Test Taking Scale*; Schutz, DiStefano, Benson, & Davis, 2004). Also, there is urgent need to develop measures assessing specific emotions as experienced by teachers, including their emotional labor in the classroom, beyond scales pertaining to omnibus constructs like teachers' burnout. Finally, measures of collective emotions and emotional climates experienced in classrooms are required.

Furthermore, measures are needed that can capture the dynamic nature of students' and teachers' emotions, and of components within these emotions. While self-report measures of emotions can be used to assess the development of emotions across situations (e.g., in experience sampling and diary methods), the value of these measures for assessing the dynamics of emotions is limited. Self-report is needed to assess the contents of emotional experiences, but it cannot render real-time estimates of emotional processes. Self-report measures are difficult to construct so that they yield interval or ratio scales needed for modeling more complex, nonlinear relationships over time. Furthermore, they are subject to response biases, and are not well suited to assess emotional processes that have limited access to consciousness.

Therefore, behavioral and neuropsychological assessment may be needed as well. In research on emotions, three groups of methods are available today: functional imaging and EEG methods analyzing cortical and subcortical affective processes (Murphy, Nimmo, & Lawrence, 2003); analysis of affect-related peripheral physiological activation; and behavioral observation of facial and postural expression of emotions and of emotion-related prosodic features of speech (Ekman & Rosenberg, 1997; Scherer, 1986). Studies on emotions in education should attempt to adapt these methodologies for the purposes of educational research. For example, there is a need for adapting observational systems of emotions such that they can be integrated into video-based classroom studies, and be used for analyzing students' and teachers' ongoing emotions in classroom discourse.

WHAT TO STUDY: FACETS, LEVELS, DYNAMICS, AND EDUCATIONAL INTERVENTION

To date, research on emotions in education focused on few emotions (above all, students' test anxiety) and correlates of emotions (such as students' academic achievement) as located on one level of analysis (the individual student), and situated in one type of sociohistorical context (North American and European educational systems). To make progress, this nascent field of research has to take more facets, levels, and contexts of students' and teachers' emotions into account than have been analyzed to date. Also, the situational, ontogenetic, and historical dynamics of emotions have to be analyzed, and there is an urgent need for carefully designed educational intervention research targeting students' and teachers' emotions.

Types and Patterns of Emotions: Facets of the Domain of Academic Emotions

The chapters of the present volume document that researchers have begun to pay attention to emotions other than test anxiety, but research on these emotions is still in its infancy. We need more studies on important unpleasant emotions like anger, frustration, shame, hopelessness, and boredom, as well as more studies on pleasant emotions like enjoyment, hope, and pride, as experienced by both students and teachers. Furthermore, to gain a better understanding of the affective climate in educational institutions, we also need research on the emotions experienced by administrators, principals, and employees of these institutions.

Beyond single emotions, patterns of emotions and of emotion components should be studied as well. Typically, emotional experiences involve more than one single emotion. Currently, however, it is unclear how exactly students' and teachers' emotions are patterned; by which individuals, and under which conditions, patterns of emotions are experienced; and how they evolve. A nonreductionist approach to emotions needs a thorough analysis of patterns of emotions and components, and of their variation across individuals and contexts, and over time (see Sansone & Thoman, 2005).

Levels and Contexts of Emotions in Education

Emotions are organized at multiple levels. Within the individual mind, component processes of emotions are organized at different levels that are related to subsystems of the central nervous system, and of peripheral physiological and motor systems. At the level of the individual student or teacher, feedback from the component processes of these different levels is merged into a holistic emotional experience that can verbally be labeled by using categories of emotion. In education, the individual, typically, is part of an educational setting such as the classroom, implying that emotions can be conceptualized at the level of social settings as well. Indeed, emotions are seen as social rather than as individual phenomena by some researchers (Ratner, 2007; Zembylas, 2007). For example, the emotional climate of classrooms is located at this level. Beyond the single classroom, relevant levels pertain to educational institutions nested within educational systems, systems nested within societies, and societies within cultures and sociohistorical macrosystems (Bronfenbrenner, 1986). Generally, from the perspective of a unit located at some specific level, the higher-level units of which it is a member can be regarded as *contexts* of the unit. For example, students are situated in the context of classrooms, classrooms in the context of institutions, and both of them in broader sociohistorical contexts.

To date, research on emotions in education has mainly addressed the individual and classroom levels of emotions, as shown in the chapters of this volume. Future research should also take into account the level of educational institutions, and of different cultural and sociohistorical contexts. Two types of studies may prove to be especially important. First, *multilevel classroom studies* are needed that analyze students' and teachers' emotions from a multilevel perspective, addressing the variation of emotions between individuals, activities, and subject domains, as well as between classrooms, schools, and educational systems. For example, in which ways can the variation of emotional experiences be explained by differences between academic activities, like studying vs. taking exams, or between different academic subjects (Goetz, Frenzel, Pekrun, & Hall, 2006)? Also, how much variance of students' emotions can be attributed to teachers and classrooms, how much of the variation between different classrooms' emotional climate can be attributed to different types of schooling, and what are the critical variables at these levels influencing students' and teachers' emotions? Answers to such questions are of fundamental importance for adequately designing educational interventions.

Second, *cross-cultural* and *sociohistorical studies* are needed that analyze the variation of emotions across different cultures that have different values and educational practices. Conducting cross-cultural studies pertaining to cultures that are present in today's world requires specifying theoretical assumptions on the nature of cross-cultural differences of emotions, constructing measures that show cross-cultural equivalence (e.g., Frenzel, Thrash, Pekrun, & Goetz, 2006), and including samples from different cultures in one investigation. Adequate sociohistorical analyses on emotions in education are even more difficult to conduct. Such analyses can be achieved by means of oral history and by an analysis of historical documents on education, including documents such as diaries giving insights into the subjective reception of educational practices.

Dynamics Over Time: Processes, Functions, and Regulation of Emotions

Most of the extant research on emotions in education provides cross-sectional snapshots describing the phenomenology of emotions, their structures and interindividual variation, or their relation with variables of learning and teaching. If progress regarding the dynamics of emotions is to be made, however, affective processes have to be studied by multiple assessments over time, instead of relying on single-shot assessments. Emotion dynamics can extend over different time frames, such as fractions of seconds in the neuropsychological dynamics of emotion components, minutes and hours in the continuing development of emotions within specific educational activity settings, years in ontogenetic development, and decades or centuries in historical development. For capturing these dynamics, process-oriented studies are best suited, including real-time physiological or observational analysis for dynamics of emotions over seconds or minutes, time interval or event sampling for processes within activity settings, and multi-wave longitudinal studies for the ontogenetic development of emotions in students' educational careers, and teachers' professional development. All of these possibilities are still underused to date. For example, even in research on test anxiety, which has been studied so often, there is a clear need for more longitudinal studies analyzing the reciprocal relations of students' anxiety and learning over the school years (Zeidner, 1998).

As part of a dynamic analysis of emotions, the effects of emotions on learning, teaching, and performance, and their antecedents need to be analyzed as well. With the exception of test anxiety, not much is known today on the specific and combined effects of different emotions on students' learning under different task conditions, and on teachers' classroom instruction and professional activities. Similarly, the individual and social antecedents of students' and teachers' emotions need to be analyzed. Finally, more research on the dynamics of students' and teachers' regulation of emotions is required (Schutz & Davis, 2000). There is a need to connect research on emotions in education to research on emotion regulation, coping, emotional competences, and emotional intelligence (Matthews, Zeidner, & Roberts, 2002).

The Need for Educational Intervention Research

Emotions are of primary practical importance in education: They affect students' and teachers' interest, engagement, and achievement, as well as their personality development, health, and well-being more generally. By implication, they can profoundly influence the productivity and quality of life in educational institutions, and in society at large. The question then arises, how can we shape education in "emotionally sound" ways (Astleitner, 2000)?

To date, practical considerations on how to foster students' and teachers' emotions can be deduced from findings of classrooms observations (e.g., Meyer & Turner, 2007), or from theoretical considerations (Pekrun et al., 2007). However, more direct evidence on educational interventions is still largely lacking, with the exception of intervention studies targeting students' test anxiety (Zeidner, 1998). Cognitive and cognitive-behavioral treatment of test anxiety is among the most successful methods of psychotherapy available to date. Even for test anxiety, however, there is a discernable lack of truly educational studies analyzing ways to prevent the development of excessive test anxiety in the first place, which would make psychological treatment unnecessary. Beyond test anxiety, what can be done to prevent or reduce anger, hopelessness, or boredom in students and teachers, and to foster their hope, pride, and enjoyment in teaching and learning? In which ways should classroom instruction, assessment practices, educational institutions, and the educational climate in our societies be shaped such that the emotions experienced by students and teachers benefit individual development as well as communities and society?

These questions are far from being trivial, as the few extant studies targeting emotions other than anxiety were partially successful at best (e.g., Glaeser-Zikuda, Fuss, Laukenmann, Metz, & Randler, 2005). Creating ways to shape educational environments in affectively sound ways will not be an easy task. The best of researchers' efforts will be required to successfully design and implement interventions targeting emotions, such that educational research on emotions can inform educational practitioners, administrators, and policy-makers on how they might be able to shape classroom instruction and educational institutions in affectively productive ways.

CONCLUSION: A CALL FOR INTERDISCIPLINARY COLLABORATION

Researchers in the field of emotions in education typically have training in psychology or education. The program of future research outlined in this chapter, however, requires using perspectives and methodologies from a number of disciplines, some of them far removed from educational research as traditionally conceived. Specifically, researching levels of emotions, from component processes within neuropsychological systems up to sociohistorical contexts, makes it necessary to integrate concepts and methods from a broad variety of disciplines. These disciplines include education and psychology, but they also include the neurosciences, sociology, economics, cultural anthropology, history, and philosophy.

Furthermore, it also seems necessary to transcend perspectives focused on emotions in education per se. For example, as noted above, structures and functions of emotions in the business world share many similarities with what is going on in education. The previously cited importance of different goal structures that involve competition and cooperation, and likely exert profound effects on participants' emotions, is but one example for the many parallels between these two worlds. In spite of the many similarities of phenomena and challenges, however, to date there has been almost no crosstalk between affective researchers in education and researchers in work psychology and economics (Pekrun & Frese, 1992). Similar arguments can be made with regard to sports, and to disciplines addressing specific institutional contexts of our youth, like family research.

More training of educational researchers by use of programs that have an interdisciplinary focus, and involve multiple strategies and methodologies of research, would seem useful to make progress along these lines. Even given extensive training and experience, however, to encompass the multitude of perspectives from all of the different disciplines mentioned would probably transcend the capacities of any single researcher. Therefore, truly collective efforts are likely needed. With such efforts, it should prove possible to make the best use of collaboration between researchers of different disciplines to make progress in the evolving field of research on emotions in education.

References

Allport, G. W. (1938). *Personality. A psychological interpretation*. London: Constable & Company.

Ainley, M. (2007). Being and feeling interested: Transient state, mood, and disposition. In P. A. Schutz and R. Pekrun, *Emotion in education* (pp. 141-157). San Diego: Elsevier Inc.

Aspinwall, L. (1998). Rethinking the role of positive affect in self-regulation. *Motivation and Emotion, 22,* 1-32.

Astleitner, H. (2000). Designing emotionally sound instruction: The FEASP-approach. *Instructional Science, 28,* 169-198.

Atkinson, J.W., & Birch, D. (1970). *A dynamic theory of action*. New York: Wiley.

Benson, J. (1998). Developing a strong program of construct validation: A test anxiety example. *Educational Measurement: Issues and Practices, 17,* 10-17.

Benson, J., & Clark, F. (1982). A guide for instrument development and validation. *American Journal of Occupational Therapy, 36*(12), 789-800.

Bronfenbrenner, U. (1986). Ecology of the family as a context for human development: Research perspectives. *Developmental Psychology, 22,* 723-742.

Brown, C.H. (1938). Emotional reactions before examinations: II. Results of a questionnaire. *Journal of Psychology, 5,* 11-26.

Clore, G. L., Schwarz, N., & Conway, M. (1994). Affective causes and consequences of social information processing. In R. S. Wyer & K. Skrull (Eds.), *Handbook of social cognition* (2nd ed.) (pp. 323-417). Hillsdale, NJ: Erlbaum.

DeCuir-Gunby, J. T., & Williams, M. R. (2007). The impact of race and racism on students' emotions: A critical race analysis. In P. A. Schutz & R. Pekrun (Eds.), *Emotion in education* (pp. 267-283). San Diego: Academic Press.

Efklides, A., & Petkaki, C. (2005). Effects of mood on students' metacognitive experiences. *Learning and Instruction, 15,* 415-431.

Efklides, A., & Volet, S. (Eds.) (2005). Feelings and emotions in the learning process [Special issue]. *Learning and Instruction, 15*(5).

Ekman, P., & Rosenberg, E. L. (Eds.). (1997). *What the face reveals: Basic and applied studies of spontaneous expression using the Facial Action Coding System (FACS)*. New York: Oxford University Press.

Fredrickson, B. L. (2001). The role of positive emotions in positive psychology: The broaden-and-build theory of positive emotions. *American Psychologist, 56,* 218-226.

Frenzel, A. C., Thrash, T. M., Pekrun, R., & Goetz, T. (in press). A cross-cultural comparison of German and Chinese emotions in the achievement context. *Journal of Cross-Cultural Psychology*.

Glaeser-Zikuda, M., Fuss, S., Laukenmann, M., Metz, K., & Randler, C. (2005). Promoting students' emotions and achievement—Instructional design and evaluation of the ECOLE-approach. *Learning and Instruction, 15, 481-495.*

Goetz, T., Frenzel, A. C., Pekrun, R., & Hall, N. C. (2006). The domain specificity of academic emotional experiences. *Journal of Experimental Education,* 75, 5-29.

Grandey, A. A. (2000). Emotional regulation in the work place: A new way to conceptualize emotional labor. *Journal of Occupational Health Psychology, 5*(1), 95-110.

Johnson, R., & Onwuegbuzie, A. J. (2004). Mixed methods research: A research paradigm whose time has come. *Educational Researcher, 33,* 14-26.

Kleinginna, P. R., & Kleinginna, A. M. (1981). A categorized list of emotion definitions, with suggestions for a consensual definition. *Motivation and Emotion, 5,* 345-379.

Lewin, K. (1935). *A dynamic theory of personality*. New York: McGraw-Hill.

Lewis, M., & Haviland-Jones, J. M. (Eds.). (2000). *Handbook of emotions* (2nd ed.). New York: Guilford Press.

Linnenbrink, E. A. (2007). The role of affect in student learning : A multidimensional approach to considering the interaction of affect, motivation and engagement. In P. A. Schutz & R. Pekrun (Eds.), *Emotion in education* (pp. 101-118). San Diego: Elsevier Inc.

Linnenbrink, E. A., & Pintrich, P. R. (2002). Achievement goal theory and affect: An asymmetrical bidirectional model. *Educational Psychologist, 37*, 69-78.

Lorenz, K. (1974). Analogy as a source of knowledge. *Science, 185*, 229-234.

Matthews, G., Zeidner, M., & Roberts, R. D. (Eds.). (2002). *Emotional intelligence: Science and myth*. Cambridge, MA: MIT Press.

Meinhardt, J., & Pekrun, R. (2003). Attentional resource allocation to emotional events: An ERP study. *Cognition and Emotion, 17*, 477-500.

Messick, S. (1989). Validity. In R. L. Linn (Ed.), *Educational measurement* (3rd ed.), (pp. 13-105). New York: Macmillan.

Meyer, D. K. & Turner, J. C. (2007). Scaffolding emotions in classrooms. In P. A. Schutz & R. Pekrun (Eds.), *Emotion in education* (pp. 235-249). San Diego: Elsevier Inc.

Morris, J. A., & Feldman, D. C. (1996). The dimensions, antecedents, and consequences of emotional labor. *Academy of Management Review, 21*(4), 986-1010.

Murphy, F. C., Nimmo, S. I., & Lawrence, A. D. (2003). Functional neuroanatomy of emotions: A meta-analysis. *Cognitive, Affective and Behavioral Neuroscience, 3*, 207-233.

Pekrun, R. (1988). *Emotion, Motivation und Persönlichkeit* [Emotion, motivation, and personality]. Munich/Weilheim: Psychologie Verlags Union.

Pekrun, R. (1992). The impact of emotions on learning and achievement: Towards a theory of cognitive/motivational mediators. *Applied Psychology, 41*, 359-376.

Pekrun, R. (2005). Progress and open problems in educational emotion research. *Learning and Instruction, 15*, 497-506.

Pekrun, R. (in press). The control-value theory of achievement emotions: Assumptions, corollaries, and implications for educational research and practice. *Educational Psychology Review*.

Pekrun, R., Elliot, A. J., & Maier, M. A. (2006). Achievement goals and discrete achievement emotions: A theoretical model and prospective test. *Journal of Educational Psychology, 98*, 583-597.

Pekrun, R., Frenzel, A., Goetz, T., & Perry, R. P. (2006, April). *Control-value theory of academic emotions: How classroom and individual factors shape students' affect*. Paper presented at the annual meeting of the American Educational Research Association, San Francisco, CA.

Pekrun, R., Frenzel, A. C., Goetz, T., & Perry R. P. (2007). The control-value theory of achievement emotions: An integrative approach to emotions in education. In P. A. Schutz & R. Pekrun (Eds.), *Emotion in education* (pp. 9-32). San Diego: Elsevier Inc.

Pekrun, R., & Frese, M. (1992). Emotions in work and achievement. In C. L. Cooper & I. T. Robertson (Eds.), *International review of industrial and organizational psychology* (Vol. 7, pp. 153-200). Chichester, UK: Wiley.

Pekrun, R., Goetz, T., & Perry, R. P. (2005). *Achievement Emotions Questionnaire (AEQ). User's manual*. Munich: Department of Psychology, University of Munich.

Pekrun, R., Goetz, T., Titz, W., & Perry, R.P. (2002). Academic emotions in students' self-regulated learning and achievement: A program of quantitative and qualitative research. *Educational Psychologist, 37*, 91-106.

Pekrun, R., & Hofmann, H. (1996, April). *Affective and motivational processes: Contrasting interindividual and intraindividual perspectives*. Paper presented at the annual meeting of the American Educational Research Association, New York.

Ratner, (2007). A macro cultural-psychological theory of emotions. In P. A. Schutz and R. Pekrun (Eds.), *Emotion in education* (pp. 85-100). San Diego: Elsevier Inc.

Robinson, W.S. (1950). Ecological correlations and the behavior of individuals. *American Sociological Review, 15*, 351-356.

Roseman, I. J., Antoniou, A. A., & Jose, P. E. (1996). Appraisal determinants of emotions: Constructing a more accurate and comprehensive theory. *Cognition and Emotion, 10*, 241-277.

Sansone, C., & Thoman, D. B. (2005). Does what we feel affect what we learn? Some answers and new questions. *Learning and Instruction, 15*, 507-515.

Schafe, G. E., Doyere, V., & LeDoux, J. E. (2005). Tracking the fear engram: The lateral amygdala is an essential locus of fear memory storage. *Journal of Neuroscience, 25*, 10010-10015.

Scherer, K. R. (1986). Vocal affect expression: A review and a model for future research. *Psychological Bulletin, 99*, 143-165.

Schiefele, U. (1991). Interest, learning, and motivation. *Educational Psychologist, 26*, 299-323.

Schmitz, B., & Skinner, E. (1993). Perceived control, effort, and academic performance: Interindividual, intraindividual, and multivariate time series analyses. *Journal of Personality and Social Psychology, 64*, 1010-1028.

Schutz, P. A., Chambless, C. B., & DeCuir, J. T. (2003). Multimethods research. In K. B deMarrais and S. D. Lapan (Eds.), *Foundations for research: Methods of inquiry in education and the social sciences* (pp. 267-282). Hillsdale, NJ: Lawrence Erlbaum.

Schutz, P. A. Cross, D. I., Hong J. Y., & Osbon, J. N. (2007). Teacher identities, beliefs, and goals related to emotions in the classroom. In P. A. Schutz & R. Pekrun (Eds.), *Emotion in education* (pp. 215-233). San Diego: Elsevier Inc.

Schutz, P. A., & Davis, H. A. (2000). Emotions and self-regulation during test taking. *Educational Psychologist, 35*, 243-256.

Schutz, P. A., & DeCuir, J. T. (2002). Inquiry on emotions in education. *Educational Psychologist, 37*, 125-134.

Schutz, P. A., DiStefano, C., Benson, J., & Davis, H. A. (2004). The development of a scale for emotional regulation during test taking. *Anxiety, Stress and Coping: An International Journal, 17*, 253-269.

Schutz, P. A. Hong J. Y., Cross, D. I., & Osbon, J. N. (in press). Reflections on investigating emotions among social-historical contexts. *Educational Psychology Review.*

Schutz, P. A., & Lanehart, S. (Eds.). (2002). Emotions in education [special issue]. *Educational Psychologist, 37*, 67-134.

Schutz, P. A., & Pekrun, R. (2007). Introduction to emotion in education. In P. A. Schutz & R. Pekrun (Eds.), *Emotion in education* (pp. 1-8). San Diego: Elsevier Inc.

Shepard, L. (1993). Evaluating test validity. *Review of Research in Education, 19*, 405-450.

Skinner, E. A. (1996). A guide to constructs of control. *Journal of Personality and Social Psychology, 71*, 549-570.

Stengel, E. (1936). Prüfungsangst und Prüfungsneurose [Exam anxiety and exam neurosis]. *Zeitschrift für Psychoanalytische Pädagogik, 10*, 300-320.

Stearns, P. N., & Stearns, C. Z. (1985). Emotionology: Clarifying the history of emotions and emotional standards. *American Historical Review, 90*, 813-836.

Turner, J. E. & Waugh, R. M. (2007). A dynamical systems perspective regarding students' learning processes: Shame reactions and emergent self-organizations. In P. A. Schutz and R. Pekrun (Eds.), *Emotion in education* (pp. 119-139). San Diego: Elsevier Inc.

Weiner, B. (1985). An attributional theory of achievement motivation and emotion. *Psychological Review, 92*, 548-573.

Weiner, B. (2007). Examining emotional diversity in the classroom: An attribution theorist considers the moral emotions. In P. A. Schutz and R. Pekrun (Eds.), *Emotion in education* (pp. 71-84). San Diego: Elsevier Inc.

Wehrle, T., & Scherer, K. R. (2001). Toward computational modeling of appraisal theories. In K. R. Scherer, A. Schorr & T. Johnstone (Eds.), *Appraisal processes in emotion* (pp. 350-365). New York: Oxford University Press.

Zeidner, M. (1998). *Test anxiety: The state of the art*. New York: Plenum.

Zeidner, M. (2007). Test anxiety in educational contexts: Concepts, findings, future directions. In P. A. Schutz & R. Pekrun (Eds.), *Emotion in education* (pp. 159-177). San Diego: Elsevier Inc.

Zembylas, M. (2007). The power and politics of emotions in teaching. In P. A. Schutz and R. Pekrun (Eds.), *Emotion in education* (pp. 285-301). San Diego: Elsevier Inc.

Index